MODERN ASPECTS OF ELECTROCHEMISTRY

No. 40

Modern Aspects of Electrochemistry

Modern Aspects of Electrochemistry, No. 39:

- Approaches to Solute-Solvent Interactions including two approaches to computational determination of solution properties, and several other procedures that establish correlations between properties of interest and certain features of the solute and/or solvent molecules
- Porous Silicon, including its morphology and formation mechanisms, as well as anodic reaction kinetics
- Modeling Electrochemical Phenomena via Markov Chains and Processes gives an introduction to Markov Theory, then discusses applications to electrochemistry, including modeling electrode surface processes, electrolyzers, the repair of failed cells, analysis of switching-circuit operations, and other electrochemical systems
- Fractal Approach to Rough Surfaces and Interfaces in Electrochemistry, from a review of Fractal Geometry to the application of Fractal Geometry to the classification of surfaces and Electrochemistry
- Phenomenology and Mechanisms of Electrochemical Treatment (ECT) of Tumors, starting from fundamentals and proceeding to electrochemical treatment of tumors in animals and then in humans

Modern Aspects of Electrochemistry, No. 38:

- Solid State Electrochemistry encompassing modern equilibria concepts, thermodymanics and kinetics of charge carriers in solids
- Electron transfer processes, with special sections devoted to hydration of the proton and its heterogeneous transfer
- Electrosorption at electrodes and its relevance to electrocatalysis and electrodeposition of metals
- The behavior of Pt and other alloy electrocatalyst crystallites used as the electrode materials for phosphoric acid electrolyte fuel-cells
- Applications of reflexology and electron microscopy to the materials science aspect of metal electrodes
- Electroplating of metal matrix composites by codeposition of suspended particles, a process that has improved physical and electrochemical properties

LIST OF CONTRIBUTORS

DAVID BLOOMFIELD
Analytical Power, LLC
2-X Gill Street
Woburn, Massachusetts 01801

VALERIE BLOOMFIELD
Analytical Power, LLC
2-X Gill Street
Woburn, Massachusetts 01801

JOHN O'M. BOCKRIS
Haile Plantation
10515 S. W. 55th Place
Gainesville, FL 32608

THOMAS Z. FAHIDY
Department of Chemical Engineering
University of Waterloo
Waterloo, Ontario
Canada N2L 3G1

BRENDA L. GARCIA
College of Engineering and Information Technology,
Swearingen Engineering Center
Department of Chemical Engineering
University of South Carolina
301 South Main Street,
Columbia, South Carolina, 29208

CHARLES R. MARTIN
Department of Chemistry
University of Florida
Gainesville, Florida 32611
crmartin@chem.ufl.edu
Fax: 352-392-8206

PARTHA P. MUKHERJEE
Electrochemical Engine Center (ECEC), and Department of Mechanical and Nuclear Engineering
The Pennsylvania State University
University Park
Pennsylvania 16802

KEITH SCOTT
School of Chemical Engineering and Advanced Materials
University of Newcastle upon Tyne,
United Kingdom

ASHOK K. SHUKLA
Solid State and Structural Chemistry Unit
Indian Institute of Science
Bangalore, India

CHARLES R. SIDES
Department of Chemistry
University of Florida
Gainesville, Florida 32611

CHAO-YANG WANG
Electrochemical Engine Center (ECEC), and Department of Mechanical and Nuclear Engineering
The Pennsylvania State University
University Park
Pennsylvania 16802
e-mail: cxw31@psu.edu,
Ph: (814) 863-4762
Fax: (814) 863-4848

GUOQING WANG
Plugpower Inc.
968 Albany-Shaker Road
Latham, New York 12110

JOHN W. WEIDNER
College of Engineering and Information Technology,
Swearingen Engineering Center
Department of Chemical Engineering,
University of South Carolina
301 South Main Street,
Columbia, South Carolina, 29208

MODERN ASPECTS OF ELECTROCHEMISTRY
No. 40

Edited by

RALPH E. WHITE
University of South Carolina
Columbia, South Carolina

C. G. VAYENAS
University of Patras
Patras, Greece

and

MARIA E. GAMBOA-ALDECO
Managing Editor
Superior, Colorado

Ralph E. White
Department of Chemical Engineering
University of South Carolina
Columbia, South Carolina
U.S.A.
white@engr.sc.edu

Library of Congress Control Number: 2005939188

ISBN-10: 0-387-46099-3   e-ISBN-10: 0-387-46106-X
ISBN-13: 978-0-387-46099-4   e-ISBN-13: 978-0-387-46106-9

Printed on acid-free paper.

© 2007 Springer Science+Business Media, LLC
All rights reserved. This work may not be translated or copied in whole or in part without the written permission of the publisher (Springer Science+Business Media, Inc., 233 Spring Street, New York, NY 10013, USA), except for brief excerpts in connection with reviews or scholarly analysis. Use in connection with any form of information storage and retrieval, electronic adaptation, computer software, or by similar or dissimilar methodology now known or hereafter developed is forbidden.
The use in this publication of trade names, trademarks, service marks, and similar terms, even if they are not identified as such, is not to be taken as an expression of opinion as to whether or not they are subject to proprietary rights.

9 8 7 6 5 4 3 2 1

springer.com

Preface

This volume begins with a tribute to Dr. Brian E. Conway by Dr. John O'M. Bockris, which is followed by six chapters. The topics covered are state of the art Polymer Electrolyte Membrane (PEM) fuel cell bipolar plates; use of graphs in electrochemical reaction networks; nano-materials in lithium ion batteries; direct methanol fuel cells (two chapters); and the last chapter presents simulation of polymer electrolyte fuel cell catalyst layers.

David and Valerie Bloomfield begin the first chapter with a discussion of the difficulties encountered when confronting bipolar plate development and state that the problems stem from the high corrosive nature of phosphoric acid. The water problems are mitigated but the oxidation problems increase. Bipolar plates are still not cheap, reliable or durable.

In Chapter 2, Thomas Z. Fahidy reviews analysis of variance (ANOVA) and includes one-way, two-way, three-way classification, and Latin squares observation methods. He moves on to a discussion of the applications of the analysis of covariance (ANCOVA) and goes over certain variables such as velocity, velocity and pressure drop, and product yields in a batch and flow electrolyzer. His conclusion is that proper statistical techniques are time savers which can save the experimenter and the process analyst considerable time and effort in trying to optimize the size of statistically meaningful experiments.

In Chapter 3 Charles R. Sides and Charles R. Martin present the effect of nanomaterials in lithium-ion battery electrode design and they describe the template-synthesis of lithium-ion battery electrode materials using polymeric templates. They present a plasma-assisted method to create a carbon replica of an alumina template membrane, which serves as a nanostructured anode for lithium-ion battery design.

Keith Scott and Ashok K. Shukla begin their discussion in Chapter 4 with a review of various types of fuel cells: phosphoric acid fuel cells, alkaline fuel cells, polymer electrolyte fuel cells, molten electrolyte fuel cells, solid oxide fuel cells, and direct methanol fuel cells. They follow this by reviewing the advances made in the performance of DMFCs since their inception. These fuel cells need to be able to stand higher temperatures than 200 °C which require more temperature stable membranes. These fuel cells have a great potential to be used in mixed reactant systems due to the simplicity of the fuel cell stack design which may result in reduced costs.

In Chapter 5, Brenda L. Garcia and John W. Weidner review direct methanol fuel cells (DMFCs). High temperature operation of DMFCs has not been achievable without pressure on the cathode side and researchers have not been able to obtain good performances at temperatures above the boiling point of water. Novel polymers that operate well with temperatures above the boiling point have not been developed yet. One solution proposed in the literature is to laminate a layer of methanol impermeable polymer between two Nafion® membranes. This solution still requires hydration at high potentials and adsorption of methanol limits the reaction rate at low potential. Two-dimensional models have been developed to explain the variation of the current along the flow path. Transient models have been proposed but not much work has been accomplished in this area.

Partha P. Mukherjee, Chao–Yang Wang, and Guoqin Wang in Chapter 6 discuss the direct numerical simulation of polymer electrolyte fuel cell catalyst layers and they present a systematic development of the direct numerical simulation model which was developed at Penn State Electrochemical Engine Center. This method is used to describe the oxygen, water, and charge transport at the pore level within 2-D and 3-D computer-generated catalyst microstructures and indicates the importance of developing high-performance catalyst layers.

R. E. White

University of South Carolina
Columbia, South Carolina

C. Vayenas

University of Patras
Patras, Greece

Memories of Brian Evans Conway
Editor 1955-2005

I remember meeting Brian Conway on the road, Queens Gate, which leads from Imperial College, London University, to the Tube Station in South Kensington, London. It was a Saturday afternoon in 1946. I was one year into a lectureship (assistant professorship in U.S. terms) at Imperial College, the technologically oriented part of London University. Brian stopped me to say that he wanted to do his graduate research with me. He was 20 and I, 23.

Brian Conway entered my research group at Imperial College in its second year. There are some who have called the years between 1946 and 1950 as the Golden Years of Electrochemistry because the development of the subject as a part of physical chemistry was pioneered in those years, primarily by the remarkable set of people who were in the group at that time and who spread out with their own students and ideas after they left me.

First among those in respect to the honors achieved was Roger Parsons. He became an FRS and President of the Faraday Division of the Chemical Society. Conway is No. 2 in this formal ranking although he published more extensively and with more original ideas than did Parsons. Then there was John Tomlinson who became a professor at the University of Wellington in New Zealand (and Vice Chancellor of the University) F. C. Potter too was in the group at that time and became in the 1980's the President of the Royal Society of New South Wales, apart from having an active career in the Council of Science and Industry of the Australian government. Harold Eagan was there too and he later achieved a grand title: *The Government Chemist*, the man in the U.K. government who is in charge of testing the purity of various elements of the country's supply.

The only woman in the group, Hanna Rosenburg, should be mentioned here because she had an extended influence on Brian Conway, based partly on her remarkable ability to discuss widely.

Martin Fleischmann, too, is relevant. He got his Ph.D. in a small group near to my group in which Conway worked. However, he joined us in various activities, particularly the discussions. He became well known internationally not only because of his contributions to physical electrochemistry, but also because in 1989 he resuscitated an idea—which had been introduced by the French and Japanese in the 1960's—that nuclear reactions could be carried out in solutions in the cold.

Conway's thesis (1949) was very original in content and contained, e.g., 3-D representations of potential energy surfaces.

After Brian Conway got his Ph.D. with me, he went to work with J. A. V. Butler at the Chester-Beatty Cancer Research Institute, a short walk from Imperial College. Butler's habit of poring over manuscripts, murmuring, and his sensitivity to noise whilst thinking, was reflected by some of Conway's habits. Butler was known to be absent minded in a way that colored the great respect with which he was regarded. Thus, according to Conway, in Butler's year at Uppsala, he was observed by a British colleague to pass two lumps of sugar to the cashier whilst thoughtfully adding the ore to his coffee.

Contact with Conway continued from 1949 to 1953 in London. We used to meet on Saturday afternoons in a Kensington coffee shop and it was there that some of our electrochemical problems were worked out.

After 1951 I decided to look towards the United States for continuation of my career. We thought of Brian as a possible post doc to take with us and wanted to get him married before the transition. Unfortunately, Brian evaluated girls principally by their intellectual abilities and the Viennese Ms. Rosenburg had given him the model. He found it difficult to meet her equal and we had eventually to point out to him that he couldn't get a visa to come to America with a woman unless he were married to her. Shortly, after this, he did invite me to Daguize, our Saturday café, to meet a former Latvian pharmacist, Nina; later I served to take Nina up the aisle at the wedding of Nina and Brian (1954).

It had seemed to me that, if we were going to bring Electrochemistry from its moribund state of the 1940s to modernity in physical chemistry, a yearly monograph would help. I therefore approached Butterworth's in London with a proposition and the first volume was published in 1954.[1] I brought Brian into Modern Aspects in Volume I as an assistant and as my co-author in a chapter on solvation. It soon seemed unfair to continue to receive Brian's help in editing unless he was made an editor too and, therefore, from the second volume on, we were both named editors of the series. We kept it that way until in 1987 we invited Ralph White to join us as the Electrochemical Engineer, thus increasing the breadth of the topics received.[2]

I moved to the University of Pennsylvania in Philadelphia in 1953 whilst Brian remained at Chester-Beatty, but I went back to London for the summer of 1954 and persuaded Brian to come over to Philadelphia at first as a post doctoral fellow.

Brian Conway's move to the U.S.A. facilitated our joint editorship in running the series. We had had many discussions in London and this encouraged us to make the composition of each volume an occasion for a meeting, lasting two to three days in a resort. This meant one to two day's hard discussion, deciding the topics we wanted as well as the authors who might write the articles, together with a reserve author in case the first one refused. The third day we spent as free time for our own discussions. Colorado Springs was a place in which we had an interesting time and then there was the Grand Hotel in White Sulfur Springs and several resorts in Canada. But there is no doubt that Bermuda was more often the locale of the Modern Aspects invitation meetings than any other. After Ralph White threw his expertise onto choice of authors, we continued to meet yearly, e.g., in Quebec City and in San Francisco, though in recent years the meetings drifted back to their original form.

Coming now to Brian's two years at the University of Pennsylvania (1954–1955), we got out of this time three papers that I think of as being foundation papers of what we now know as physical electrochemistry. The first concerned the mechanism of the high mobility of the proton in water. The second arose out of the interpretation of plots of log i_o against the strength of the metal-hydrogen bond and found two groups of metals, one involving

---

[1] Plenum Press purchased Butterworth's in 1955 after the second volume for Modern Aspects of Electrochemistry had been published.
[2] Our policy throughout—as I believe it is with the present editors—was to make in Modern Aspects of Electrochemistry a truly *broad* presentation of the field although we always avoided analytical topics for which there was already so much presentation.

proton discharge onto planar sites and the other proton discharge (rate determining) onto adsorbed H. The last of the three concerned the early stages of metal deposition, a subject that raises many fundamental questions in electrochemical theory (e.g., partial charge transfer). The first two of these works were published in the Journal of Chemical Physics and the last one in the Proceedings of the Royal Society. To have papers accepted in these prestigious journals is unusual, particularly for electrochemical topics so often mired in lowly technological issues.

Conway transferred to the University of Ottawa in 1955 and remained there for the rest of his career. He published near to 400 papers whilst in Ottawa and it is interesting to mention a few of them.

In Ottawa I would pick out the work with Currie in 1978. The topic was on pressure dependence of electrode reactions and gave a new mechanism criterion.

Agar had suggested in 1947 that there might be a temperature dependence of the electrode kinetic parameter. Conway took this up in 1982 and showed experimentally that in certain reactions this was the case.

The most well known work that Conway and his colleagues completed in Ottawa was on the analysis of potential sweep curves. I had been critical of the application of potential sweep theory to reactions that involved intermediates on the electrode surface. Working with Gilaedi and then with Halina Kozlowska, and to some extent with Paul Stonehart, Conway developed an analysis of the effect of intermediate radicals on the shape and properties of potential sweep showing how interesting electrode kinetic parameters could be thereby obtained.

This is the point to stress the part played by Halina Angerstein-Kozlowska in Brian's work. She played an important role in administration of the co-workers and supervision of the work, particularly in Brian's absences at meetings, many in Europe. Her position was similar to that of the late Srinivasan in respect to John Appleby electrochemical research group at Texas A&M. It is more creative and directional than her formal position as a senior researcher indicates.

I continued an active discussion relationship with Brian Conway for more than fifty years. This was made possible not only by the yearly meetings related to Modern Aspects but by letters, most of which concerned discussions of topics apart from electrochemistry. I remember those on the inheritance of acquired characteristics with Hanna Rosenburg in London in the fifties, the discussions of the position and number of the galaxies, and then, later, consideration of methods by aggressive groups could be restrained from war. In more recent years we have been concerned with the speed of electrons in telephone lines. Many people think that electrons there must travel very quickly but in fact their movement during the passage of messages is measured in a few centimeters per second.

Although the correspondence with me concerned topics discussed in terms of the science of the day, Conway's interests were broad and he had interests in the Arts, stimulated by his wife in this direction (his house was decorated with her paintings). In this respect, Conway exemplifies the Scientist as a Great Man, a man of Knowledge in all and any direction. One is reminded of Frumkin in Moscow who kept abreast of modern literature from Anglo-America, France and Germany and Hinshelwood at Oxford, a physical chemist whose alternative interests were in Japanese Poetry.

The long lasting scientific correspondence with Conway started to ebb in the mid nineties and I assumed that this was due to normal processes of aging but it turned out that it was the beginning of the end—he was frequently in the hospital.

One day in this Summer of 2005, Nina called to tell me that the doctors declared the prostate cancer that he had suffered metastasized and that they could no longer control it. I placed myself at Brian's disposal in respect to anything that he wanted—I worked here through the well-known Barry Macdougall, an ex-Conway student. Through him I passed to Brian photographs from the London days, and these contained, among others, the one he most wanted—that of the student he had known 55 years earlier, Hanna Rosenburg.

Brian leaves behind him Nina (later a professional artist and musician) who had contributed so much to his career by her help and fortitude, and a son, Adrian, who is a competent theorist in telephone networks.

Brian chose to have his body cremated.

John O'M. Bockris
Editor 1953–2003

Contents

Chapter 1

PEM FUEL CELL BIPOLAR PLATES

D. Bloomfield and V. Bloomfield

I.	Introduction	1
	1. NAFION MEA Based Bipolar Plate Problems ...	2
	2. Polybenzimidazole/H_3PO_4	3
II.	Definition	3
	1. Separator Plate	4
	2. Flow Field	5
	3. Port and Port Bridges	6
	4. Seals	7
	5. Frame	8
III.	Bipolar Plate Features	8
	1. Tolerances	9
	2. Thermal Management	9
	3. Electrical Conduction	10
	4. Water Management	11
	5. Low Cost	13
	6. Stable, Free from Corrosion Products	13
	(*i*) Galvanic Corrosion	13
IV.	Materials and Processes	15
	1. Comparison of Carbon and Metal	15
	(*i*) Operational	16
	(*ii*) Forming Cost	16
	2. Carbon	18
	(*i*) Molded Graphite	18
	(*ii*) Paper	19
	(*iii*) Stamped Exfoliated Graphite (Grafoil, Graflex)	19

3. Metal 19
   (*i*) Forming Metal Bipolar Plates 20
   (*ii*) Intrinsically Corrosion Resistant Metals ... 21
   (*iii*) Direct Coatings 21
   (*iv*) Conductive Polymer Grafting 24
References 33

Chapter 2

BASIC APPLICATIONS OF THE ANALYSIS OF VARIANCE AND COVARIANCE IN ELECTROCHEMICAL SCIENCE AND ENGINEERING

Thomas Z. Fahidy

I. Introduction 37
II. Basic Principles and Notions 38
III. ANOVA: One-Way Classification 40
  1. Completely Randomized Experiment (CRE) 42
  2. Randomized Block Experiment (RBE) 42
     (*i*) Example 1: A Historical Perspective of Caustic Soda Production 43
     (*ii*) Example 2: Metallic Corrosion 45
IV. ANOVA: Two-Way Classification 46
  1. Null and Alternative Hypotheses 46
  2. Illustration of Two-Way Classification: Specific Energy Requirement for an Electrolytic Process 47
V. ANOVA: Three-Way Classifications 49
VI. ANOVA: Latin Squares (LS) 51
VII. Applications of the Analysis of Covariance (ANCOVA) 53
  1. ANCOVA with Velocity as Single Concomitant Variable 53
     (*i*) Pattern A(CRE) 53
     (*ii*) Pattern B(RBE) 56
  2. ANCOVA with Velocity and Pressure Drop Acting as Two Concomitant Variables 58
  3. Two Covariate-Based ANCOVA of Product Yields in a Batch and in a Flow Electrolyzer 58
  4. Covariance Analysis for a Two-Factor, Single Cofactor CRE 60

VIII.	Miscellaneous Topics	62
	1. Estimation of the Type II Error in ANOVA	62
	2. Hierarchical Classification	64
	3. ANOVA-Related Random Effects	66
	4. Introductory Concepts of Contrasts Analysis	69
IX.	Final Remarks	71
	Acknowledgments	72
	List of Principal Symbols	72
	References	73

Chapter 3

NANOMATERIALS IN LI-ION BATTERY ELECTRODE DESIGN

Charles R. Sides and Charles R. Martin

I.	Introduction	75
II.	Templates Used	78
	1. Track-Etch Membranes	78
	2. Alumina Membranes	80
	3. Other Templates	81
II.	Nanostructured Cathodic Electrode Materials	83
	1. Electrode Fabrication	84
	(*i*) Nanostructured Electrode	84
	(*ii*) Control Electrodes	85
	2. Structural Investigations	86
	3. Electrochemical Characterization	87
	(*i*) Cyclic Voltammetry	87
	(*ii*) Rate Capabilities	89
III.	Nanostructured Anodic Electrodes	91
	1. Electrode Fabrication	92
	(*i*) Nanostructured Electrodes	92
	(*ii*) Control Electrodes	92
	2. Structural Investigations	93
	3. Electrochemical Investigations	95
V.	Nanoelectrode Applications	97
	1. Low-Temperature Performance	97
	(*i*) Electrode Fabrication	97
	(*ii*) Strategy	98
	(*iii*) Electrochemical Results	99

	(*iv*) Electronic Conductivity	101
	(*v*) Cycle Life	102
2.	Variations on a Synthetic Theme	102
	(*i*) Nanocomposite of LiFePO$_4$/Carbon	102
	(*ii*) Improving Volumetric Capacity	109
VI.	Carbon Honeycomb	117
1.	Preparation of Honeycomb Carbon	118
2.	Electrochemical Characterization	121
VII.	Conclusions	123
	Acknowledgements	123
	References	124

Chapter 4

DIRECT METHANOL FUEL CELLS: FUNDAMENTALS, PROBLEMS AND PERSPECTIVES

Keith Scott and Ashok K. Shukla

I.	Introduction	127
II.	Operating Principle of the SPE-DMFC	128
III.	Electrode Reaction Mechanisms in SPE-DMFCs	132
	1. Anodic Oxidation of Methanol	132
	2. Cathodic Reduction of Oxygen	139
IV.	Materials for SPE-DMFCS	140
	1. Catalyst Materials	140
	(*i*) Anode Catalysts	140
	(*ii*) Oxygen Reduction Catalysts	149
	(*iii*) Membrane Materials	156
V.	Direct Methanol Fuel Cell Performance	163
	1. DMFC Stack Performance	175
	2. Alternative Catalysts and Membranes in the DMFC	178
	3. Alkaline Conducting Membrane and Alternative Oxidants	183
VI.	Conventional vs. Mixed-Reactant SPE-DMFCs	185
VII.	Mathematical Modelling of the DMFC	192
	1. Methanol Oxidation	195
	2. Empirical Models for Cell Voltage Behaviour	198
	3. Membrane Transport	202

		4. Effect of Methanol Crossover on Fuel Cell Performance	204
		5. Mass Transport and Gas Evolution	205
		6. DMFC Electrode Modelling	209
		7. Cell Models	210
		8. Single Phase Flow	212
		9. Two- and Three-Dimensional Modelling	213
		10. Dynamics and Modelling	215
		11. Stack Hydraulic and Thermal Models	215
VIII.	Conclusions		216
	List of Symbols		217
	References		218

Chapter 5

REVIEW OF DIRECT METHANOL FUEL CELLS

Brenda L. García and John W. Weidner

I.	Introduction		229
II.	Anode Kinetics		232
	1. Reaction Mechanism		232
	2. Methanol Oxidation Catalysts		233
		(*i*) Platinum and Platinum Catalyst Structure	233
		(*ii*) Platinum and Platinum Alloy Catalyst Performance	240
III.	Oxygen Reduction Reaction Catalysts		247
IV.	High Temperature Membranes		248
V.	Methanol Crossover		253
	1. Magnitude of Crossover		253
	2. Effect of CO_2 Crossover		258
	3. Mixed-Potential Effects		260
	4. Novel Membranes to Reduce Methanol Crossover		261
VI.	DMFC Modeling Review		264
	1. One-Dimensional Models		265
	2. Two-Dimensional and Three-Dimensional Models		273
VII.	Summary		278
	References		280

Chapter 6

DIRECT NUMERICAL SIMULATION OF POLYMER ELECTROLYTE FUEL CELL CATALYST LAYERS

Partha P. Mukherjee, Guoqing Wang, and
Chao-Yang Wang

I.	Introduction	285
II.	Direct Numerical Simulation (DNS) Approach	288
	1. Advantages and Objectives of the DNS Approach	289
	2. DNS Model - Idealized 2-D Microstructure..........	290
	3. Three-Dimensional Regular Microstructure	293
	4. Results and Discussion	299
	(*i*) 2-D Model: Kinetics- vs. Transport-Limited Regimes	299
	(*ii*) Comparison of the Polarization Curves between 2-D and 3-D Simulations	304
III.	Three-Dimensional Random Microstructure	305
	1. Random Structure	306
	2. Structural Analysis and Identification	307
	3. Governing Equations	311
	4. Boundary Conditions	314
	5. Results and Discussion	316
IV.	DNS Model – Water Transport	320
	1. Water Transport Mechanism	321
	2. Mathematical Description	323
	3. Results and Discussion	327
	(*i*) Inlet-Air Humidity Effect	327
	(*ii*) Water Crossover Effect	330
	(*iii*) Optimization of Catalyst Layer Compositions	331
V.	3-D Correlated Microstructure	333
	1. Stochastic Generation Method	333
	2. Governing Equations, Boundary Conditions and Numerical Procedure	334
	3. Results and Discussion	337
VI.	Conclusions	340
	Acknowledgements	340
	References	341
INDEX	....	343

1

PEM Fuel Cell Bipolar Plates

D. Bloomfield and V. Bloomfield

Analytical Power, LLC, Woburn, MA

I. INTRODUCTION

The objective of this paper is to describe the state-of-the-art of PEM fuel cell bipolar plate technology. We will present the problems confronting bipolar plate development along with several potential paths to solution.

The paper first describes the functions, properties, and attributes of bipolar plates. The balance of the paper is devoted to the materials and fabrication processes that have been explored and are currently under investigation. One of the materials is carbon, in a variety of forms from molded graphite, to paper, to exfoliated stamped graphite. The other set of materials are metals which can be intrinsically corrosion resistant, or coated with corrosion protection materials. Metal bipolar plates are generally composite structures with different materials performing different functions. Sometimes the only metal part in the metal bipolar plate is the flat sheet separator. The recent problems with PEM MEA: durability, cathode catalyst and carbon loss, have place more emphasis on the development of bipolar plates with high *in plane* conductivity, namely metal. The high in plane conductivity enables a potential systems solution to this problem. A patent is pending by Howard Baker.[25]

The properties of bipolar plates are more or less defined by the membrane electrode assemblies (MEAs). The set of problems associated with low temperature membrane MEAs such as Nafion are due to water and, more specifically, the liquid product water that leaches out of the membrane and finds its way into the bipolar plate flow fields. The other major problem is that of carbon. While carbon is a base material, its oxidation kinetics are so slow and its hydrogen evolution overpotential so low that it often behaves as if it was more noble than platinum, see Sedricks.[23]

Recently there has been a push to develop higher temperature MEAs such as polybenzimidazole gels with phosphoric acid. The problems for bipolar plates stem from the high temperature corrosive nature of phosphoric acid.

"Things go wrong, and things go wrong faster at high temperature."

Murphy

While the water problem is mitigated, in the high temperature MEAs, the oxidation problems are exacerbated. These problems were extensively studied in the 1960s and 1970s when the conventional PAFC was developed. The PAFC led to the present day o-molded graphite bipolar plate, as well as carbon supported platinum catalyst and carbon paper GDLs.

1. NAFION MEA Based Bipolar Plate Problems

The problem with PEM Fuel Cells is that after 40 years of R&D, they are not cheap, reliable, or durable. (Gemini 3 flew on March 23, 1965.) Most of the problems lie in the MEA, but the source of many problems can be found in the MEA/bipolar plate interaction; in particular, the initial degradation of the polymer membrane. This is due to the presence of short chain sulfonic acid groups in Nafion membranes. These short chain groups have a relatively high solubility in water and leach out of the fuel cell when it is initially started.[18] The process is simply demonstrated by measuring the pH of water extracted from the reactant cavities of a fuel cell that has run for about 100 hours. Water is always found in the reactant cavities of PEM cells. As the cell ages, imperfectly terminated polymers will also release HF and increased amounts of sulfonic acid. Peroxides are formed at the cell anode. The membrane attack by peroxides is well known[20,21] and has been well know for a long time. Peroxides are stable at anode potentials, and they attack the thin ionomer films surrounding the anode catalyst. Polyvalent cations, especially Fe, Co, and Ni catalyze peroxide formation.

The low pH of the cell water and the use of carbon paper gas diffusion layers (GDLs) make metal bipolar plates problematic. Carbon has a low hydrogen evolution potential, lower than all but the precious metals, and it is the most widely used GDL material. Most metal bipolar plates in electrical contact with carbon GDLs become anodic with respect to the GDL and corrode. Carbon in the catalyst support and in bipolar plates is often a source of polyvalent cations which form in the low pH water surrounding the bipolar plates diffuse back to the MEA and catalyze the formation of peroxide.

2. Polybenzimidazole/H_3PO_4

A great deal of relevant work was done by UTC, Westinghouse and Fuel Cell Energy in the 1960's and 1970's on phosphoric acid fuel cells. None of it was encouraging for metal bipolar plates. The elevated operating temperature, 190C, of the PBI membranes currently limits bipolar plate material choices to graphite. No liquid water is found in operating cells but condensate can be found on the surfaces of plates in shutdown cell stacks. The condensate has a low pH from the phosphoric acid it dissolves.

II. DEFINITION

Bipolar plates are used in PEM fuel cell stacks because no one has successfully engineered an alternative to the planar, gas diffusion electrode. The function of the plates is to ensure that gaseous reactants are uniformly supplied to the cells in a stack and uniformly distributed over the surfaces of the cell or cooler. The bipolar plate separates the gases and coolants and provides sealing between the cell reactant chambers and the environment. The bipolar plate provides for the continuous conduction of electrons from one cell anode to the next fuel cell cathode. In the case of internally manifolded stacks, the manifolds are formed by the stacking of plates. In the case of internal manifolds, port design is of great importance. In some designs, the bipolar plate manages condensed product water to allow unimpeded flow of fuel and air. Condensate is undesirable in the cell stack because it facilitates corrosion, blocks reactant flow, and increases the heat load on the cooling system.

Bipolar plates constitute everything in the fuel cell stack that is not membrane electrode assembly (MEA) or follow up system (tie rods and end plates). We generally consider MEAs everything between the gas diffusion layers (GDLs), including the membrane and catalyst layers. Bipolar plates are comprised of several functionally distinct parts including:

- separator
- flow field
- port and port bridges
- seals
- frame

1. Separator Plate

The separator plate separates the anode and cathode sides of the bipolar plate so that gases don't mix. The two sides of the plate are electrically connected so that they are at the same potential. The separator can be a separate part, such as a metal sheet, or integral with the material, such as the web in a molded graphite plate. Graphite plates functionally separate the gases but do not usually have a discrete separator plate part. In the case of a metal separator plate, only one side of the plate is exposed to low pH water. The other side is exposed to the coolant cavity. This is an advantage in some cases such as using a clad metal separator where cladding is not needed on both sides. Type of coolant used influences the materials of construction. Aluminum, which has a cost advantage, cannot be used with a water coolant but can be used with air, silicone fluid, or hydrocarbon-ether based coolants. In some designs the separator is merely a plastic sheet or membrane

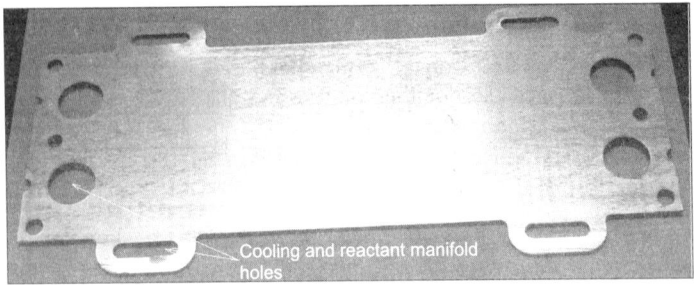

Figure 1. TiN-coated stainless-steel separator plate showing manifold holes and alignment holes.

that seals gases between the two flow fields but allows electrical connections through the sheet by hot pressing or other means. An example of a separator plate with manifold and alignment holes is shown in Figure 1.

2. Flow Field

The purpose of the flow field is to distribute the reactants across the active region (catalyzed area) of the cell (Figure 2). The flow field can be a separate part from the plate or integral with the plate. Combining parts reduces assembly time. The gas distribution channels are optimized to limit partial pressure gradients and temperature gradients that can lead to performance loss. A long channel has a higher frictional pressure drop than a short channel. A smaller cross sectional area will result in a higher reactant gas pressure loss. Deep channels are disadvantageous since the extra flow field thickness to accommodate the depth results in longer stacks requiring more material and bigger manifolds. Shallow channels are preferred but must be wider to maintain sufficient cross sectional area.

The reactant partial pressure gradients exist across any one cell from inlet to exit as well as from the bottom of the channel to the surface of the membrane. The gas must diffuse perpendicular to the flow direction, through the gas diffusion layer, in order to get at the catalyst. If the channels are spaced far apart, the gas must diffuse in the plane of the GDL as well as through the plane of the

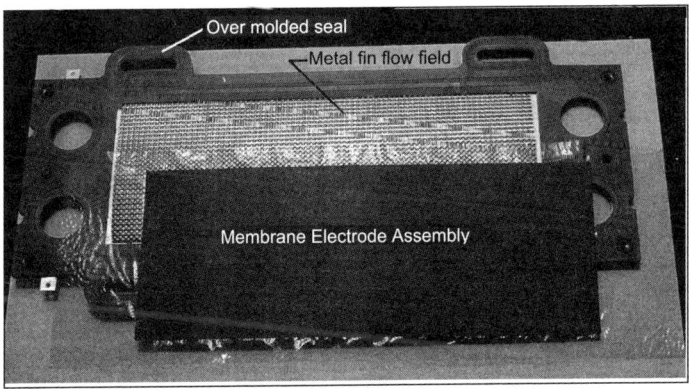

Figure 2. Metal bipolar plate assembly with lanced and offset roll fin flow field.

GDL to access the full catalyst area. Since the in-plane diffusion is a longer path length the catalyst above the lands of the flow field is called shadowed and is not as easily accessible to reactants. The rule of thumb for channel spacing is many shallow channels very closely spaced. The spacing of channel area to land area must be balanced with electrical conduction requirements through the stack.

Losses can also exist across multiple cells in the stack depending on manifold length and size. The rule of thumb here is that the cell should have an order of magnitude higher pressure drop than the manifold. The cells act like a shower head to the manifold ensuring uniform distribution to all cells. The bulk of the cell pressure drop takes place in the ports of an internally manifolded cell.

The flow field may include water management structures such as wicks. Flow channel cross sectional area changes are also used to optimize velocity gradients and entrain water droplets. The inclusion of extra components to manage water generally leads to higher assembly costs, and pressure drop (read parasite power). These considerations constitute the water management trade off.

3. Port and Port Bridges

The port is the transition from manifold to flow field. It is usually comprised of small flow channels that distribute the gas from the manifold to each flow channel or set of flow channels. The port is generally designed to be the highest pressure drop in the manifold/port/cell flow path. This limits the pressure variation across the active area.

However, this is not the case with long serpentine channels where the pressure loss is significant from inlet to exit across the face of the active area. A separate port bridge is sometimes necessary to support the thin cell membrane. The term *bridge* refers to the flat top of the port bridge with a grooved underside (Figure 3). There are at least as many port designs as there are PEM fuel cell companies. Molded plates with grooves on both sides often use a bridge structure with holes drilled in the plate (Figure 4). On one side of the plate the holes communicate with the manifold. On the opposite side, the holes communicate with the flow field grooves. The flat bridge support is needed not just to prevent the membrane from expanding into the grooves but as a sealing surface as well. It is usually not a good idea to use the GDL

PEM Fuel Cell Bipolar Plates

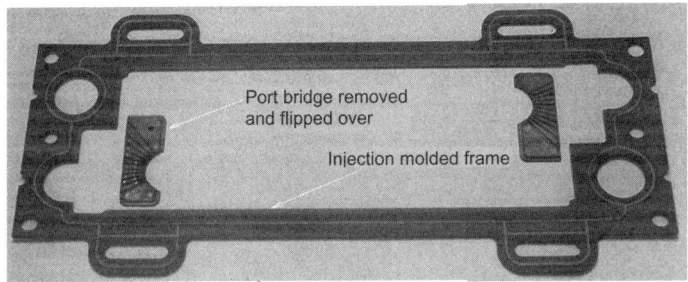

Figure 3. Plastic frame for metal bipolar plate showing removable ports and overmolded seals.

as the bridge material. It limits the seal compression since high sealing forces will crush the GDL.

4. Seals

Nothing leaks like hydrogen (except, perhaps, aqueous KOH). Seals can be flat, punched gaskets, or O-rings integral to the bipolar plate, or as separate pieces. Seal materials can be over-molded onto a grooved cell frame or molded plate. This saves insertion labor and prevents seal misalignment or movement during assembly. Seals prevent the gases and cooling fluid from mixing and keep the reactants from leaking to the surroundings. Seals can be made part of the bipolar plate or part of the MEA. In some designs, the seal is injected into a sealing manifold after the entire stack is built up. As we will show, this feature greatly improves stack assembly. The sealing load will compress rubber seals about 30%. Part tolerances can affect the compression causing crushing or gaps.

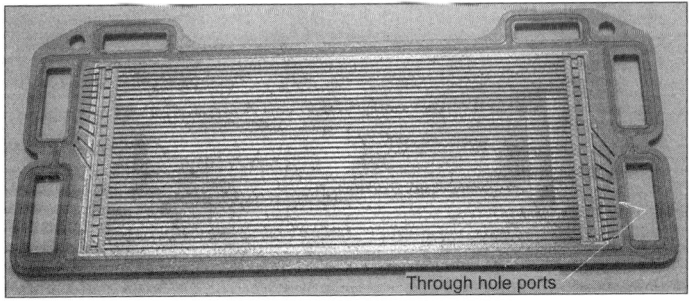

Figure 4. Molded graphite bipolar plate with *through hole* ports.

The requirement to compress seals incurs assembly problems. The uncompressed stack is 30% longer and must be uniformly compressed in order to get a low contact resistance. Misalignment of parts and seals during compression is a common problem. The problem is not trivial because it causes variations in cell to cell pressure drop resulting in cell reactant flow maldistribution.

5. Frame

A frame is required if the flow field is a separately formed part. A conductive molded material such as resin filled graphite can form the entire bipolar plate as a unit part except for the seal. Its advantage is the reduced number of parts required to assemble a stack. The disadvantage is in cost, weight and volume. Resin molded graphite requires a thick web on the order of 0.05"– 0.06". This is the minimum thickness of the plate at the depth of the grooves on both sides. This web thickness is about an order of magnitude larger than metal designs. Recent advances in molding materials have reduced this thickness. Thin stamped or formed metal flow fields often require a separate frame and/or seal. The frames are typically injection molded. A separate part requires additional seals not only to join the MEA to the plate but to join the plate to the frame.

The ratio of frame area to active area should be minimized to reduce the cost of the membrane electrode assembly (MEA). If the frame includes the manifold (called *internal manifolding*) then the manifold cross section area and frame is constant for all plates in the stack. The cost of the added material, seal, and additional membrane area is multiplied by the number of bipolar plates in the stack. This added cost can be significant. In some cases, external manifold are preferred. External manifolds are generally *boxes* fitted with an inlet or exit pipe. The *boxes* are simply held by compression against the sides of the stack where they form a seal. The shape of the external manifold can vary over the stack length to balance pressure gradients. Externally manifolded stack cost is typically much lower as well.

III. BIPOLAR PLATE FEATURES

To utilize all of the catalytic area of the cell the MEA must be well hydrated, have free access to reactants and a very good electrical contact with the bipolar plate. The heat of reaction must be removed to prevent temperature gradients and loss of water from the membrane hot spots.

1. Tolerances

Tolerances are critical in a fuel cell stack. The higher the stack voltage (number of cells) the more important tolerances are. A tolerance of +/– 0.002 inch on all bipolar plate components can add up to 0.408 inches in a 48 volt (nominal) stack if all parts are at the maximum plus tolerance (assuming three parts to each bipolar plate). Part flatness is just as important as overall thickness. If one portion of the frame is consistently thicker than the rest, the seal must make up the difference. In an injection over-molded seal, 2 mils (50 microns) can be 15% of the seal thickness. The result is under-compression in the thin parts of the frame. Applying higher compression loads to improve seal only results in crushing of the frames and often the MEA and GDL as well. In some cases, the thin frames will crack. Separate flow field and frame assemblies make this problem even more complex. An advantage of unitized plates is that the plate can be surface ground to a uniform thickness. Good plate tolerances will not only ensure sealing but will improve conductivity.

2. Thermal Management

The operating temperature of the cell is set by choice of MEA. In the case of Nafion cells, the MEA conductivity is intimately associated with the water content of the membrane. Since there is no vapor pressure suppression of membrane water by the electrolyte (which is covalently bonded to the membrane solid phase), the amount of water in the membrane is a function of the cell temperature, pressure, reactant water concentrations and velocity, and the current density. The relationship is difficult to analyze and even more difficult to control. To remove the cell waste heat, a single-phase coolant must absorb the waste heat as sensible heat. Two-phase coolants can absorb the waste heat as latent heat. Of course as the reactant gas streams gather moisture traversing the cell, a temperature gradient is helpful for removing waste heat and moisture. This can be closely coupled to the bipolar plate material. If the material has a high thermal conductivity then the inlet to exit plate temperature is minimized by axial conduction. One can readily see how complex thermal management becomes when one realizes that the flow through the cells in a stack is only approximately equal and that the transients in power and the performance characteristics of the fluid movers are coupled into the problem.

The in plane thermal conductivity of metals allows a small temperature difference between the inlet and the outlet of a (rectangular cell). This axial conduction can be used advantageously.

3. Electrical Conduction

In addition to the functions described above, the bipolar plate transmits the electrons from the cell anodes to the adjacent cell cathodes perpendicular to the plane of the plate, or *through plane*. Monopolar cells transmit electrons *in the plane* of the cell requiring a minimum bipolar plate thickness and zero conductivity. This type of cell is rarely used because the current that can be collected from a monopolar cell is only about half that of a conventional bipolar cell for a given cell voltage. The intrinsic conductivity in a bipolar cell is generally not a concern. The interfacial conductivity between cell components is a function of the stack contact pressure and contact area and is very important. High thickness and flatness tolerances are needed to maximize contact area since the parts are not soft and deflection of carbon can lead to crushing and increased diffusion losses. Intrinsic conductivity in bipolar arrangements is only important when implementing performance enhancement techniques that require in-plane conduction to promote cathode catalyst reduction.

Electronic conduction in the GDL and the flow field influences flow field design. Just as the reactant gas must diffuse in the plane of the GDL to get to the shadowed catalyst, the electrons must move in the plane of the GDL from the lands of the bipolar plate to the region over the grooves in order to supply the catalyst with electrons. Several design features can enhance the in-plane conduction. Narrow channels and lands reduce the conduction path length. Increased surface area contact between the land and the gas diffusion layer can be done by micro-texturing the material in contact with the GDL. Improving the in-plane conductivity of the GDL using micro-layers, or conductive fillers is now common. The micro-layers are either a separate thin layer placed between GDL and MEA or carbon/Teflon filler material that is printed onto the carbon paper GDL and sintered.

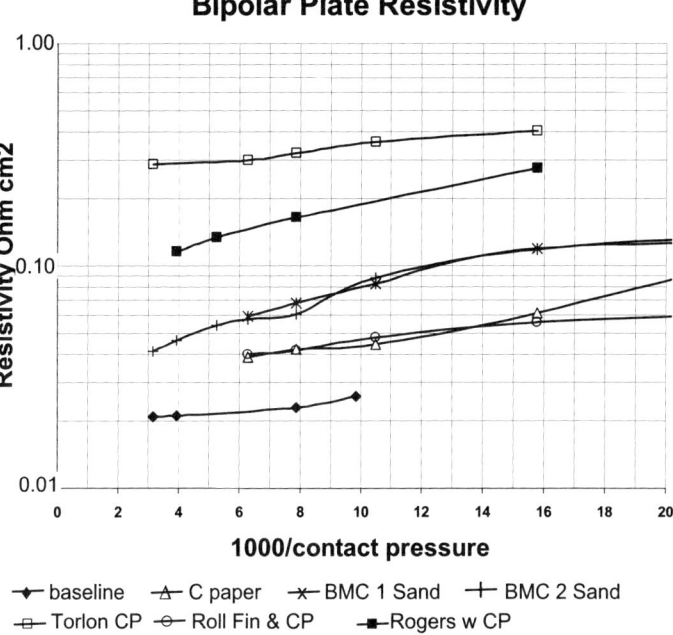

Figure 5. Comparison of bipolar plate resistivity.

Figure 5 shows the resistivity of several bipolar plate materials. The lowest resistivity on the chart is the baseline which represents the test hardware. All other curves include the baseline hardware resistance. The BMC compounds are moldable graphite and show moderate resistance. The highest resistivity was shown by Torlon, a composite TiB filler embedded in a plastic resin. Roll fin is a lanced offset metal flow field and CP means carbon paper. Samples are often tested as a sandwich with carbon paper on each side to representing the bipolar plate arrangement.

4. Water Management

The bipolar plate design is an integral part of the PEM cell water management. Product water vapor that is not carried out with the cathode exhaust will condense, forming droplets in the gas channels and adversely affecting reactant distribution. The droplets are often not swept out because the Reynolds number in the flow field is ~16. In the case of condensation, flow distribution is dynamically unstable and CFD analysis on flow distribution becomes moot. When a reactant gas channel becomes clogged with a water droplet, the gas flow preferentially bypasses the clogged channel and diverts into unclogged channels. However, the same amount of current is passing through the clogged and unclogged regions of the cell. The reactant gas supply in the clogged region is reduced meaning that the clogged region of the cell operates at a higher utilization and the polarization increases. This causes the clogged region of the cell to run hotter which may clear the channel. Hot, dry inlet gases which could carry away additional product water will only dry out the inlet region of the membrane leading to gas crossover and failure. Liquid product water at the exit is undesirable yet dry gases at the inlet are also undesirable. All the water that is needed is produced by the cell yet it is in the wrong location. Some designs recycle this water internally by passively absorbing the liquid at the exit and providing humidification at the inlet. A porous bipolar plate which is in liquid communication with cooling water fluid can accomplish this.[26,27] Care must be taken to balance the pressure between the two cavities to maintain a capillary seal. This is one of the most reliable methods since the liquid will be absorbed wherever it may condense. Other designs recycle the water externally but the water must first exit the bipolar plate by other means. This typically requires additional energy and cost. For example, the enthalpy wheel[28] recycles the moisture from the cathode exit back to the cathode inlet without recycling the spent gases. It is a regenerative heat and mass transport device. Moisture transfer membranes are also used for this purpose.

There are six basic methods for water removal. They are concentration gradient, capillary forces, electro-osmosis, purging (causing high gas velocity for droplet entrainment), thermal gradient (evaporation) and surface tension gradient. These methods fall into two categories: active and passive. An actively managed design removes water from the cathode gas space using controls such as increased air flow, gas purging, decreased humidity (a form of concentration gradient), reduced coolant flow (thermal gradient), etc. While most concentration gradient methods focus on humidity and temperature, a concentration gradient can be set up between the cooling fluid and the cell reactants. A membrane separator is required to limit gas exchange but allow liquid exchange. A separate circulation loop is needed to control the concentration of the circulating fluid. A passively managed design removes water without the need for system components such as described above.

There are many approaches to removing liquid from the flow field region besides the porous bipolar plate. One approach removes the water in droplet form using a hydrophobic surface. Another approach evaporates the water and spreading the liquid film out becomes desirable. For film spreading, some techniques apply micro-texturing[34] to the base of the groove to help spread the water and enhance evaporation. Others include application of polymers to absorb the liquid. Moving the water in droplet form can be accomplished both passively and actively. In most applications gas pressure is used to propel the droplets out the exhaust. There is another approach that applies a surface tension gradient. This method alters the contact angle on the leading edge of the droplet which propels it forward. Thus the droplet moves without any additional energy or system controls or components. In any event, removing the water as condensate increases the cooling load on the cell stack since the water latent heat is released within the envelop of the stack.

5. Low Cost

Bipolar plate cost is a function of material and forming process. The forming process cost is a function of production rate and number of process steps, as well as the tooling costs (dies, etc.). In the case of large production volumes, one must consider the cost of forming equipment such as presses, extruders, and real estate. The cost is also affected by the number of parts, especially in the case of composite plates. Often the process that is optimum with respect to small production volumes is not optimum in the case of a large production volume.

6. Stable, Free from Corrosion Products

(i) Galvanic Corrosion

The relatively low hydrogen evolution over voltage of carbon causes most metals in contact with carbon and water, especially low pH water, to corrode anodically. Shunt currents in the coolant manifolds of carbon bipolar plates are problematic and, if used, the coolant water must be carefully deionized.

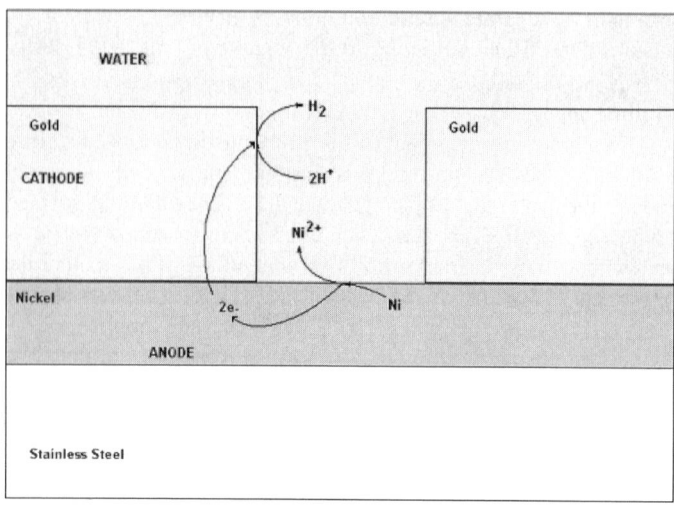

Figure 6. Galvanic corrosion mechanism.

The problem of galvanic corrosion is shown in Figure 6. Here a gold plate has been deposited on a nickel *strike layer*. The noble metal (gold) becomes cathodic and the base metal (nickel) becomes anodic. The figure shows a pore in the gold plating that is filled with water. The mechanism shown is peculiar to acidic media. Neutral or basic media mechanisms are somewhat different, but nickel is more resistant to corrosion in alkaline or basic media. Both gold and nickel have a low hydrogen evolution overpotential. When two different metals are immersed in an electrolyte, each establishes its own corrosion potential where the rates of the anodic and cathodic reactions are equal. If the corrosion potential of the materials are significantly different, and they are electronically connected, the metal with the more noble potential will become cathodic and the other metal will become anodic. A corrosion current flows between the anode and the cathode. The anode corrosion rate will be increased and cathode corrosion rate decreased or entirely stopped. The extent of galvanic corrosion depends on the area ratio of the two metals, the cathodic efficiency of the more noble metal, and the difference in corrosion potential between the two uncoupled metals. For a given current flow in a galvanic cell the current density is greater on a small electrode than on a larger one, a small anode will have a greater current density and hence a greater corrosion rate than a large anode. The principle of galvanic corrosion is used effectively in the use of *sacrificial anodes* and is applied in protecting pipelines and ship propellers from corroding. In fuel cell bipolar plates, galvanic corrosion is the principal reason why metal bipolar plates are problematic. The other is oxidation.

IV. MATERIALS AND PROCESSES

1. Comparison of Carbon and Metal

There are two material approaches to provide the conductive portion of the bipolar plates, namely metal and carbon. Graphite has a distinct advantage because it doesn't corrode. However, it causes the corrosion of other materials. Metal plates have taken on a renewed significance because of the work of Uribe and Zawodzinski[10,11] who describe an operational alternative to 40 years of unsuccessful electrocatalyst research. Their method is to lower the potential of platinum catalyst for a brief period. Platinum forms an oxide at high potential.[11,29,30] Lowering the cathode potential reduces the oxide to the metal, which has a much higher activity for oxygen reduction. Bernie Baker, the founder of FuelCell Energy, often said that

> "…platinum alloy catalysts start out as binary or ternary alloys, but they all wind up as platinum."

It is a prescient comment in more ways than one. The system solution suggested by Zawodzinski and Uribe can be implemented in a manner invisible to the load – but it requires a high in-plane conductivity in the bipolar plate. The method, proposed by H. Baker in work at Analytic Energy Systems, requires high conductivity, metal bipolar plates. The advantage is operating the cells at potentially higher efficiency because the cathode may operate at a (net) higher potential. Meyer and Darling[29,30] show that the platinum oxide formed is not restricted to the surface. Periodically cycling the catalyst may be used to stem corrosion of cathode catalysts by restricting the oxidation of the Pt to a thin surface layer.

In most other respects carbon and metal bipolar plates are roughly equivalent. Carbon bipolar plates are often bulkier than metal, but flow fields and current collection incur more design problems than material problems.

(i) Operational

Operationally, metals can provide a more uniform cell temperature because of high in plane thermal conductivity. There is also a reduced thermal gradient through the plane of the bipolar plate due to reduced web thickness between coolant and reactant cavities. During operation a graphite stack which is generally solid graphite in the central and perimeter region will experience a higher heat loss from the surface than a metal stack, which usually has insulating material frames. The overall effect is that the reactant temperature in the graphite plate manifold can vary, altering the reactant temperature to each cell, and affecting performance and water management. To limit surface heat loss the graphite bipolar plate stack must be insulated.

(ii) Forming Cost

One of the reasons for pursuing polymer coated metal bipolar plates is low manufacturing cost of metal bipolar plates. Table 1 shows a comparison of costs for manufacturing stamped steel plates, stamped and extruded aluminum plates, molded graphite plates and injection molded graphite plates.

The stamped metal bipolar plates have the lowest projected cost. The molded graphite plates have a high cost because of the low production rate. The dramatic improvement in the injection molded graphite bipolar plate is shown to be due to the much reduced amount of material required as well as the greatly reduced mold time which has dropped from 8 minutes to 45 seconds. This still compares unfavorably with the stamped metal approach that can produce about 30–40 plates per minute. Modular stamping dies for a 5 in x 11 in plate cost about $25,000. When production cost is low and production volume is high, if the capital cost of the forming machinery is high it is more economical to contract the production rather than purchase the press. The production rates of stampings are very high and would not justify the capital cost of a press unless production volumes were extraordinarily high. This is also true of extruded aluminum. The production rates would have to be much higher than 10,000 stacks per year. Extruded aluminum bipolar plates are not competitive in cost because they require twice as much aluminum as stamped plates.

Injection molded plate frames and O-ring seals are exceptionally expensive. All the plates were costed on the same

Table 1
Cost Comparison for Bipolar Plate forming Technologies.[1]

Weight and volume	316 SS Stamping	Aluminum stamping	Aluminum stamping	Molded graphite	Injection molded	Units
Total weight	0.231	0.099	0.202	0.481	0.077	lb/plate
Material cost $/lb	$ 1.80	$ 0.78	$ 0.78	$ 3.50	$ 3.50	$/lb
Subtotal material	$ 0.42	$ 0.08	$ 0.16	$ 1.68	$ 0.27	$/plate
Frame and seal or "O" ring	$ 1.30	$ 1.30	$ 0.05	$ 0.50	$ 1.50	$/plate
Material cost per plate	$ 1.72	$ 1.38	$ 0.21	$ 2.18	$ 1.77	$/plate
Forming cost 10,000 units/year						
Total time	0.53	0.53		8.95	1.12	min
Labor rate	50.00	50.00		50.00	50.00	$/hr
Press capital cost	$ 0.01	$ 0.01		$ 0.43	$ 0.51	$/plate
Per plate labor	$ 0.44	$ 0.44		$ 7.46	$ 0.93	$/plate
Subtotal forming cost	$ 0.45	$ 0.45	$ 4.45	$ 7.89	$ 1.44	$/plate
Plate finishing cost						
Coating materials	$ 0.90	$ 0.90	$ 0.90			$/ft^2
Corrosion coating cost materials	$ 0.34	$ 0.34	$ 0.20			$/plate
Coating application cost	$ 0.05	$ 0.05	$ 0.05			$/plate
Planarization cost				$ 0.35		$/plate
Subtotal	$ 0.39	$ 0.39	$ 0.25	$ 0.35		$/plate
Total manufactured cost	$ 2.56	$ 2.22	$ 4.91	$ 10.42	$ 3.21	$/plate
Stack production cost						
Plates/stack	136	136	136	136	136	
Cost/stack	$ 348	$ 301	$ 668	$ 1,417	$ 440	
Weight/stack	31.42	13.46	27.47	65.45	10.55	

basis using an internal manifold approach, with the exception of the extruded aluminum plate which can only be used with external manifolds. If external manifolds are used then the material costs of stamped aluminum plates drops from $1.38 to $.08 each. There are other reasons for using external manifolds.

2. Carbon

(i) *Molded Graphite*

Carbon is often chosen as a bipolar plate material because it can survive in the PEM fuel cell environment where most alternatives corrode. There are several approaches to fabrication (Figure 7). The earliest was to use compression molding of a mixture of graphitic powder and thermoset resins, usually phenolic or polybutadiene. More recently injection molded mixtures with thermoplastic polymer binders have been used. While the injection molded types have a much faster fabrication cycle than the compression molded units, the electronic conductivity is much lower. The compression-molded plates sometimes have an excess of plastic at the surfaces. This must be ground away, but the process is inexpensive. The molding process uses an inexpensive

Figure 7. 2-kW molded graphite plate stack.

starting material. However the process of compression molding and subsequent cooling is a time consuming process.

(ii) Paper

More recently, carbon paper plates have been fabricated. These plates are formed using a mixture of carbon fibers either pitch or polyacrilonitrile (PAN) based fibers and thermoset resin. The materials are graphitized, and because of the expensive curing process and the shrinkage of material, the starting material cost is not cheap. Fairly thick carbon paper, on the order of 50 mils, costs roughly 25$/$m^2$ in quantities of about 1000m^2. The flow field grooves can be compression molded prior to pyrolysis of the resin or machined (Gang Sawed, milled, etc.). after the graphitization step. The plate/flow field is made in two or three parts, one each for the anode, cathode and coolant cavity. Often it is useful to render the surfaces of the carbon paper hydrophilic by evolving oxygen from the surface of the plate. The edges of the plates may be sealed by polymers or elastomers impregnation or can they can be used in conjunction with a solid frame.

(iii) Stamped Exfoliated Graphite (Grafoil, Graflex)

Another method of forming bipolar plates uses stamped Grafoil or Graflex exfoliated graphite. The materials and processes are inexpensive enough for exfoliated graphite to be used as a head gasket material in automobiles. This material combines the ease of fabrication of metal with the corrosion resistance of graphite.

3. Metal

General Electric ran three PEM fuel cells for 15 years.[32] The cells used Niobium bipolar plates. Metal bipolar plates (Figure 8) offer the potential advantages of very low fabrication cost and high in plane conductivity. Metal bipolar plates, especially low cost metals such as aluminum, are attractive because of intrinsic low cost, high electrical and thermal conductivity, and high speed manufacture (stamping). Many stack fabricators, including Analytic Power, have investigated intrinsically corrosion resistant metals.[33] These are the valve metals (Ti, Nb, Hf, Zr, Ta) and stainless steels. The valve metals, with little or no aqueous chemistry are ideal for fuel cells, such as titanium, but all tend to form oxides. Niobium is

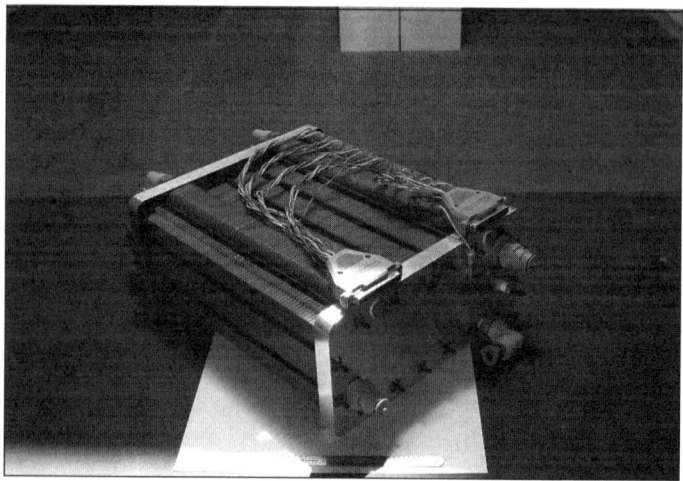

Figure 8. 2-kW fuel cell stack with metal bipolar plates.

attractive because its oxide is conductive, but it is expensive. Many corrosion resistant coatings have been tried. In stainless the chromium forms a conductive oxide. The coating processes most frequently discussed are: plating, cladding, thermal deposition and in situ polymerization of composite coatings.

The uncoated metals used are Niobium, stainless steels and high performance alloys for intrinsic corrosion resistance. The coated metals are generally aluminum or copper. To date coating appears to be on the verge of success. Many of the major automotive fuel cell manufacturers have patents on metal coatings. Coatings are applied directly or as mixtures of polymers with conductive particles. There is a large patent literature on the subject of fuel cell metal bipolar plates.

(i) *Forming Metal Bipolar Plates*

The metal forming processes most frequently used for forming the plate are punching and stamping. More rarely used approaches are injection molding of metal filled plastic composites. Aluminum can also be readily extruded (i.e., ladders). The insulating frames used with metal bipolar plates are injection molded and seals can be injection molded or elastomer bonded to the metal surfaces. Flow fields used with metal bipolar plates can be rolled or folded fin, generally in a lanced and offset pattern.

(ii) Intrinsically Corrosion Resistant Metals

PAFC studies conducted since ~1967 examined many high nickel alloys. A great deal of work has been done in the past at LANL and by Allegheney Ludlum on ferritic steels. None of these materials have shown promise with sulfonic acids at low temperature or with high temperature PBI/H_3PO_4 cells. Several metals are being tested in conjunction with Polymer/Phosphoric Acid based MEAs. Some are reported to be corrosion resistant. Usually one metal is corrosion resistant at the anode and another at the cathode. The more promising materials are:

- GenCell Corp stamped material - 310 stainless
- Hastelloy C-276
- Super Alloy HASTELLOY® C22® alloy
  - Inconel 622
  - Thyssen Krup metals Nicrofer 5621 hMoW (tm)
- Allegheny Ludlum Alloy 59
  - Thyssen Krup metals NiCrofer 5923
- E-Brite Ferritic Stainless Steels

(iii) Direct Coatings

As we mentioned several nitrides, carbides, oxides, silicides and borides of titanium, tantalum, chromium, lanthanum, zirconium, tungsten and niobium are the most frequently examined coatings. Some of these materials have high conductivity and are corrosion resistant. The structure of most of these materials is cubic, except the borides that are hexagonal and have the highest conductivity. Titanium carbide (TiC) coating is a well-developed technology and is widely used for coating cutting tools, ball bearings, etc. In MCFC's, TiC and TiN coatings on separator plates have been used for their corrosion protection properties. Some of the methodologies adopted for metal bipolar plate coating are listed next.

(a) Plating

Plating is generally unsuccessful because the plating is invariably porous. Hole free coatings are so thick that they are prohibitively expensive. Since the anodic current density (exposed holes) is much larger than the cathodic (plated portion) current density in a porous plated metal bipolar plate, the substrate metal rapidly develops pits. Electroless plating is occasionally used. The coatings are not generally as conductive because of phosphorus inclusions (up to 12%) in the coating.

(*b*) *Cladding*

Metal cladding approach costs about 14\$/ft^2. Niobium cladding does provide excellent corrosion resistance. The material cannot be cut after cladding without exposing the interior metal. The cut regions have to be protected from corrosion that limits design options.

(*c*) *Vapor deposition plasma, jet spray, CVD, laser ablation, LAFAD*

The latest of these processes is large area filtered arc deposition (LAFAD) methods. The process deposits dense coatings at low rates and then can put down high coating rates of medium density. Some of these coatings are NiCrN on aluminum (MER Corp.). Some have been nitriding NiCr alloy substrates (ORNL) with early success. This is not a coating process but a thermal nitriding process which reacts the surface NiCr alloy with nitrogen forming NiCrN. In general Ni should be avoided. Ni and many other polyvalent cations will catalyze the formation of peroxide at PEM anodes. These materials which generally slowly leach from alloys have very little effect on the MEA resistance until they accumulate in relatively large concentrations. The formation of peroxide has a steady degrading effect on the cell limiting current density. The peroxides destroy the thin membrane layers formed around catalyst particles.

(*d*) *Flame spraying*

TiC has been successfully sprayed on relatively thick metal parts. In our experience it makes a moderately dense coating. It is difficult to apply to thin metal plates without warping because of the very high application temperature.

(*e*) *Thermal vapor deposition*

Thermal vapor deposition of TiN has been a well known method of coating titanium, and stainless steel tools for many years. It has not, to our knowledge, been successfully employed as a corrosion proof coating. Analytic Power extensively explored this metal in the mid 1990's. The corrosion results obtained at UMinn[35,36] were not encouraging.

(f) Thermally activated CVD

$TiCl_4$ is used with CH_4 as precursor to synthesize TiC in a CVD chamber. Deposition temperatures of over 900 °C are required for the process. Austenitic steels (SS 304) do not show phase transformation between room temperature and 1000 °C, hence can be used as substrates for the above process. However, the process temperature is too high for most steel substrates.

(g) Fast CVD

In this method, acrylic acid and $TiCl_4$ are used as precursors for the formation of TiC. Use of acrylic acid (mp = 140 °C) helps in reducing the temperature of operation to 800 °C. High deposition rates are possible with this method (5 μm in 5 min) but the method often leaves graphitic inclusions in the deposit.

(h) Thermal laser assisted CVD (LCVD)

In this process a non-focused stationary CO_2 laser beam is used to deposit thin films of TiC from a gaseous mixture of $TiCl_4$, CH_4 and H_2. Hydrogen gas promotes the reduction reaction of $TiCl_4$. The P_{TiCl4}/P_{CH4} controls the stoichiometry of the TiC film.

(i) Reactive ion beam-assisted electron beam-physical vapor deposition

In RIBA, EB-PVD titanium rods are evaporated by electron beam and the chamber is maintained under a constant acetylene/argon flow. The temperature of the TiC synthesis is relatively low in this process, compared to the CVD process. But this technique does require sophisticated equipment to achieve accurate stoichiometry.

(j) Reactive plasma spraying (RPS)

Titanium powders are used as starting material in the plasma and nitrogen is used as carrier gas. The chamber has the ability to spray various reactive gases (CH₄ for TiC films). Coating porosity is an issue with this technique.

(k) *Pulsed laser deposition (PLD)*

Laser assisted deposition has attracted attention for obtaining tailored coatings. There exist several types of PLD methodologies to achieve TiC coatings. The process is expensive and is used in the electronics industry.

(iv) *Conductive Polymer Grafting*

Conductive, corrosion proof conductive polymer coatings have been explored by several companies, notably Ineos Chlor. The use of composite polymer coatings has been studied at Analytic Power for several years. Chemical grafting either alone or combined with siloxane bonding has, potentially, the lowest coating cost. If it can be made durable, then it will be the coating of choice in PEM cells. Analytic's coatings, while not yet providing complete corrosion resistance, cost about 32 ¢/ft². Figure 9 shows the early results of testing an aluminum sample submerged in pH 2.5 methane sulfonic acid for two weeks. The test shows the effects of permeability in the plastic coating.

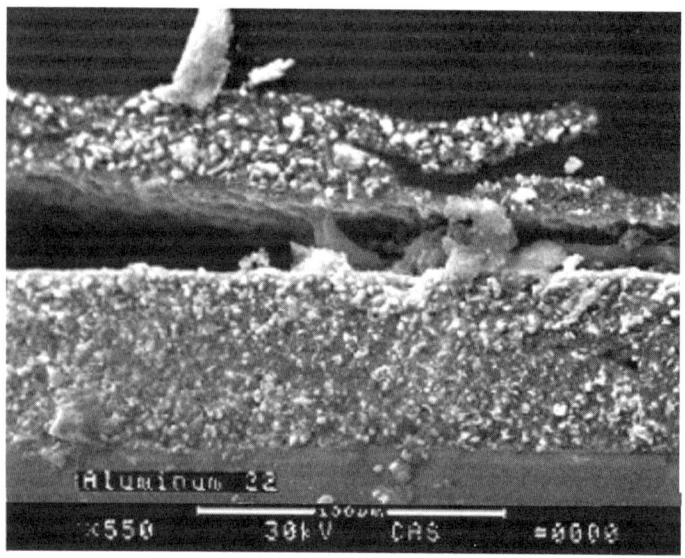

Figure 9. Early TiC/polymer coating on Al post test.

The aluminum bipolar plates coatings that Analytic Power prepared and tested were highly conductive and tough. Their principal weaknesses were that they are too permeable to water and the bond between the polymer and the aluminum substrate and the conductive particles is weak. The problems are interrelated. The coatings use:

- aluminum coupling agent: silanes – precursors to the siloxane aluminum bond
- polymer matrix: epoxy/acrylic, and imide monomer and pre-polymer blends
- accelerators: accelerate the reaction between catalyst and a resin A
- crosslinker: hardens polymer matrix films
- graft initiator: silver ion links monomers and prepolymers to metal substrate (and filler) via chemical graft.
- catalyst regenerator: peroxides (silver ion catalyst)
- supporting solvents: methyl pyrrolidone, MEK, butanol, butyl acetate, water
- conductive filler: titanium carbides and borides

While the number of components in the coating seems large, the coatings are relatively inexpensive ~$0.32/ft^2. The bulk of the cost is invariably in the conductive filler. The filler can cost 70% to 82% of the total coating cost.

(*a*) *Aluminum coupling*

Two processes for covalently bonding polymers into inorganic solids and metals are used. The first treats the surface with a silane and the second is chemical grafting with free radical initiated polymerization. Chemical grafting can be used with bare aluminum, with a silane surface coating, or both. The two processes provide covalent bonds to conductive particles which are then linked into a polymer matrix. Bonding to hydrated or hydroxyl coated surfaces generally yield superior coatings, and is well known. Bonding to carbides and borides are not well characterized. This bond is an important problem to be resolved, probably by increasing the number of polymer links into the inorganic/metal surface.

Silanes. The silane coupling agent forms a link between the inorganic, aluminum substrate and the polymeric structure.[7] It forms the binder of the conductive coating. Unlike simple physical bonding, silanes enable an adhesive or sealant to chemically bond to a metal. The covalent bond is less susceptible to the effects of moisture and temperature than physical bonds.

The nature of the substrate surface plays an important role in achieving good adhesion. The more chemically active sites the substrate has—preferably hydroxyl groups—the better the adhesion will be. The use of a potassium hydroxide treatment of the aluminum substrate to enhance the surface reactivity and number of sites is discussed in the section on fillers. Further improvements in adhesion may be achieved by applying a silane primer prior to adhesive or sealant application.

A potential problem area with the use of silanes is described by van Ooij:[14]

"As the Si–O–Al bonds are known to hydrolyze, the corrosion protection is related to the hydrophobicity of the siloxane films formed on the metal substrate."

Others[15,16,17] have also discussed acid catalyzed hyrolysis of silanes and how the dimethoxy silanes are more stable than the trimethoxy silanes. One of our concerns is that when the silanes are used in conjunction with the other coating components, the kinetics of reaction between the silane alkoxy groups and these components might be faster than the reaction rate between alkoxy groups and the metal surface.

Analytic Power uses a commercial phenylaminosilane (Siliquest Y9669), shown in Figure 10. It combines phenyl and amino functionalities which are reactive with resins such as epoxies, acrylates, isocyanates, phenolics and silicones. For optimal adhesion, the silane functionalities must be matched to the type of polymer matrix used and the physical properties desired.

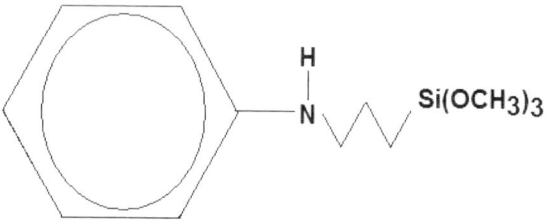

Figure 10. Silquest Y9669 silane (phenylaminosilane).

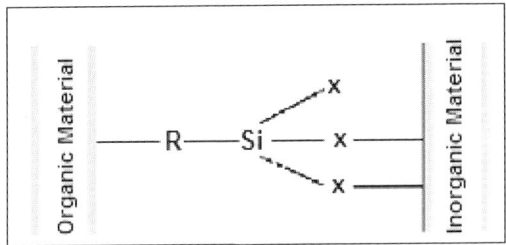

Figure 11. Silane bonding mechanism.

The amount required is determined empirically, but one percent (1%) by weight is the usual starting point.

Figure 11 illustrates the bonding mechanism characteristic of Silanes. The "R" is the amino group shown in Figure 10. The amino group reacts with the epoxy and acrylate groups. The "x" corresponds to the three methoxy groups in Figure 10. These groups bond to the inorganic material, in our case, aluminum and particulate fillers. The methoxy groups hydrolyze in the presence of water to form metal hydroxide, siloxane bonds. It is known that silanes bond well with silica, alumina, glass quartz and aluminosilicates. It is less effectively bonded to talc, hydrated clay, and iron powder. It appears to bond with asbestos, titanium dioxide and zinc oxide. Its bonding activity with titanium carbide or boride is not described in the literature.

Analytic Power has also used a commercial β-(3,4-epoxycyclohexyl) ethyltriethoxysilane, commercially, CoatOSil 1770. It is both an aluminum coupler and an epoxy functional crosslinker. It is useful with epoxy films since it can link directly with the Bisphenol A pre-polymer. This silane is used principally with water borne resins and the toughness of the film can be enhanced by lowering the pH of the solution.

Chemical grafting. Polymer grafting is also a well-known technology that produces covalent bonds between monomers or pre-polymers and dissimilar surfaces. Frequently it involves activating monomers with plasma, corona discharge, x radiation, or sputtering. Dissimilar monomers can be polymerized by this process. For example, it is used in Ballard's BAM[9] membrane and by RAI Pall for membranes of general industrial interest. Analytic's chemical grafting uses silver ions and free radicals to initiate bonding to a surface. About ten patents cover the chemical grafting process. The process of grafting to aluminum surfaces consists of the following steps:

- Surface activation. The graft initiator, generally a silver nitrate or perchlorate, removes hydrogen from the surface to be treated, in our case a hydrous aluminum oxide film on an aluminum bipolar plate. The hydrated aluminum (R) oxide in (1) reacts with a silver ion in solution to form an aluminum oxide radical, to produce a reduced silver and a proton. The aluminum shown in (2) reacts with a vinyl to produce an ether link between the surface of the aluminum and a monomer; in the case of reaction (2), a vinyl group:

$$ROH + Ag^+ \rightarrow RO^* + H^+ + Ag \quad (1)$$

$$RO^* + n(CH\ X = CH_2) \rightarrow$$
$$RO(CH_2 = CHX)-(CH_2 = CHX^*)_{n-1} \quad (2)$$

Grafting and polymerization The grafting reaction with vinyl groups in Figure 12 is for illustrative purposes. This monomer may be replaced by epoxy, isocyanate or other groups. In our case the important group is epoxy. The (X) in the reaction (2) represents pendant groups. The polymerization reaction can be terminated by coupling with a monomer of its own type or with a radical formed by reaction with silver ion. As shown in Figure 13

Small amounts of peroxide in solution act to oxidize the silver (4) and regenerate silver ion providing more free radicals to initiate polymerization:

$$ROOH + Ag \rightarrow RO^* + OH^- + AG^+ \quad (3)$$

$$RO^* + nCH_2 - CHX$$
$$\rightarrow RO\ (CH_2\text{---}CHX)_{n-1} - C^*H_2\text{-}CHX \quad (4)$$

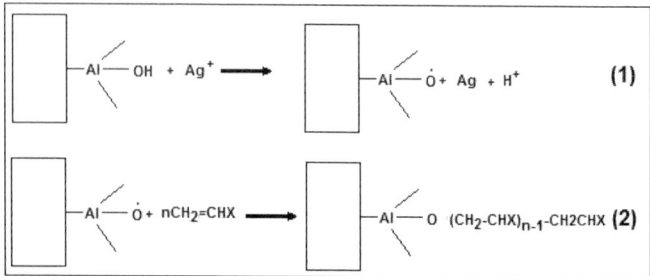

Figure 12. Substrate activation and radical initiated polymerization.

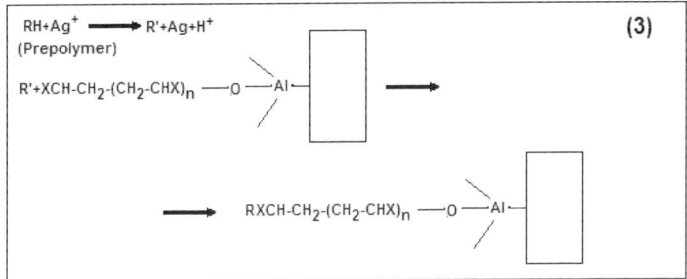

Figure 13. Polymer termination.

This approach was patented by Carl Horowitz et al.[2,3] for chemically grafting polymers into aluminum surfaces and non-metallic surfaces.

(b) Polymer matrix

The second component of the coating is its polymer matrix. The function of the polymer matrix is to bind the conductive filler particles and link to the silane and chemically grafted surface. Analytic Power has used thermosetting plastics for the polymer matrix because they are amenable to chemical grafting and free radical polymerization. The coating is polymerized in situ from a solution of monomers and prepolymers. The solution can contain imides, a blend of epoxy and acrylic, or all three.

Epoxy.[4] Questions have been raised concerning the hydrolytic stability of epoxies. This is a concern since it may be the cause of the relatively poor bond between the polymer coating. Professor Wicks[6] provided some comments in this area:

> "Epoxy based coatings (are) pretty much uniformly accepted as having the best hydrolytic stability provided they do not have a polyester component to them. The bonds formed in preparation of epoxies from amine curing agents are - alkyl amines, not something prone to hydrolysis and the backbone of the epoxy only presents ethers and carbon-carbon bonds, again not something prone to hydrolysis. Bisphenol A epoxide systems are found in the most aggressive environments for hydrolysis—concrete coatings, marine coatings, water treatment plant coatings, cathodic electrodeposition systems, etc."

Others[12] indicate that electrostatically sprayed epoxy coatings are not chemically stable enough for use as protective metal coatings. Water transport is a complex process in epoxy composite films and is characterized by activated diffusion rather than a Fickian process.[13]

In polyester and epoxy, absorbed water causes a depression of the T_g of plastics and a concomitant a loss of other performance properties.[5] DGEBA (diglycidyl ether of Bisphenol A) cured with TETA (triethylene tetramine) is sensitive to absorbed water. A promising approach uses thermally cured bisphenol A resin, a high molecular weight pre-polymer (Araldite GZ488 N40) with high hydroxyl content that permits crosslinking with melamine and phenolic formaldehyde. Degradation of amine cured epoxy resin occurs by the formation of microcavities and is discussed in several references. The amount of damage depends on temperature and humidity levels.

Based on the literature, and discussions with investigators, indicate the blending of acrylics with epoxies imparts hydrolysis resistance to the coating. The epoxy coatings bond well to the particles in the coating but are weakly bonded to the substrate as evidenced by their delamination.[8] This problem may be part of the silane hydrolysis problem previously discussed.

Polyimide. Polyimides are potentially useful polymer matrix materials. Like the epoxy films, the imide is polymerized by evaporating the solvent. It is crosslinked with a hexamethoxy melamine (Cymel 303) and a butyl etherified phenolic. Polyimide has higher water absorptivity than epoxy which may render it unsuitable for long term applications. It also shows good bonding to the particles but a poor bond between the coating and aluminum substrate as evidenced by pitting corrosion.

(*c*) *Accelerators*

Accelerators are used to speed the reaction between catalysts and resins. They are also used in conjunction with the hardeners to aid the curing process. Epoxy hardeners can be amines or phenolics. We have used a strong aromatic sulfonic acid (Cycat 4040) to accelerate the curing of melamine-formaldehyde crosslinkers and to reduce the required cure temperature.

(d) Crosslinkers

Several crosslinkers are employed and these are used to harden the films and make them less permeable to water. Melamine prepolymer crosslinkers are used with epoxy films and a phenolic prepolymer, with polyimide films. Trifunctional methacrylate ester can be used in epoxy and polyimide film. While this methacrylate monomer also provides additional metal adhesion, it's use might be counter productive in view of Prof. Wicks[6] comments. The β-(3,4-epoxycyclohexyl) ethyltriethoxysilane (CoatOSil) serves as a crosslinker when heat is applied.

(e) Graft initiators and regenerators

The graft initiator initiates the coupling of the polymer to the metal. The initiator is a silver ion (a nitrate or perchlorate in origin). We have listed the graft initiator separately since it is also instrumental in linking the epoxy or polyimide film into the silane treated metal or conductive particle surface. In this case, the silane we are discussing is the phenyamino silane, Silquest Y9669. The graft initiator can link prepolymers (both epoxy and polyamide) into the amine group on the silane. As discussed in the aluminum coupling section, the metallic silver is reoxidized (regenerated) to its active state by small amounts of peroxide in solution.

(f) Solvents

The supporting solvent can be: an organic, MEK, Xylene, toluene; an alcohol or water; or an intermediate such as N methyl pyrrolidone. The function of the solvent is to suppress film formation and reactions between the film forming constituents and act as a carrier in the coating application process. Film formation is initiated by the evaporation of solvent and completed with the application of heat. In spray application, the evaporation of the solvent starts with the application of the coating. Fast evaporating solvents, such as butyl acetate, are often included in the film, to speed drying. The coating is next subjected to a drying operation at about 100 °C for about 5 minutes. The final heat treatment takes place at about 200 °C for 30 minutes.

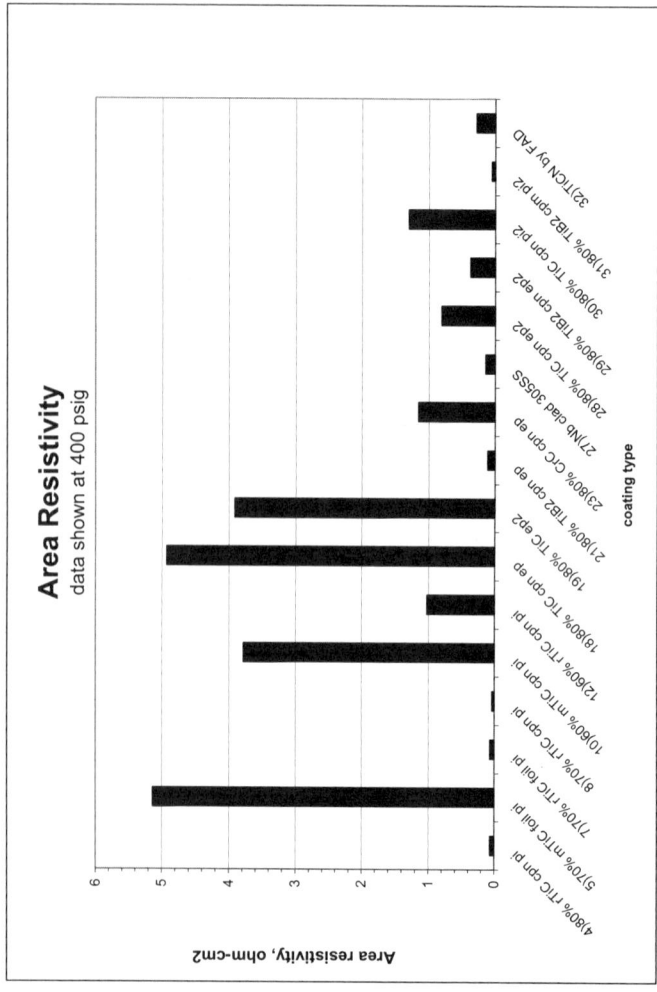

Figure 14. Area resistivity of conductive polymer coatings.

(g) Conductive fillers

While several conductive fillers are available, Analytic has had the most success with TiC and TiB_2. Size –325 mesh particles are ball milled with the polymer solution for about 3 days in order to produce films that are smooth enough to pass testing with a Hegman No. 7 gauge. Very fine particle sizes, (~ 50nm TiC) produce much higher resistivity coatings than –325 mesh particles (44 microns). Since surface area is proportional to the inverse square of the particle diameter, the surface area of the films is increased by almost 800 as the particle diameter dropped from 44 microns to 50 nm. If the film thickness is roughly constant, irrespective of the particle diameter, then the amount of nonconductive polymer film in the coating is increased by the same factor. Experimentally we have found that larger particle sizes yield much higher conductivity, as shown by comparison in Figure 14 samples 5 and 7.

The chemical grafting and the silane processes suggest that fillers with surface hydroxyl groups are more effectively bound to the polymer. Such a metal might be nickel, and its carbides, nitrides, borides and silicides. Nickel forms a hydroxide readily between 8 < pH < 12. Unfortunately, nickel hydroxide dissolves readily in acid. Nickel, or other polyvalent cations which escape into the fuel cell PEM can catalyze the formation of peroxide.

The extent to which carbide or boride particles have been covalently or only physically bonded into the polymer matrix, is currently not known. If the latter is true, then the wetable surfaces of the carbides and borides will provide water permeation pathways to the surface of the aluminum and, promote corrosion. There is an extensive literature on the subject of water permeability in epoxy and composite (glass) epoxy films.[9]

REFERENCES

[1] V. Bloomfield et al., *Cost effective, chemically grafted metal bipolar plate coating*, Fuel Cell Seminar 2004, Poster 95, Analytic Power LLC, 2-X Gill Street, Woburn, MA 01801.
[2] C. Horowitz et al., Method of protectively coating metallic aluminum containing substrate, U. S. Patent 4,105,811 (August 1978).
[3] C. Horowitz et al., Method of grafting polymerizable monomers onto substrates, U. S. Patent 5,232,748 (August 3, 1993).
[4] Department of Polymer Science University of Southern Mississippi, Polycondensation and curing of epoxy polymers, (1996), http://www.psrc.usm.edu/macrog/lab/epoxy.htm.

[5] *Engineering Materials Handbook,* Vol. 2, Engineering Plastics, ASM Intl, Metals Park, Ohio, 1988.
[6] Prof. Douglas Wicks, personal communication, University of Southern Mississippi.
[7] Prof. Wim van Ooij, personal communication, University of Cincinnati.
[8] L. L. Marsh, R. Lasky, D. P. Seraphim, and G. S. Springer, Moisture solubility and diffusion in epoxy and epoxy-glass composites, *IBM J. Res. Develop.* **28**(6) (1984).
[9] J. Wei, C. Stone, A. Steck, Trifluorostyrene and substituted trifluorostyrene copolymeric compositions and ion-exchange membranes formed therefrom, U. S. Patent 5422411, (June 6, 1995).
[10] F. A. Uribe and T. Zawoddzinski, Method for improving fuel cell performance, U. S. Patent A 2001/0044040A1, (November 22, 2001).
[11] F. A. Uribe and T. A. Zawodzinski, A study of polymer electrolyte fuel cell performance at high voltages.dependence on cathode catalyst layer composition and on voltage conditioning, J. Electrochimica Acta, **47** (2002) 3799.
[12] G. Ya. Vyaseleva, A. A. Konopleva, D. M. Torsuev, R. A. Kop'ev, and V. P. Barabanov, On the protective properties of electrostatically sprayed fluoroplastic and epoxy coatings, *Protection of Metals* **39**(4) (2003) 389.
[13] L. L. Marsh, R. Lasky, D. P. Seraphim, and G. S. Springer, Moisture solubility and diffusion in epoxy and epoxy-glass composites, *IBM I. RES. DEVELOP.* **28**(6), November (1984).
[14] J. Song and W. J. Van Ooij, Bonding and corrosion protection mechanisms of γ-APS and BTSE silane films on aluminum substrates, *J. Adhesion Science and Technology* **17**(16) (2003) 2191.
[15] Z. Zhang and S. Sakka, Hydrolysis and polymerization of dimethyldiethoxysilane, methyltrimethoxysilane and tetramethoxysilane in presence of aluminum acetylacetonate. a complex catalyst for the formation of siloxanes, *J. Sol-Gel Science and Technology* **16**(3) (1999) 209.
[16] H. Ishida, *Introduction to Polymer Composite Processing*, Ch. 5, NSF Center for Molecular and Microstructure of Composites (CMMC), Department of Mascromolecular Science, Case Western Reserve University, Cleveland, Ohio, p.p. 9-15.
[17] N. B. Madsen, Modification and characterization of the interface in polymer/inorganic composites, PhD thesis, Risø National Laboratory, Roskilde, Denmark, 1999.
[18] D. Bloomfield and I. Kriksunova, *Low Cost Composite Bipolar Plate for Proton Exchange Membrane Fuel Cells,* NSF Phase I SBIR Program, July 15, 1997, Award 9661078.
[19] M. Pourbaix, *Atlas of Electrochemical Equilibria in Aqueous Solutions*, National Assn. of Corrosion Engineers, Houston, TX, 1974.
[20] A. B. LaConti, M. Hamdan and R. C. McDonald, Mechanisms of membrane degradation for PEMFCs, in *Handbook for Fuel Cells – Fundamentals Technology and Applications*, Vol. 3, Ch. 55, Ed. by W. Vielstich, et al., John Wiley and Sons. LTD, 2003.
[21] D. A. Blom, J. R.Dunlap, T. A. Noonan and L. F. Allard, Preparation of cross-sectional samples of proton exchange membrane by ultramicrotomy for TEM, *J. Electrochem. Soc.* **150**(4) (2003) A414.
[22] W. van Ooij, et al., Method of preventing corrosion of metals using silanes, U. S. Patent 6,261,638 (July 17, 2001).

[23] A. J. Sedricks, *Corrosion of Stainless Steels*, Wiley Interscience, John Wiley & Sons, New York.
[24] H. Colón, H. Kim, and B. Popov, *Evaluation of Durability and Catalytic Activity of Different Pt Oxygen Electrocatalysts,* Fourth International Symposium on Proton Conducting Membrane Fuel Cells, Electrochemcial Society, Department of Chemical Engineering University of South Carolina, Columbia, SC, October 3-8, 2004.
[25] H. Baker, Cell maintenance device for fuel cell stacks, Patent application pending.
[26] C. Reiser, and R. Sawyer, Solid Polymer Electrolyte Fuel Cell Stack Water Management System, U. S. Patent 4,769,297 (September 6, 1988).
[27] C. Reiser, Ion exchange membrane fuel cell power plant with water management pressure differentials, U. S. Patent 5,853,909 (December 29, 1998).
[28] R. A.DuBose, Fuel cell gas management system, U. S. Patent 6,013,385 (January 11, 2000).
[29] R. M. Darling and J. P. Meyers, Mathematical model of platinum movement in PEM fuel cells, J. Electrochem. Soc. **152**(1) **(**2004) A242.
[30] R. M. Darling and J. P. Meyers, Kinetic model of platinum dissolution in PEMFCs, J. Electrochem. Soc. **150**(11) (2003), A1523.
[31] *Laminated Fluid Flow Field Assembly for Electrochemical Fuel Cells*, SP 5,300,370, Washington, Wilkinson, Voss, (Ballard).
[32] J. McElroy, personal communication.
[33] G. Faita, et al., Electrochemical cell provided with ion exchange membranes and bipolar metal plates, U. S. Patent 5,482,792 (January 9,1996).
[34] O. Adlhart, Fuel cell system utilizing ion exchange membranes and bipolar plates, U. S. Patent 4175165 (November 20 1979).
[35] D. Shores et al., Corrosion of current collector materials, Multidisciplinary Research Program of the University Research Initiative Army Research Office (MURI/ARO), presentation (August 1998).
[36] L. Ma, S. Warthesen, and D. Shores, *Evaluation of Materials for Bipolar Plates in PEMFC's,* Research Center, University of Minn., August 1999.
[37] J. Wei, C. Stone, and A. Steck, Trifluorostyrene and substituted trifluorostyrene copolymeric compositions and ion-exchange membranes formed therefrom, U. S. Patent 5422411(June 6, 1995).

2

Basic Applications of the Analysis of Variance and Covariance in Electrochemical Science and Engineering

Thomas Z. Fahidy

Department of Chemical Engineering, University of Waterloo, Waterloo, Ontario Canada

I. INTRODUCTION

An electrochemical scientist specializing in corrosion studies wishes to decide whether tensile and compressive stress measured at various temperatures in samples of corroding metals differ significantly, and if the influence of temperature is statistically important. An electrochemical engineer specializing in process performance/quality control wishes to decide if specific energy requirements of an electrolytic product, varying from plant to plant, are significantly different. The analysis of variance, ANOVA, a *body of statistics*[1] answers such questions. It is devoted to the study of the variability of factors influencing experimental observations, involving simple (one- and two-factor), and complex (multiple-factor) experiments and designs.

If the relative importance of a priori hidden factors is to be determined, the analysis of covariance, ANCOVA, is included in the overall statistical exercise. The scientist may worry about the pretreatment (if any) of the metal samples prior to corrosion, and the engineer may suspect a possible surface quality effect of the anodes used in the cells. The introduction of such concomitant[2] variables or covariates[2,3] renders, when necessary, a more sophisticated picture at the expense of modest additional numerical work.

The purpose of this chapter is to illustrate the application of ANOVA and ANCOVA via selected problems of interest to the electrochemical scientist and engineer. ANOVA categories discussed include one-way classification, latin squares, 2- and 3-way classification involving completely randomized experiments (CRE), randomized block experiments (RBE) and hierarchical arrangements. ANCOVA techniques are limited to one or two concomitant variables, and the effect of their neglect in the overall analysis.

An unfortunate impediment to the application of statistical analysis to experimental data published in the electrochemical literature is the paucity of sufficiently detailed numerical observation sets, including replicate measurements. It is, therefore, inevitable that numerical data in the Tables have to be considered as if they had been obtained in accordance with ANOVA principles. For the sake of illustration of the statistical techniques, it is immaterial whether this imposition is strictly justified, inasmuch as the ultimate objective of tests: the attainment of appropriate decisions, is not compromised.

There is an impressive amount of textbooks (and Internet-based material) devoted to the subject matter in general, and it would be impossible as well as impractical to supply a full list in this chapter. In addition to the comprehensive References [1-3], the text by Snedecor and Cochran[4] is an outstanding source on account of its clarity and organization. Readers interested in pursuing the subject matter at an advanced level will find ample material in the literature to satisfy their needs.

The objective here is not to turn electrochemical scientists and engineers into statisticians, but to promote the appreciation of certain statistical principles and techniques useful to electrochemical process analysis and design.

II. BASIC PRINCIPLES AND NOTIONS

The fundamental role of statistics is to draw inferences about population parameters from parameters of a single sample, or several samples. In fixed-effect experiments the population (or treatment) parameters are considered to be constant and the observations to belong to normal populations with unknown means, and constant variance. The errors associated with the observations are assumed to belong independently to a normal population of mean zero and variance σ^2. A fundamental tenet of ANOVA is the null hypothesis that the population/treatment means are equal, i.e.,

$$H_0 : \mu_1 = \mu_2 = ... = \mu_r \qquad (1)$$

if r populations provide the samples. A perhaps more thoughtful way to state Eq. (1) would be that there exists no statistically distinguishable difference between the population means. The alternative (or counter) hypothesis:

$$H_a\text{: at least two means are different} \qquad (2)$$

negates H_0. All the tests seen in the sequel have to do with Eqs. (1) and (2), at various levels of complexity.

On what basis can it be decided that H_0 can be rejected in a statistically meaningful manner? Phrasing it otherwise, how much error is committed if H_0 is rejected in favour of H_a, although the latter is false? In ANOVA the magnitude of this (so-called) Type I error depends on the numerical value f of the F-statistic; the Type I error is the probability that in a given test, a numerical value of F would be equal to or exceed f. In classical statistics the 5% Type I error is called *significant*, and the 1% Type I error is called *highly significant*. The less categorical modern statistician would not put the 5% and 1% Type I error on a high pedestal, but would prefer to think in terms of the *P-value*, which is the Type I error committed when the null hypothesis is rejected in face of the observation-based F-statistic. This is a more flexible point of view, where the influence of *outside* (i.e., not necessarily statistical factors) is also taken into account.

Critical values of the F-statistic are determined by the size of the Type I error and the two degrees of freedom, as shown in the partial F-table (Table 1). Extensive tabulations are widely available in statistical textbooks and handbooks.

At very small values of v_2, the second degree of freedom, the F-test is weak in the sense that the rejection of a null hypothesis will be virtually impossible. As the degrees of freedom increase, rejection of H_0 becomes increasingly possible, regardless of the size of the admitted Type I error α. When the degrees of freedom are fixed, the critical F-values increase with a decrease in the Type I error. If, e.g., a computed F-statistic related to $v_1 = 3$ and $v_2 = 10$

is $f = 5.32$, the Type I error is between 1% and 5%; linear interpolation between the critical values yields $\alpha \approx 0.027$, i.e., rejection of H_0 carries an about 2.7% Type I error.

What about the error committed in failing to reject H_0 when it is false? This is the Type II error β (the symbol α is also used for denoting the Type I error). The *power of test*, $(1 - β)$, is difficult to compute for the F-distribution, but facilitated by charts e.g., Table 6, pp. 188–193 in Ref. [1].

The computation of the F-statistic can conveniently be followed in ANOVA and ANCOVA tables. An F statistic is obtained as the ratio of two mean squares; the mean squares are themselves obtained from sums of squares divided by their degree of freedom. The degrees of freedom depend on the size of the observation sets, the number of treatments and the type of statistical experiment. ANOVA tables are also a standard component of software packages for computers.

When only two treatments are compared, the F-test can be replaced by a test based on Student's T-statistic, on account of the $f_\alpha(1, v_2) = t_{\alpha/2}(v_2)$ equivalence.

If H_0 is rejected, a statistical Pandora's box is opened in the sense that various *derivative* null hypotheses can be set up to test if differences in certain means are significant with respect to differences in some other means. A linear combination of treatment means is called a *contrast*, if the sum of the coefficients related to the treatment means is zero. Contrasts are at the heart of multiple comparison methods,[5] but their detailed discussion is outside the scope of this chapter, beyond Section VIII.4.

Notation in the sequel follows closely Guenther's[1] (which agrees to a large extent with a large number of statistics textbooks). Specifically, a random variable or distribution is denoted by a capital letter, and its numerical value by a lower case letter.

III. ANOVA: ONE-WAY CLASSIFICATION

One-way classification (or single-factor) ANOVA deals with differences between means of observations *of a kind* pertaining to treatments. Current efficiencies measured repeatedly at various current densities (e.g., at a certain type of electrode) constitute a proper electrochemical example, where each current density is a treatment, but no other factor is assumed to influence current efficiency.

Table 1
Selected Critical Values of the F-Statistic at Three Magnitudes of the Critical Region

v_2 \ v_1	$\alpha = 0.05$ (or 5%)			$\alpha = 0.01$ (or 1%)			$\alpha = 0.001$ (or 0.1%)		
	1	2	3	1	2	3	1	2	3
1	161.4	199.5	215.7	4052	4999.5	5403	4×10^5	5×10^5	5.4×10^5
2	18.51	19.00	19.16	98.50	99.00	99.17	998.5	999.0	999.2
5	6.61	5.79	5.41	16.26	13.27	12.06	47.18	37.12	33.20
10	4.96	4.10	3.71	10.04	7.56	6.55	21.04	14.91	12.55
20	4.35	3.49	3.10	8.10	5.85	4.94	14.82	9.95	8.10

1. Completely Randomized Experiment (CRE)

In a statistical experiment subjects (or units) may be assigned randomly to any one of the r treatments. The null hypothesis

$$H_0 : \mu_1 = \mu_2 = ... = \mu_r \tag{3}$$

is tested against H_a: at least two treatment means are unequal. In terms of the observation set

$$x_{ij} = \mu + \beta_j + e_{ij}; \quad i = 1,2,...,n_j; \quad j = 1,2,...,r \tag{4}$$

the null hypothesis rewritten as

$$H_0 : \beta_j = 0; \quad j = 1,2,...,r \tag{5}$$

with

$$n_1\beta_1 + n_2\beta_2 + ... + n_r\beta_r = 0 \tag{6}$$

is more conducive to ANOVA manipulations. The errors e_{ij} are postulated to belong to the normal distribution $N(0;\sigma^2)$. As indicated in Section II, this is a fixed-effect formulation, where $\{\beta_j\}_r$ is a set of constant, albeit unknown, measures of deviation from an average of the r population means.

2. Randomized Block Experiment (RBE)

In a statistical experiment, subjects (or units) may also be assigned randomly to treatments after having been first placed into groups with certain *homogeneous* characteristics. The purpose of RBE is either to remove a nuisance or a believed-to-be unimportant variable, or to identify a variable to which the subjects might be particularly sensitive. Defining

$$\alpha_i \equiv \mu_{i.} - \mu \tag{7}$$

and

$$\beta_j \equiv \mu_{.j} - \mu \tag{8}$$

where $\mu_{.j}$ are the treatment means and $\mu_{i.}$ are the block means, the null hypothesis to test is

$$H_0 : \beta_j = 0; \quad j = 1, 2, ..., r \tag{9}$$

with alternative hypothesis H_a: at least one β_j is nonzero.

(i) Example 1: A Historical Perspective of Caustic Soda Production

Table 2 summarizes calculations based on information by Hine,[6] transformed by normalizing the production figures with respect to production in country A observed in 1970. The eight countries cited by Hine are denoted by letters in this Section.

The CRE approach disregards the countries as separate entities, and utilizes only the column totals for ANOVA. The symbol N denotes the total number of observations. The sums of squares are computed as

$$SS_T = \sum_{i=1}^{n_j} \sum_{j=1}^{r} x_{ij}^2 - \frac{T_{..}^2}{N} = 26.2747 \tag{10}$$

$$SS_A = \sum_{j=1}^{r} \frac{T_{.j}^2}{n_j} - \frac{T_{..}^2}{N} = 0.2737 \tag{11}$$

and

$$SS_W = SS_T - SS_A = 26.0010 \tag{12}$$

The degrees of freedom are $(N-1)$ for SS_T; $(r-1)$ for SS_A, and $(N-r)$ for SS_W, respectively.

The RBE approach considers the eight countries as blocks. The missing entry for country H in 1970 is estimated[7] to be

$$x_{ij} = \frac{rT_{.j}' + nT_{i.}' - T_{..}'}{(r-1)(n-1)} = 0.241 \tag{13}$$

where the primes denote the totals with one entry missing. Accordingly,

Table 2
Production of Caustic Soda in Coded Units and Countries in Certain Years

Country	1970	1975	1979	$T_{i.}$	$x_{mi.}$
A	1.00	1.12	1.12	3.24	1.080
B	0.35	0.31	0.42	1.08	0.360
C	0.42	0.42	0.54	1.38	0.460
D	0.39	0.39	0.42	1.20	0.400
E	0.69	0.92	1.15	2.76	0.920
F	0.65	0.96	1.31	2.92	0.973
G	3.54	3.23	4.27	11.04	3.680
H	–	0.38	0.46	0.84	0.420
$T_{.j}$	7.04	7.73	9.69	$T_{..} = 24.46$;	$x_{m..} = 1.063$
$x_{m.j}$	1.0057	0.9663	1.2113	$r = 3$; $n_1 = 7$;	$n_2 = n_3 = 8$

$$\sum_{i=1}^{n_j}\sum_{j=1}^{r} x_{ij}^2 = 52.2874$$

ANOVA/CRE				
Source	SS	DF	MS	F	P-value
Among columns	0.2737	2	0.1368	0.105	> 0.25
Within columns	26.001	20	1.3		
Total	26.275	22			
$f_{2,20}$ ($\alpha = 0.25$) = 1.49					
ANOVA/RBE					
Treatments	0.3891	2	0.1945	2.09	0.18
Blocks	25.234	7	3.6048	38.80	< 0.001
Error	1.3003	14	0.0929		
Total	26.923	23			
$f_{2,14}(\alpha = 0.1) = 2.73$; $f_{2,14}(\alpha = 0.25) = 1.53$; $f_{7,14}(\alpha = 0.001) = 7.08$					

$$SS_T = \sum_{i,j} x_{ij}^2 - \frac{T_{...}^2}{rn} = 26.9233 \tag{14}$$

$$SS_{Tr} = \sum_{j=1}^{r} \frac{T_{.j}^2}{n} - \frac{T_{...}^2}{rn} = 0.41025 \tag{15}$$

$$SS_{Tr}^{'} = SS_{Tr} - \frac{[T_{i.}^{'} - (r-1)x_{ij}]^2}{r(r-1)} = 0.3891 \tag{16}$$

$$SS_B = \sum_{i=1}^{n} \frac{T_{i.}^2}{r} - \frac{T_{..}^2}{rn} = 25.234 \tag{17}$$

and

$$SS_E = SS_T - SS_B - SS'_{Tr} = 1.3003 \tag{18}$$

The degrees of freedom are $(nr - 1)$ for SS_T, $(r - 1)$ for treatments, $(n - 1)$ for blocks, and $(n - 1)(r - 1)$ for error, respectively. The RBE-based analysis yields the more realistic result that the block effect is extremely significant, while the treatment effect is not significant. The difference between the eight countries is much more important than the years of observation.

(ii) Example 2: Metallic Corrosion

The partial molal volume of hydrogen gas in Armco iron metal corroded under tensile and compressive stress at various temperatures has been reported[8] as 2.60, 2.69, 2.62, 2.52 cm^3 (tensile stress) and 2.73, 2.74, 2.60, 2.61 cm^3 (compressive stress) at temperatures 27, 40, 50, 60 ^{0}C, respectively. If the volumes are considered to be four random readings per test, i.e., temperature as a factor is ignored, a CRE- based calculation scheme similar to the one shown in Table 2, yields $f = 1.89$ with a Type I error of about 23%. It would be, therefore, rather hazardous to reject the null hypothesis of no difference between the mean partial volumes due to tensile and compression stress.

An RBE-based analysis, with temperature as the second factor, yields a similar result. In this configuration,

$$SS_T = \sum_{i=1}^{n_j} \sum_{j=1}^{r} x_{ij}^2 - \frac{T_{..}^{'2}}{N} = 0.03949 \tag{19}$$

$$SS_{Tr} = \sum_{j=1}^{r} \frac{T_{.j}^2}{n_j} - \frac{T_{..}^2}{N} = 0.007825 \tag{20}$$

$$SS_B = \sum_{i=1}^{n_j} \frac{T_{i.}^2}{r} - \frac{T_{..}^2}{N} = 0.025 \tag{21}$$

and

$$SS_E = SS_T - SS_R - SS_B = 0.006663 \tag{22}$$

The F-statistics can, in effect, be obtained without necessarily constructing an ANOVA table, viz.

for treatments: $f_{Tr} = \dfrac{SS_{Tr}}{SS_E} \cdot \dfrac{(n-1)(r-1)}{r-1} = 3.52$ (23)

for blocks: $f_{BL} = \dfrac{SS_B}{SS_E} \cdot \dfrac{(n-1)(r-1)}{n-1} = 3.75$ (24)

Since $f_{1,3} = 5.54$ ($\alpha = 0.10$), $f_{1,3} = 2.02$ ($\alpha = 0.25$), $f_{3,3} = 5.39$ ($\alpha = 0.10$), and $f_{3,3} = 2.36$ ($\alpha = 0.25$), neither effect is significant with P-values roughly equal to 0.2, for the mean values of the molal volume of H_2. The result is not surprising in view of the extremely large Type I error that would be committed if the null hypothesis of zero slope in a linear regression on volume versus temperature (independently and combined) were rejected.

IV. ANOVA: TWO-WAY CLASSIFICATION

In two-way (or two-factor) classification the subjects to be measured under each treatment may be grouped according to properties which are expected to influence observations due to each treatment, or the two factors may be treated as sources of equally random observations. This is the process analyst's choice, especially if experiments are to be a priori designed. Accordingly, the evaluation process follows either the CRE or the RBE path.

1. Null and Alternative Hypotheses

The general model can be written as

$$\begin{aligned} x_{ijk} &= \mu + \alpha_i + \beta_j + (\alpha\beta)_{ij} + e_{ijk}; \\ & i = 1,2,\ldots,a;\ j = 1,2,\ldots,b;\ k = 1,2,\ldots,n \end{aligned} \tag{25}$$

where the e_{ijk} errors belong independently to normal population $N(0;\ \sigma^2)$, and

$$\sum_i \alpha_i = \sum_j \beta_j = \sum_{i,j}(\alpha\beta)_{ij} = 0 \qquad (26)$$

The three null hypotheses can be stated as $\alpha_i = 0$; $\beta_j = 0$; $(\alpha\beta)_{ij} = 0$ for all pertinent indices. The alternative hypotheses state that not all α_i, not all β_j, and not all $(\alpha\beta)_{ij}$ are zero. The presence of the $(\alpha\beta)_{ij}$ terms assumes interaction between the two factors; if the hypothesis that the sum of these terms is zero, cannot be rejected, the two factor effects (or factor/treatment and block effects) are simply additive.

2. Illustration of Two-Way Classification: Specific Energy Requirement for an Electrolytic Process

Selecting three anode types to be tested in electrolyzers at four plant locations, an electrochemical engineer intends to determine the relative importance of these two factors on the specific energy requirement (SER) for a certain electrolytic product. In each case three replicate measurements have been made as shown in Table 3. For the sake of comparison via identical experimental data, it is assumed that the measurements were obtained either by a CRE-based, or an RBE-based experimental protocol.

In the CRE-based approach, the pertinent sums of squares are

$$SS_T = \sum_{i=1}^{a}\sum_{j=1}^{b}\sum_{k=1}^{n} x_{ijk}^2 - \frac{T_{...}^2}{abn} = 250.06 \qquad (27)$$

$$SS_{Tr} = \frac{\sum_{i=1}^{a}\sum_{j=1}^{b} T_{ij.}^2}{n} - \frac{T_{...}^2}{abn} = 185.72 \qquad (28)$$

$$SS_E = SS_T - SS_{Tr} = 64.33 \qquad (29)$$

$$SS_A = \frac{\sum_{j=1}^{b} T_{i..}^2}{bn} - \frac{T_{...}^2}{abn} = 117.06 \qquad (30)$$

$$SS_B = \frac{\sum_{j=1}^{b} T_{.j.}^2}{an} - \frac{T_{...}^2}{abn} = 49.06 \tag{31}$$

$$SS_{AB} = SS_{Tr} - SS_A - SS_B = 19.61 \tag{32}$$

Table 3
Analysis of SER Observations (kWh/metric ton) via Two-Factor Experiment

Plant location (A levels)	Anode type (B levels)			$T_{i..}$
	1	2	3	
1	7.5	10.0	11.0	
	9.5	12.0	8.5	80.5
	6.0	9.0	7.0	
2	8.5	12.0	13.0	
	5.0	9.0	9.5	85.0
	6.5	11.0	10.5	
3	4.5	6.0	5.0	
	6.0	7.5	2.5	43.5
	3.0	5.0	4.0	
4	7.0	10.5	9.5	
	4.0	8.0	7.5	65.0
	5.5	7.0	6.0	
$T_{.j.}$	73.0	107.0	94.0	$T_{...} = 274.0$

ANOVA/CRE					
Source	SS	DF	MS	F	P-value
Treatment A	117.06	3	39.02	14.56	< 0.001
Treatment B	49.06	2	24.53	9.15	< 0.0013
Treatment AB (interaction)	19.61	6	3.27	1.22	> 0.25
Error	64.33	24	2.68		
Total	250.06	35			

$f_{2,24}(\alpha = 0.001) = 9.34$; $f_{2,24}(\alpha = 0.005) = 6.66$; $f_{3,24}(\alpha = 0.001) = 7.55$; $f_{6,24}(\alpha = 0.25) = 1.41$

ANOVA/RBE					
Treatment A	117.06	3	39.02	23.42	< 0.001
Treatment B	49.06	2	24.53	14.72	< 0.001
Treatment AB	19.61	6	3.27	1.96	0.123
Block	24.68	2	12.34	7.41	0.004
Error	39.65	22	1.67		
Total	250.06	35			

$f_{2,22}(\alpha = 0.005) = 6.81$; $f_{2,22}(\alpha = 0.001) = 9.61$; $f_{3,22}(\alpha = 0.001) = 7.71$; $f_{6,22}(\alpha = 0.1) = 2.06$; $f_{6,22}(\alpha = 0.25) = 1.42$

The degrees of freedom are $(abn - 1)$ for SS_T, $(a-1)$ for SS_A, $(b - 1)$ for SS_B, $(a - 1)(b - 1)$ for SS_{AB}, and $(ab[n - 1])$ for SS_E, respectively. The P-values indicate that

1. the effect of plant location on SER is extremely important,
2. the effect of anode type used is extremely important, and
3. interaction between plant location and anode type is negligible, i.e., the *behaviour* of each anode type is independent of plant location.

In the RBE-based approach, the sum of squares due to blocking is

$$SS_{BL} = \frac{\sum_{i=1}^{a} T_{..k}^2}{ab} - \frac{T_{...}^2}{abn} = 24.68 \qquad (33)$$

and the error sum of squares is accordingly modified to

$$SS_E = SS_T - SS_{BL} - SS_{Tr} = 39.65 \qquad (34)$$

The rest of the sums of squares remain unchanged. The degrees of freedom are $(abn - 1)$ for SS_T, $(a - 1)$ for SS_A, $(b - 1)$ for SS_B, $(a - 1)(b - 1)$ for SS_{AB}, and $(n - 1)(ab - 1)$ for SS_E, respectively. Consequently, plant location and anode type are extremely significant, the replication effect is at least highly significant, but interaction between plant location and anode type is insignificant. The CRE and RBE lead to essentially identical statistical conclusions.

V. ANOVA: THREE-WAY CLASSIFICATIONS

The mathematical framework required for the analysis of three treatment categories and their possible interactions is cumbersome, and since it has been described at length in the textbook literature, only a numerical summary of analysis pertaining to a specific illustration is discussed in this Section.

In an electrolytic process the concentration of an undesirable byproduct content in the exit stream from a flow electrolyzer is considered to depend on electrolyzer type, temperature and current density. Table 4 shows measured values of byproduct

Table 4
Analysis of Byproduct Content Levels (ppm) via Completely Randomized Three-Factor Experiment

Current density (Level C)	Single tank electrolyzer (Level A) Temperature (Level B)			Electrolyzer/mixer cascade (Level A) Temperature (Level B)			Totals
	1	2	3	1	2	3	
1	140	160	182	150	160	191	983
2	139	145	172	98	115	148	817
3	94	108	141	71	80	119	613
Totals	1281			1132			2413

ANOVA

Source	SS	DF	MS	F	P-value
Treatment A	1233.4	$a - 1 = 1$	1233.4	67.06	0.0017
Treatment B	6066.8	$b - 1 = 2$	3033.4	164.86	< 0.001
Treatment C	11448.5	$c - 1 = 2$	5724.3	311.23	< 0.001
Treatment AB	41.43	$(a - 1)(b - 1) = 2$	20.72	1.13	> 0.25
Treatment AC	1219.1	$(a - 1)(c - 1) = 2$	609.5	33.14	0.0042
Treatment BC	52.2	$(b - 1)(c - 1) = 4$	13.1	0.71	> 0.25
Treatment ABC (error)	73.6	$(a - 1)(b - 1)(c - 1) = 4$	18.4		
Total	20075.0	$abcn - 1 = 17$			

$f_{1,4}(\alpha = 0.005) = 31.33$; $f_{1,4}(\alpha = 0.001) = 74.14$; $f_{2,4}(\alpha = 0.001) = 61.25$;
$f_{2,4}(\alpha = 0.005) = 26.28$; $f_{2,4}(\alpha = 0.25) = 2.00$

concentration by randomization among treatments A, B and C. Treatment A has two levels, representing a simple tank flow electrolyzer, and an electrolyzer-mixer cascade with recycle to the electrolyzer. Treatment B consists of three temperature levels, and Treatment C carries three current density levels. In view of the lack of replicate observations, the mean square of the ABC treatment interaction serves for the computation of the F-statistics.[9] The electrolyzer/temperature and temperature/current density effects are additive, but all other treatments are strongly significant. They indicate the acute sensitiveness of the undesired byproduct concentration to all operating factors considered in the analysis; this may be good news for the process analyst in providing a wide choice for controlling its presence in the effluent.

VI. ANOVA: LATIN SQUARES (LS)

The LS technique aims to reduce the number of observations required to extract the desired information from data, and it can be a good choice for removing (unwanted) sources of variation. In this respect, it can be regarded as an alternative to randomized blocks.

Assume that the performance of an electrolyzer is determined by the size of the corroded fraction of an electrode surface upon a reference duration of electrolysis. Postulating three levels of operating conditions, three levels of current density, and three electrode types A, B and C, the LS-based experimental protocol, illustrated in Table 5, requires only nine observations instead of running twenty seven experiments to deal with every possible set of conditions. This is achieved via specific assignment schemes of the electrode types. The null hypotheses can be written as

$$H_0: \mu_{..1} = \mu_{..2} = \mu_{..3} = \mu \tag{35}$$

$$H_0': \mu_{1..} = \mu_{2..} = \mu_{3..} = \mu \tag{36}$$

$$H_0'': \mu_{.1.} = \mu_{.2.} = \mu_{.3.} = \mu \tag{37}$$

The pertinent sums of squares are computed as

$$SS_T = \sum_{i,j,k} x_{ijk}^2 - \frac{T_{...}^2}{r^2} = 934.00 \tag{38}$$

$$SS_R = \frac{\sum_{i=1}^{r} T_{i..}^2}{r} - \frac{T_{...}^2}{r^2} = 484.67 \tag{39}$$

$$SS_C = \frac{\sum_{j=1}^{r} T_{.j.}^2}{r} - \frac{T_{...}^2}{r^2} = 206.00 \tag{40}$$

Table 5
Latin Square Experiment Applied to the Electrode Corrosion Problem

Current density	Electrolyzer 1	Electrolyzer 2	Electrolyzer 3	$T_{i..}$
1	A: x_{111} = 41	B: x_{122} = 39	C: x_{133} = 55	135
2	B: x_{212} = 66	C: x_{223} = 55	A: x_{231} = 63	184
3	C: x_{313} = 59	A: x_{321} = 39	B: x_{332} = 42	140
$T_{.j.}$	166	133	160	$T_{...}$ = 459

Randomly assigned order to electrode types: A → 1; B → 2; C → 3

$$\sum_{ijk} x_{ijk}^2 = 24343; \quad \frac{T_{...}^2}{N} = 23409; \quad T_{..1} = 143; \quad T_{..2} = 147; \quad T_{..3} = 169$$

ANOVA

Source	SS	DF	MS	F	P-value
Rows	484.67	2	242.34	4.30	0.217
Columns	206	2	103	1.83	> 0.25
Treatments	130.67	2	65.33	1.16	> 0.25
Error	112.66	2	56.33		
Total	934	8			

$f_{2,2}$ (α = 0.25) = 3.00; $f_{2,2}$ (α = 0.10) = 9.00

$$SS_{Tr} = \frac{\sum_{k=1}^{r} T_{..k}^2}{r} - \frac{T_{...}^2}{r^2} = 130.67 \tag{41}$$

$$SS_E = SS_T - SS_R - SS_C - SS_{Tr} = 112.66 \tag{42}$$

The degrees of freedom are $(r^2 - 1)$ for SS_T, $(r - 1)$ for SS_R, SS_C, SS_{Tr}, and $(r - 1)(r - 2)$ for SS_E, respectively. The ANOVA table shows that neither null hypothesis can be rejected without committing a serious Type I error.

The array in Table 5 is one of many possible random assignments. For instance, the arrangement 2:1:3 for rows, 3:1:2 for columns and A→2; B→3; C→ 1 would finally produce the array

	Electrolyzer 1	Electrolyzer 2	Electrolyzer 3
Current density 1	x_{112}	x_{123}	x_{131}
Current density 2	x_{213}	x_{221}	x_{231}
Current density 3	x_{311}	x_{322}	x_{333}

and the test would require conditions different from those in Table 5. Consequently, ANOVA may well lead to different conclusions. In any event, Table 5 would suggest to a process analyst that variability of electrode corrosion is most likely due to other (possibly undetected) physical reasons. This kind of result is one important benefit of statistical analysis.

VII. APPLICATIONS OF THE ANALYSIS OF COVARIANCE (ANCOVA)

The analysis of covariance accomplishes the same task of eliminating variables of no interest as do randomized blocks and Latin squares, but in a conceptually different manner. It employs concomitant variables, also known as cofactors, to *sharpen* the testing of null hypotheses. The desired outcome is the reduction of the mean square error with respect to treatment, block, etc. mean squares, i.e., the attainment of sufficiently large F-statistics for the rejection of H_0 at the expense of a (very) small Type I error.

A case in point is a statistical interpretation of the results of the design calculation for a single-pass high conversion electrochemical reactor presented by Rode et al.[10] For the sake of discussion, model-computed velocity, pressure drop, and average current density data in Table 4[10] are considered *as if they had been measured* in both non-segmented and segmented cell. The two cells are defined as the treatments, and velocity and pressure drop as concomitant-variable/cofactor candidates.

1. ANCOVA with Velocity as Single Concomitant Variable

The array in Table 6 provides quantities required for ANCOVA in conjunction with CRE (Pattern A) and RBE (Pattern B).

(i) Pattern A(CRE)

The postulates (1) y_{ij} belong independently to normal population $N(\mu_{ij}, \sigma^2)$, $i = 1,2,\ldots,n_j$; $j = 1,2,\ldots,r$; (2) $\mu_{ij} = \mu + \beta_j + \gamma(x_{ij} - x_{m..})$; (3) $n_1\beta_1 + n_2\beta_2 + \ldots + n_r\beta_r = 0$ lead to hypotheses

$$H_0 : \beta_j = 0; \quad j = 1,\ldots,r \tag{43}$$

$$H_a : \text{not all } \beta_j \text{ are zero} \tag{44}$$

Table 6
Data Array for ANCOVA with Velocity as a Single Concomitant Variable (Section VI.1)
$r = 2$; $n_j = n = 6$; $N = 12$

	Treatment			
	$j = 1$ (non-segmented cell)		$j = 2$ (segmented cell)	
	x_{i1} (cm/s)	y_{i1} (kA/m^2)	x_{i2} (cm/s)	y_{i2} (kA/m^2)
	16.1	8.206	17.8	9.091
	6.04	6.154	7.68	7.828
	2.68	4.103	4.20	6.413
	1.01	2.051	2.31	4.699
	0.447	1.026	1.55	3.565
	0.212	0.513	1.15	2.790
$T_{x.j}$	26.489		34.69	
$T_{y.j}$		22.053		34.386
34.386	4.4148		5.782	
$x_{m.j}$		3.6755		5.7310
$y_{m.j}$				

$T_{ix.}$ ($i = 1,\ldots,6$): 33.90; 13.720; 6.880; 3.320; 1.997; 1.362
$x_{mi.}$ ($i = 1,\ldots,6$): 16.95; 6.86; 3.44; 1.66; 0.999; 0.681
$T_{iy.}$ ($i = 1,\ldots,6$): 17.300; 13.980; 10.516; 6.750; 4.591; 3.303
$y_{mi.}$ ($i = 1,\ldots,6$): 8.650; 6.990; 5.258; 3.375; 2.296; 1.652

$$\sum_{i=1}^{6}\sum_{j=1}^{2} x_{ij}^2 = 706.6624; \quad \sum_{i=1}^{6}\sum_{j=1}^{2} y_{ij}^2 = 355.1916; \quad \sum_{i=1}^{6}\sum_{j=1}^{2} x_{ij}y_{ij} = 451.3841$$

and

$$H_0^*: \gamma = 0 \tag{45}$$

$$H_a^*: \gamma \neq 0 \tag{46}$$

Equation (43) states that all modified (*corrected*) treatments effects, due to the concomitant variable, are equal, and Eq. (45) states that the concomitant variable is not required. From Table 6, the sums of squares are computed as

$$SS_{xT} = \sum_{i=1}^{n_j}\sum_{j=1}^{r} x_{ij}^2 - \frac{T_{x..}^2}{N} = 394.76 \tag{47}$$

$$SS_{xTr} = \sum_{j=1}^{r} \frac{T_{x.j}^2}{n_j} - \frac{T_{x..}^2}{N} = 5.60 \tag{48}$$

$$SS_{xE} = SS_{xT} - SS_{xTr} = 389.15 \tag{49}$$

$$SS_{yT} = \sum_{i=1}^{n_j}\sum_{j=1}^{r} y_{ij}^2 - \frac{T_{y..}^2}{N} = 89.74 \tag{50}$$

$$SS_{yTr} = \sum_{j=1}^{r} \frac{T_{y.j}^2}{n_j} - \frac{T_{y..}^2}{N} = 12.66 \tag{51}$$

$$SS_{yE} = SS_{yT} - SS_{yTr} = 77.07 \tag{52}$$

For ANCOVA three sums of products are also needed:

$$SP_T = \sum_{i=1}^{n_j}\sum_{j=1}^{r} x_{ij} y_{ij} - \frac{T_{x..} T_{y..}}{N} = 163.64 \tag{53}$$

$$SP_{Tr} = \sum_{j=1}^{r} \frac{T_{x.j} T_{y.j}}{n_j} - \frac{T_{x..} T_{y..}}{N} = 8.43 \tag{54}$$

$$SP_E = SP_T - SP_{Tr} = 155.21 \tag{55}$$

In order to calculate pertinent F-statistics, three modified sums of squares are obtained as

$$SS'_{yT} = SS_{yT} - \frac{(SP_T)^2}{SS_{xT}} = 21.91 \tag{56}$$

$$SS'_{yE} = SS_{yE} - \frac{(SP_E)^2}{SS_{xE}} = 15.16 \tag{57}$$

$$SS'_{yTr} = SS'_{yT} - SS'_{yE} = 6.75 \tag{58}$$

Bypassing the ANOVA table, the F-statistic related to Eqs. (43) and (44) are computed as

$$f_{r-1,N-r-1} = \frac{SS'_{yTr}}{SS'_{yEr}} \cdot \frac{N-r-1}{r-1} = 4.05 \tag{59}$$

Since $f_{1,9}$ = 5.12 (α = 0.05) and $f_{1,9}$ = 3.36 (α = 0.10), H_0 in Eq.(43) may be rejected (with a Type I error of about 8%), indicating that the modified/corrected treatments are not significantly different.

To test H_0 in Eq. (45), the corresponding F-statistic, computed as

$$f_{1,N-r-1} = \frac{(SP_E)^2}{(SS_{xE})(SS'_{yE})} \cdot \frac{N-r-1}{r-1} = 36.75 \tag{60}$$

strongly favours H_a^*, i.e., velocity most definitely qualifies as a concomitant variable.

(ii) Pattern B(RBE)

In this configuration there are six block elements for each treatment and (1) $\mu_{ij} = \mu + \alpha_i + \beta_j + \gamma(x_{ij} - x_{m..})$; (2) $\alpha_1 + \alpha_2 + \ldots + \alpha_n = 0$; $\beta_1 + \beta_2 + \ldots + \beta_r = 0$ are postulated. In other words,

$$H_0 : \beta_j = 0; j = 1,\ldots,r \tag{61}$$

and

$$H_a : \text{not all } \beta_j \text{ are zero} \tag{62}$$

The sums of squares for the blocks are determined as

$$SS_{xB} = \sum_{j=1}^{r} \frac{T_{ix.}^2}{r} - \frac{T_{x..}^2}{N} = 388.92 \tag{63}$$

$$SS_{yB} = \sum_{j=1}^{r} \frac{T_{iy.}^2}{r} - \frac{T_{y..}^2}{N} = 75.99 \tag{64}$$

Consequently, the error squares are found to be

$$SS_{xE} = SS_{xT} - SS_{xB} - SS_{xTr} = 0.234 \tag{65}$$

$$SS_{yE} = SS_{yT} - SS_{yB} - SS_{yTr} = 1.083 \tag{66}$$

The product sums arising from the RBE approach are

$$SP_B = \sum_{i=1}^{n} \frac{T_{ix.}T_{iy.}}{r} - \frac{T_{x..}T_{y..}}{N} = 155.61 \tag{67}$$

$$SP_E = SP_T - SP_B - SP_{Tr} = -0.396 \tag{68}$$

For the calculation of the F-statistics, further sums of squares are needed, namely:

$$SS'_{yE} = SS_{yE} - \frac{(SP_E)^2}{SS_{xE}} = 0.413 \tag{69}$$

$$SS'_{y(Tr+E)} = SS_{yTr} + SS_{yE} - \frac{(SP_{Tr} + SP_E)^2}{SS_{xTr} + SS_{xE}} = 2.706 \tag{70}$$

$$SS'_{yTr} = SS'_{y(Tr+E)} - SS'_{yE} = 2.293 \tag{71}$$

Testing the validity of H_0 in Eq. (61), the F-statistic

$$f_{r-1,[(n-1)(r-1)-1]} = \frac{SS'_{yTr}}{SS'_{yE}} \cdot \frac{[(n-1)(r-1)-1]}{r-1} = 22.22 \tag{72}$$

rejects H_0 at a highly significant level in favour of H_a in Eq. (62), in as much as $f_{1,4}(\alpha = 0.01) = 21.20$. The Type I error of this rejection is about 0.95% ($f_{1,4} = 31.33$ at $\alpha = 0.005$), and it can be stated that the modified/corrected treatment effects are highly significantly different.

The validity of H_0^* in Eq. (45) is determined by

$$f_{1,[(n-1)(r-1)-1]} = \frac{(SP_E)^2}{(SS_{xE})(SS'_{yE})} \cdot \frac{[(n-1)(r-1)-1]}{r-1} = 6.55 \tag{73}$$

and it follows that rejection of H_0^* in favour of H_a^* in Eq. (46) carries a Type I error of about 6% [$f_{1,4} = 4.54$ and 7.71, at $\alpha = 0.1$, and 0.05, respectively). The necessity of retaining the concomitant variable may well be an admissible conclusion, since 6% is reasonably close to the significant level. If H_0^* is not rejected from a traditional viewpoint, ANOVA is recommended[11] for testing H_0. In that instance, both treatment effects and block effects are more than highly significant with F-statistics 58.41 (treatments) and 70.03 (blocks), inasmuch as the critical F-values are $f_{1,5} = 47.18$ and $f_{2,5} = 37.12$ ($\alpha = 0.001$). The final result is (not surprisingly) the same. A similar procedure would apply if the pressure drop alone were considered to be the sole concomitant variable.

2. ANCOVA with Velocity and Pressure Drop Acting as Two Concomitant Variables

With two covariates, analysis requires a complex calculation procedure[12-14] in order to test the null hypothesis of no differences between the adjusted treatment means. Extension to several covariates is relatively straightforward[15] as an elegant application of matrix algebra. Table 7.65[16] provides a useful summary of the origin of corrected sums of squares, their degrees of freedom and associated computing formulae.

The principal result of ANCOVA applied to the high-conversion chemical reactor summarized in Table 7 is that rejection of the null hypothesis carries a Type I error of approximately 7.8%. Recalling the a priori postulate that x, y and z are measurements in randomly arranged experiments (instead of model-based numbers), a process analyst may conclude on the basis of a comprehensive understanding of the electrochemical process, that even an almost 8% Type I error may be judged to be sufficiently small to accept H_a. If so, the segmented and the non-segmented reactors are believed to behave differently. A process analyst strictly obeying conventional statistics, on the other hand, would conclude that the two reactors are not significantly different.

3. Two Covariate-Based ANCOVA of Product Yields in a Batch and in a Flow Electrolyzer

Table 8 contains hypothetical product yield data obtained via randomly arranged measurements in a batch electrolyzer, and in a flow electrolyzer with a fixed flow rate. The concentration of a contaminant ionic species and current density at the working electrode are considered to be the concomitant variables/cofactors. The ANOVA table signals no significant difference between the

Table 7
Data Array for ANCOVA with Velocity and Pressure Drop as Concomitant Variables (Section VI.2)

Treatment					
$j = 1$ (non-segmented cell)			$j = 2$ (segmented cell)		
x_{i1} (cm/s)	z_{i1} (kPa)	y_{i1} (kA/m^2)	x_{i2} (cm/s)	z_{i2} (kPa)	y_{i2} (kA/m^2)
16.1	29.0	8.206	17.8	32.0	9.091
6.04	10.9	6.154	7.68	14.0	7.828
2.68	4.83	4.103	4.20	7.6	6.413
1.01	1.81	2.051	2.31	4.1	4.699
0.447	0.805	1.026	1.55	2.8	3.565
0.212	0.381	0.513	1.15	2.1	2.790
Total: 26.489	47.726	22.053	34.69	62.6	34.386
Mean: 4.4148	7.954	3.6755	5.7820	10.4330	5.7310

ANOVA					
{degrees of freedom: $(rn - 3)$ for SS_T; $(r - 1)$ for SS_{Tr}; $[r(n - 1) - 2]$ for SS_E}					
Source	SS	DF	MS	F	P-value
Treatments	7.7456	1	7.7456	4.29	0.078
Error	14.4560	8	1.8070		
Total	22.2016	9			

$f_{1,8} (\alpha = 0.10) = 3.46$; $f_{1,8} (\alpha = 0.05) = 5.32$

Table 8
Data Array for ANCOVA with Containment Impurity and Current Density as Concomitant Variables (Section IV.3)

Treatment					
$j = 1$ (batch electrolyzer)			$j = 2$ (flow electrolyzer with fixed flowrate)		
x_{i1} (mass%)	z_{i1} (kA/m^2)	y_{i1} (%)	x_{i2} (mass%)	z_{i2} (kA/m^2)	y_{i2} (%)
1.5	1.1	90	1.9	3.0	87
2.0	2.8	82	2.1	4.0	84
2.5	3.5	85	2.2	1.5	87
2.8	3.0	87	2.4	5.0	83
3.5	0.9	90	2.6	6.0	82
Total: 12.30	11.3	434	11.2	19.5	423
Mean: 2.46	2.26	86.8	2.24	3.90	84.6

ANOVA					
Source	SS	DF	MS	F	P-value
Treatments	9.7513	1	9.7513	1.61	~ 0.25
Error	36.2753	6	6.0459		
Total	46.0266	7			

$f_{1,6} (\alpha = 0.25) = 1.62$

two kinds of electrolyzer, inasmuch as the opposite conclusion can be reached only at an inadmissibly high 25% error. If the cofactors are suppressed, a conventional ANOVA test yields a similar result with $f = 1.42$. Since $f_{1,18} = 1.54$ ($\alpha = 0.25$), rejecting H_0 would be highly unadvisable.

4. Covariance Analysis for a Two-Factor, Single Cofactor CRE

The performance of a cathodic process is analyzed in the following manner. As shown in Table 9, two levels of electrolyte composition (A_1 and A_2) and three different cathodes (levels B_1, B_2, B_3) are employed to determine the fraction of current wasted on hydrogen evolution at various settings of the cathode current density. The latter is taken to be the single concomitant variable.

The underlying postulate states that the y_{ijk} measurements (%-wasted current) belong to a normal population with mean μ_{ijk} and variance σ^2, and that in the expression

$$\mu_{ijk} = \mu + \alpha_i + \beta_j + (\alpha\beta)_{ij} + \gamma(x_{ijk} - x_{m..}) \qquad (74)$$

the parameters obey the condition

$$\sum_{i=1}^{a}\alpha_i = \sum_{j=1}^{b}\beta_j = \sum_{i=1}^{a}(\alpha\beta)_{ij} = \sum_{j=1}^{b}(\alpha\beta)_{ij} = 0 \qquad (75)$$

The null hypothesis set may be written as

$$H_0' : \alpha_i = 0; i = 1,2,...,a; H_a' : \text{not all } \alpha_i \text{ are zero} \qquad (76)$$

$$H_0'' : \beta_j = 0; j = 1,2,...,b; H_a'' : \text{not all } \beta_j \text{ are zero} \qquad (77)$$

$$H_0''' : (\alpha\beta)_{ij} = 0; i = 1,2,...,a; j = 1,2,...,b; H_a''' : \text{not all } (\alpha\beta)_{ij} \text{ are zero} \qquad (78)$$

and, in addition,

$$H_0^* : \gamma = 0; H_a''' : \gamma \neq 0 \qquad (79)$$

Table 9
Data Array for ANCOVA in Section IV.4 with Current Density as Single Concomitant Variable (Two-Factor CRE)

Level B_1 Cathode 1		Level B_2 Cathode 2		Level B_3 Cathode 3	
x (A/dm^2)	y (%)	x (A/dm^2)	y (%)	x (A/dm^2)	y (%)
Level A_1					
4.0	9.5	3.0	8.5	5.0	9.0
3.5	8.0	4.0	10.0	4.0	8.5
4.0	9.5	4.5	8.5	4.0	9.0
5.0	10.5	4.0	9.0	3.0	8.0
4.5	10.0	4.0	9.0	4.0	8.5
Level A_2					
5.0	10.0	5.0	10.0	4.5	9.5
3.0	9.5	3.0	9.0	3.0	8.5
3.5	9.5	4.0	9.5	2.5	7.5
4.5	11.0	4.5	9.0	5.0	10.5
3.0	8.8	4.0	9.5	3.5	8.5
Totals		X		Y	
$T_{1..}$		60.5		135.5	
$x_{m1..}$		58.0		140.3	
$T_{.1.}$		40.0		96.3	
$T_{.2.}$		40.0		92.0	
$T_{.3.}$		38.5		87.5	
$T_{...}$		118.5		275.8	
Overall mean		3.950		9.193	
Q		$\sum_{i=1}^{a}\sum_{j=1}^{b}\sum_{k=1}^{n} Q_{ijk}$			
x^2		483.25			
y^2		2554.44			
xy		1100.65			

$a = 2; b = 3; n = 5; N = 30$

Table 10
ANCOVA Array for the Data Shown in Table 9

Source[a]	SS_x	SP	SS_y	SS'_y	DF	MS'_y	F	P-value
A	0.208	−0.4	0.745	1.452	1	1.452	5.56	0.028
B	0.150	0.665	3.873	2.928	2	1.464	5.61	~0.01
AB	0.517	0.115	0.0487	0.167	2	0.0834	0.31	>0.25
Error	14.3	10.86	14.252	6.005	23	0.2611		
Total	15.18	11.24	18.919					

$f_{1,23}(\alpha = 0.025) = 5.75;\ f_{1,23}(\alpha = 0.05) = 4.28;\ f_{2,23}(\alpha = 0.01) = 5.66$

[a] Degrees of freedom: $(a-1)$ for Treatment A; $(b-1)$ for Treatment B; $(a-1)(b-1)$ for Treatment AB; $[ab(n-1)-1]$ for Error; $(abn-2)$ for Total.

The involved computation scheme, presented in sufficient detail by e.g., Guenther,[17] yields the numerical values assembled in Table 10. The electrolyte composition effects are significant, but not highly significant, and rejection of H_0' carries an about 3% Type I error. The cathode type effects are essentially highly significant, since rejection of H_0'' carries at most a 1% Type I error. The combined effects of the two treatments are additive, since H_0''' cannot be rejected. Finally, H_0^* is rejected at an extremely high level of confidence, because the computed F-statistic

$$f = \frac{(SP_E)^2}{(SS_{xE})(MS'_{yE})} = \frac{(10.86)^2}{(14.3)(0.2611)} = 31.59 \tag{80}$$

is considerably larger than the critical value $f_{1,23} = 14.19$ ($\alpha = 0.001$). Consequently, the current density is rightly considered to be the concomitant variable.

VIII. MISCELLANEOUS TOPICS

1. Estimation of the Type II Error in ANOVA

In general, a Type II error is committed when a false hypothesis is not rejected. More specifically, the experimenter or process analyst not rejecting H_0 when H_a is true risks this error. The usual symbol for Type II error is β, and the power of the test is $(1 - β)$. Since the latter is the probability of rejecting the false null hypothesis when the alternative hypothesis is true, it follows that if $H_a = H_0$, the power is the same as α, i.e., the Type I error.

The computation of power is straightforward if α and β can readily be computed on the basis of normal, binomial, Poisson, etc. cumulative probability distributions. For instance, an analyst may be given the statement that one half of a particular type of low-temperature non-aqueous cells will retain its design potential at a 2.1 Ah capacity under a set load and temperature during discharge. The analyst might then decide that if out of 100 randomly chosen cells 57 (or more) cells are still operational after the passage of 2.1 Ah, the null hypothesis $H_0 : p = 0.5$ will be rejected in favour of $H_a : p > 0.5$. Using the normal distribution as a close approximation to the rigorous binomial, the Type I error is about 8.1%. If the true fraction (unknown to the process analyst) of the surviving cells were, however, 0.6, then the Type II error and power would be about 27% and 73%, respectively. If the true fraction were 0.7, β = 0.0024 (i.e., 0.24%) and $(1 - \beta)$ would be about 98.8%. However, if the true fraction were 0.55, the Type II error would be a rather large 66%, but of very small importance, since the two postulated means are very close. Such calculations, because of their relative simplicity, are routine subject matter in elementary textbooks on statistics.

The computation of power in ANOVA is considerably more complicated, involving numerical integration arising from the theory of the non-central F-distribution. To avoid this encumbrance, the Pearson-Hartley graphs[18-20] established for α = 0.05 and α = 0.01 can be employed. The power is given in terms of the second degree of freedom (v_2) and a non-centrality parameter ψ as a function of the first degree of freedom (v_1) of the F-distribution.

The results of a one-way classification test, shown in Table 11, are employed for illustration. Electrolysis is assumed to have been carried out at three different current densities in eight identical cells assigned to each current density, with current efficiency measured in each cell. The P-value is about 0.08 and it follows that the null hypothesis of insignificant current density effects cannot be conventionally rejected. In other words, the cell-to-cell variations are too large to assign variability to current density, at least by a one-way classification. The question of Type II error arises when at least one of the current efficiency sets pertaining to a particular current density differs from the rest, but this is unknown to the process analyst. Since the ψ parameter depends on the true variance σ^2, it is convenient to set this deviation in terms of σ-multiples. The analyst may postulate e.g., a $\beta_1 = \beta_3 = \beta$; $\beta_2 = \beta + k\sigma$ deviation scheme, where k is an arbitrary constant. In a one-way classification test the non-centrality parameter can be written for treatment sets of equal size as[21]

$$\psi = \frac{1}{\sigma}(\frac{n}{r}\sum_{j=1}^{r}\beta_j^2)^{1/2} \qquad (81)$$

hence, in the case under discussion,

$$\psi = \frac{1}{3}\sqrt{2nk} \qquad (82)$$

Table 11
Analysis of Type II Error in One-Way Classification Employing Three Current Density Levels as Treatment in Eight Cells Each

Source	SS	DF	MS	F	P-value
Among treatments	169.00	2	84.50	2.91	0.08
Within Treatments	610.00	21	29.05		
Total	779.00	23			

$f_{22,21}$ ($\alpha = 0.05$) = 3.47; $f_{2,21}$ ($\alpha = 0.1$) = 2.58

	k	ψ	($1-\beta$)	
			$\alpha = 0.05$	$\alpha = 0.01$
$n = 8; r = 3$	1.0	4/3	~ 0.45	~ 0.20
	1.5	2	~ 0.81	~ 0.58
	2.0	8/3	~ 0.98	~ 0.90

is the final form of the parameter. Table 11 demonstrates the sensitiveness of power to ψ, at a fixed degree of freedom. For a one-sigma deviation of the mean ($k = 1$) the power of the test is expectably low.

A major role of power lies in determining the size of observations required for a pre-set value of errors. With the data in Table 11, if the required power is at least 0.90 when the Type I error is 0.05, a quick successive iteration involving $v_1 = 2$, $v_2 = 3n - 3$, Eq. (82), and the Pearson-Hartley graphs leads to $n = 21$ when $k = 1$. This means that at least 21 identical cells should be operated at each current density, in order to ensure a Type II error not exceeding 10%. Such an arrangement may, of course, not be practical, and the process analyst would have to settle for a lower power in order to keep down the number of experimental cells.

2. Hierarchical Classification

Consider electrodes delivered to a company operating several electrolytic plants using a large number of multiple-electrode cells. The electrodes contain an electro-catalyst whose percentage in the electrodes varies randomly (but within close limits). The process analyst may wish to determine if plant-to-plant variations, and/or cell-to-cell variations within a plant are significant by using the hierarchical classification procedure. In its fixed-effect version, electrode samples are taken randomly from b number of cells in a plant chosen from a number of plants, and n number of electrodes per cell. It is postulated that each population is normal, having the same variance. The observations of catalyst content in an electrode can then be written as

$$x_{ijk} = \mu + \alpha_i + B_{ij} + e_{ijk} \tag{83}$$

where e_{ijk} are normally distributed with zero mean and variance σ^2. Also, $\alpha_1 + \alpha_2 + \ldots + \alpha_a = 0$, and $B_{i1} + B_{i2} + \ldots + B_{ib} = 0$; $i = 1, 2, \ldots, a$. The hypotheses may then be expressed as

$$H_0 : \alpha_i = 0; i = 1, 2, \ldots, a \tag{84}$$

$$H_a : \text{not all } \alpha_i \text{ are zero} \tag{85}$$

$$H_0' : B_{ij} = 0; \quad i = 1, 2, \ldots, a; \quad j = 1, 2, \ldots, b \tag{86}$$

$$H_a' : \text{not all } B_{ij} \text{ are zero} \tag{87}$$

In this classification, SS_T and SS_A have essentially the same role as in two-factor analysis, and $SS_{B(A)}$ is the sum of squares for variation levels of factor B *within* factor A (i.e., cells within plants). Consequently, SS_E is the same sum of squares for

Table 12
Hierarchical Classification Table for Coded Catalyst Content with a = 3 Electrolytic Plants, b = 4 Cells per Plant, and n = 5 Electrodes Taken Randomly from Each Cell

$T_{11.} = 15.6$;	$T_{21.} = 23.2$;	$T_{31.} = 35.3$;	$T_{12.} = 18.7$;	$T_{22.} = 18.1$;	$T_{32.} = 29.6$;
$T_{13.} = 18.3$;	$T_{23.} = 22.1$;	$T_{33.} = 36.3$;	$T_{14.} = 19.7$;	$T_{24.} = 23.8$;	$T_{34.} = 33.3$
$T_{1..} = 72.3$;	$T_{2..} = 87.2$;	$T_{3..} = 134.5$;	$T_{...} = 294$;	$\sum_{ijk} x_{ijk}^2 = 1586.46$	

ANOVA					
Source	SS	DF	MS	F	P-value
Levels of A	105.47	2	52.73	86.30	< 0.001
Levels of B within A	11.04	9	1.23	2.01	0.063
Error	29.35	48	0.611		
Total	145.86	59			

$f_{2,48}(\alpha = 0.001) = 8.06$; $f_{9,48}(\alpha = 0.05) = 2.09$; $f_{9,48}(\alpha = 0.10) = 1.77$

variations within the same B-level within the same A-level (i.e., electrodes from the same cell). Table 12 shows a numerical illustration with $a = 3$, $b = 4$, and $n = 5$. The sums of squares are computed as

$$SS_T = \sum_{i=1}^{a}\sum_{j=1}^{b}\sum_{k=1}^{n} x_{ijk}^2 - \frac{T_{...}^2}{abn} = 145.86 \tag{88}$$

$$SS_A = \frac{\sum_{i=1}^{a} T_{i..}^2}{bn} - \frac{T_{...}^2}{abn} = 105.47 \tag{89}$$

$$SS_{B(A)} = \frac{\sum_{i=1}^{a}\sum_{j=1}^{b} T_{ij.}^2}{n} - \frac{\sum_{i=1}^{a} T_{i..}^2}{bn} = 11.04 \tag{90}$$

$$SS_E = SS_T - SS_A - SS_{B(A)} = 29.35 \tag{91}$$

The degrees of freedom are $(abn - 1)$ for SS_T, $(a - 1)$ for SS_A, $a(b - 1)$ for $SS_{B(A)}$, and $ab(n - 1)$ for SS_E, respectively.

The results indicate that the catalyst content of the electrodes varies sharply from plant to plant, but variations between cells within plants are not significant. This finding would suggest the possibility of a faulty distribution scheme to the process analyst.

Hierarchical classification with mixed-model and random-effect model formulations, the estimation of the Type II error,[22] and related analysis of covariance have been discussed elsewhere.[23]

3. ANOVA-Related Random Effects

The basic tenet of fixed-effect based analysis is that the treatments are self-contained populations. If, however, treatments are considered to be samples of a large treatment population, then the β_j parameters are no longer (unknown) constants, but random variables belonging to the normal distribution $N(0;\sigma^2)$. In fact, the means μ_1, $\mu_2, \ldots, \mu_r$ are themselves random. Under such circumstances the null and alternative hypotheses are:

$$H_0 : \sigma_\beta^2 \leq K_0\sigma^2 \tag{92}$$

$$H_a : \sigma_\beta^2 \geq K_0\sigma^2 \tag{93}$$

where K_0 is an arbitrary constant. These hypotheses are more practical than the perhaps more obvious hypotheses of the β-variance being zero or non-zero. Alternatively, the confidence interval of the σ_β^2/σ^2 ratio may be established by the inequality

$$\frac{1}{n}(\frac{MS_A}{MS_W} \frac{1}{F_{r-1;N-r;1-0.5\alpha}} - 1) < \\ \sigma_\beta^2/\sigma^2 < \frac{1}{n}(\frac{MS_A}{MS_W} \frac{1}{F_{r-1;N-r;0.5\alpha}} - 1) \tag{94}$$

may be employed in lieu of Eqs. (92) and (93). It is worth noting that the sample-based estimate of the variance, s_β^2, is given by the expression

$$s_\beta^2 = \frac{MS_A - MS_W}{n} \tag{95}$$

for treatment observations of equal size.

In the case of randomized blocks Eq. (86) is slightly modified; the MS_A/MS_W ratio is replaced by the MS_{Tr}/MS_E ratio, and the second degree of freedom of the F-statistic becomes $v_2 = (n-1)(r-1)$.

When treatment observations are of unequal size, the parameter

$$n' = \frac{(\sum_{j=1}^{r} n_j)^2 - \sum_{j=1}^{r} n_j^2}{(r-1)\sum_{j=1}^{r} n_j} \tag{96}$$

replaces n in Eq. (94).

Table 13 contains assumed membrane deterioration data observed with four different membranes employed over a three-month period in three different electrolyzers. Fixed-effect based

Table 13
RBE Applied to the Membrane Deterioration Problem in Section VIII.2[a]

Blocks	Treatment				T_i	$x_{mi.}$
	Membrane 1	Membrane 2	Membrane 3	Membrane 4		
Electrolyzer1	7.4	9.3	11.1	10.5	38.3	9.58
Electrolyzer2	3.9	8.7	8.1	9.9	30.6	7.65
Electrolyzer3	10.2	9.4	8.5	12.1	40.2	10.05
T_j	21.5	27.4	27.7	32.5	Overall sum: 109.1	
$x_{m.j}$	7.17	9.13	9.23	10.83	Overall mean: 9.09	

ANOVA

Source	SS	DF	MS	F	P-value
Treatments	20.2833	2	10.1417	4.01	0.084
Blocks	12.9225	3	4.3075	1.70	0.154
Error	15.1834	6	2.5306		
Total	43.3892	11			

$v_1 = 3; v_2 = 6$				
Type I error, α	0.10	0.05	0.01	0.002
$f_{\alpha/2}$	4.76	6.60	12.92	23.70
$f_{1-\alpha/2}$	0.1119	0.0678	0.0223	0.00753

[a]The entries are on a dimensionless scale 1 (best) – 15 (worst)

ANOVA fails to distinguish between electrolyzers (blocks) and membranes (treatments). If, however, the electrolyzers are taken to be a sample of a large number of similar electrolyzers slated for the same process with the same membrane types, the confidence intervals of the σ_β^2/σ^2 variance ratio, shown in Table 14, indicate a large domain of possible β-variance values.

The effect of treatment population size on this confidence interval is illustrated in Table 15. The first ANOVA array is established by considering the first three current density (25, 50, 75 mA/cm^2) treatments of anodes constructed of eight different alloys in a primary alkaline battery whose anode efficiency was studied by Paramasivan et al.[24] The second array refers to the entire experimental set with three additional treatments (100,125,150 mA/cm^2). Since the magnitude of the F-statistics indicates an extremely small Type I error in both cases, the conclusion that both current density and alloy composition exhibit extremely important effects is the same for the subset and the full set. The major difference is in the considerable reduction in the width of the confidence interval for the β-variance, owing to the larger amount of information rendered by the six-treatment set.

Table 14
The Effect of the Type I Error on the Confidence Interval for the Variance Ratio via Equation (94)

α	Lower bound of variance ratio[*]	Upper bound of variance ratio
0.10	0	11.56
0.05	0	19.27
0.01	0	59.33
0.002	0	176.38

[*] negative lower limits are replaced by zero

4. Introductory Concepts of Contrasts Analysis

As mentioned briefly in Section II, contrasts analysis is a means of deciding what treatment means are statistically different, if ANOVA rejects the null hypothesis of treatment-mean equality. An especially thorough discussion[25] indicates the variety of techniques available for analysis; in this Section only the method by Scheffé[5,26,27] is illustrated, on account of its generality and the convenience of using F-distribution (instead of special distributions required by other methods, e.g., Duncan's, Tukey's, etc.)

Table 15
ANOVA of Anode Efficiency Data in Table 6[22]

Treatments: current density; blocks: alloy composition					
(a) Subset of 3 treatments ($n = 8$; $r = 3$)					
Source	SS	DF	MS	F	P-value
Treatments	995.25	2	497.63	101.33	< 0.001
Blocks	418.50	7	59.79	12.17	< 0.001
Error	68.75	14	4.91		
Total	1482.5	23			
$f_{2,14}(\alpha = 0.001) = 11.78$; $f_{7,14}(\alpha = 0.001) = 7.08$					
(b) Total set of 6 treatments ($n = 8$; $r = 6$)					
Source	SS	DF	MS	F	P-value
Treatments	2732.50	5	546.50	102.74	< 0.001
Blocks	362.33	7	51.76	9.73	< 0.001
Error	186.17	35			
Total	3281.00	47			
$f_{5,35}(\alpha = 0.001) = 5.30$; $f_{7,35}(\alpha = 0.001) = 4.60$					
95% confidence interval of σ_β^2/σ^2					
Number of treatments	Lower limit		Upper limit		
3	2.48		499.4		
6	4.20		79.6		

Table 16
Contrasts Analysis Related to a One-Way Classification ANOVA of SER Data

Treatments[a]			
Plant 1	Plant 2	Plant 3	Plant 4
4.714	4.757	4.843	5.057
4.671	4.971	4.628	5.271
4.500	4.671	4.671	5.357
T_j : 13.885	14.399	14.142	15.685
$x_{m.j}$: 4.628	4.800	4.714	5.228

$$\sum_{ij} x_{ij}^2 = 282.1935$$

ANOVA				
Source	SS	DF	MS	F	P-value
Among treatments	0.6393	3	0.2131	11.60	0.0037
Within treatments	0.1469	8	0.0184		
Total	0.7862	11			

$f_{3,8}(\alpha = 0.005) = 9.60; f_{3,8}(\alpha = 0.001) = 15.83$

Differences in observation means		
4 – 1: 0.6	4 – 3: 0.514	4 – 2: 0.428
2 - 1: 0.172	2 – 3: 0.086	
3 – 1: 0.086		

[a]The entries are in units of MWh/metric ton of sodium chlorate

The gist of the method is to ascertain if (selected) sample-mean differences are larger or not than the right hand side of Eq. (97). Denoting an arbitrary sample mean difference as Δ_{ij}, if the condition

$$\Delta_{ij} > [(r-1)F_{\alpha, r-1, N-r}(MS_W \sum_{j=1}^{r} \frac{c_j^2}{n_j})]^{1/2} \qquad (97)$$

is satisfied, the treatment means in question are different at an α-level of significance. Eq. (97) applies to one-way classification only, each classification having its own set of conditions. Specifically, in the case of randomized blocks, MS_W in Eq. (97) is replaced by MS_E and the second degree of freedom is $v_2 = (n-1)(r-1)$.

Table 16 contains observations of SER from four plants producing electrolytic sodium chlorate. The null hypothesis of no plant-to-plant difference in SER is rejected with a Type I error of about 0.4%, hence at least two means are highly significantly different. At α = 0.01, the critical F-statistic is $f_{0.01,3,8}$ = 7.59, and considering differences between only two means, Eq. (97) yields the condition Δ_{ij} > 0.528, which is satisfied only by (5.228 − 4.628) = 0.6, all other differences being smaller than 0.528. It then follows, that the method identifies only the ($\mu_4 - \mu_1$) difference as highly significant. If α = 0.05 is chosen, $f_{0.05,3,8}$ being 4.07, the significance condition Δ_{ij} > 0.387 identifies the mean-differences ($\mu_4 - \mu_1$), ($\mu_4 - \mu_3$) and ($\mu_4 - \mu_2$) as significant. Other contrast-analytic methods do not necessarily lead to the same results; it is to be noted that Scheffé's test is the "most conservative with respect to Type I error and will lead to the smallest number of significant differences."[28]

IX. FINAL REMARKS

The fundamental postulate of ANOVA, that observations belong to normal distributions with a constant (albeit usually unknown) variance cannot be *ab ovo* guaranteed. It is known, however, that the F-test for the equality of means is insensitive to deviations from normality,[29] and in the case of large observations, normality is a good approximation to the real distribution of data.

To query the postulate of constant variance (i.e., to test the homogeneity of the observations), the Bartlett test is commonly used in ANOVA. The test can be carried out either by using tables of Bartlett's probability distribution,[30] or by chi-square distribution tables of a statistic whose sampling distribution is nearly chi-square.[31] The test is particularly recommended for treatments carrying observations of unequal size. For equal-size observations, a quick test[32] consists of comparing the ratio of the largest to the smallest treatment variance to critical values of the F_{max}-statistic in appropriate tables.[33]

The major relevance of material presented in this chapter, best viewed as an *appetite-whetter*, resides in economy. Proper statistical techniques can save considerable time and effort to the experimenter and the process analyst in trying to extract the largest possible amount of information from available data, and to optimize the size of statistically meaningful experiments.

ACKNOWLEDGMENTS

The author is grateful to the Natural Sciences and Engineering Research Council of Canada (NSERC) and the University of Waterloo for facilities provided to prepare this chapter.

LIST OF PRINCIPAL SYMBOLS

a, b, r	factor dimension
CRE	completely randomized experiment
DF	degree of freedom
e	error
F	F-statistic; f its numerical value
H_0	null hypothesis
H_a	alternative (or counter) hypothesis
L	contrast
MS_Q	means square related to quantity Q
n	observation size per treatment
N	total observation size
$N(\mu;\sigma^2)$	normal distribution with mean μ and variance σ^2
RBE	randomized block experiment
SER	specific energy requirement
SP_Q	sum of products related to quantity Q
SSQ	sum of squares related to quantity Q
$SS_Q^{\sim}$	modified sum of square related to quantity Q
T	sum of observations
x, y	observed quantity

Subscripts

A	among groups (or treatments)
A, B, C	treatment (factor)
AB, BC, ABC	interaction
BL	block
C	column
E	error
R	row
T	total
Tr	treatment
x, y	observations
W	within groups (or treatments)

Greek Symbols

α	deviation variable, and size of Type I error
β	deviation variable, and size of Type II error
μ	treatment mean
σ^2	variance
ν	degree of freedom

REFERENCES

[1] W. C. Guenther, *Analysis of Variance*, Prentice Hall, Englewood Cliffs, 1964.
[2] B. J. Winer, *Statistical Principles in Experimental Design*, 2nd Ed., McGraw Hill, New York, 1971, Section 10.1, p.752.
[3] R. G. D. Steel and J. H. Torrie, *Principles and Procedures of Statistics*, 2nd Ed., McGraw Hill, New York, 1980, Section 17.3, p.405.
[4] G. W. Snedecor and W. G. Cochran, *Statistical Methods*, 8th Ed., Iowa State Univ. Press, Ames, 1989, Chapters 11 – 16,18.
[5] H. Scheffé, *The Analysis of Variance*, Wiley and Sons, New York, 1959.
[6] F. Hine, *Electrode Processes and Electrochemical Engineering*, Plenum, New York, 1985, Table 7.10, p.140.
[7] W. C. Guenther, loc. cit., Eq. (3.20), p.77.
[8] J. O'M. Bockris and S. U. M. Khan, *Surface Electrochemistry*, Plenum, New York, 1993, Table 8.10, p.819.
[9] R. E. Walpole, R. H. Myers, S.L. Myers and K. Ye, *Probability and Statistics for Engineers and Scientists*, 7th Ed., Prentice Hall, Upper Saddle River, 2002, p.536.
[10] S. Rode, S. Altmeyer and M. Matlosz, *J. Appl. Electrochem.* **34** (2004) 671.
[11] W. C. Guenther, loc. cit., p.146.
[12] B. J. Winer, loc. cit., Section 10.7, p.809; 1st edn., 1962, Section 11.7, p. 618.
[13] R. G. D. Steel and J. H. Torrie, loc. cit., Section 17.12, p.428.
[14] G. W. Snedecor and W. G. Cochran, loc. cit., Section 18.9, p.393.
[15] B. J. Winer, loc. cit., p.812 (2nd edn.),p.621(1st edn.)
[16] C. A. Bennett and N. L. Franklin, *Statistical Analysis in Chemistry and the Chemical Industry*, Wiley and Sons, New York, 1963, p. 456.
[17] W. C. Guenther, loc. ci.t, Section 6.7, p.156.
[18] E. S. Pearson and H. O. Hartley, *Biometrika*, **38** (1951) 112.
[19] W. C. Guenther, loc. cit., Table 6, pp. 186–193.
[20] *CRC Handbook of tables for Probability and Statistics,* 2nd Ed., Ed. by W. H. Beyer, CRC Press, Boca Raton, 1968, p.313;1st edn., 1966, p.105.
[21] W. C. Guenther, loc. cit., p.47.
[22] W.C. Guenther, loc. cit., Section 5.14, p.136.
[23] W. C. Guenther, loc. cit., Section 6-8, p.161.
[24] M. Paramasivan, M. Jayachandran and S. Venkatakrishna Iyer, *J. Appl. Electrochem.* **33** (2003) 303.
[25] R. G. Steel and J. H. Torrie, loc. cit., Chapter 8, p.172.
[26] R. G. Steel and J. H. Torrie, loc. cit., Section 8.5, p.183.
[27] W. C. Guenther, loc. cit., Section 2-14, p.57
[28] B. J. Winer, loc. cit., Section 3-9, p.196.

[29] W. C. Guenther, loc. cit., p.63.
[30] R. E. Walpole et al., loc. cit., Section 13.4, p.469.
[31] R. E. Walpole, *Introduction to Statistics*, Collier-Macmillan, London, 1968, Section 12.3, p.299.
[32] J. L. Bruning and B. L. Kintz, *Computational Handbook of Statistics*, 2nd Ed., Scoot, Foreman and Co., Glenview, Illinois, 1977, Section 3.3, p.112.
[33] J. L. Bruning and B. L. Kintz, loc. cit., Appendix I, p.259.

3

Nanomaterials in Li-Ion Battery Electrode Design

Charles R. Sides and Charles R. Martin*

Department of Chemistry, University of Florida, Gainesville, FL
*Contact author

I. INTRODUCTION

Li-ion batteries have generated great interest as lightweight, portable, rechargeable power sources over the last decade. Their introduction in 1990 by T. Nagaura and K. Tozawa of SonyTec Inc. fueled the explosion of personal electronic devices.[1] Li-ion batteries are now the power source of choice for laptops, cell phones, and digital cameras. The public has quickly embraced this technology, which accounts for an approximately $3 billion annual market.[2] Despite (or perhaps as a result of) the commercial success of these batteries, a global research initiative exists to improve the existing design. The goal of which is to apply this technology to more demanding and exotic uses, such as the electric component of hybrid vehicles, low-temperature applications, and power supplies for MEMs. However, the current design cannot adequately satisfy the power requirements of such systems, due to the inability to deliver a sufficient quantity of charge at high discharge currents.[3] This chapter will detail the efforts of laboratories, ours in particular, to incorporate the field of nanomaterials to improve upon Li-ion batteries.

Li-ion batteries operate by reversibly intercalating charge in each of two electrodes. *Intercalation* is the process by which a guest species (Li^+) is able to reversibly enter/exit a host structure, causing little or no difference to the lattice of the host. These electrodes are separated by an ion-conductive electrolyte. Upon discharge, the Li-ions deintercalate from the low-potential electrode, migrate through the electrolyte, and insert into the high-potential electrode. The ions then rely on solid-state diffusion to fill the non-surface intercalation sites. Obeying the governing laws of charge neutrality, electrons compensate for the movement of the ions. If current flow is reversed (from cathode to anode), Li-ions insert into the low-potential electrode and the system is charged. The low-potential electrode is the *anode* and the high-potential electrode is the *cathode*. This convention (adopted from the discharge process) is obeyed regardless of the direction of current flow. In the analysis of a battery system, both the ionic conductivity and electronic conductivity must be considered. Nanomaterials are advantageous in both regards.

The Martin research group has pioneered the nanofabrication strategy of template synthesis.[4] This general method has been used to synthesize nanostructures of a variety of materials such as gold,[5-8] carbon,[9-11] semiconductors,[12,13] polymers,[14,15] and Li-ion battery electrodes,[11,13,16-28] our focus here. In general, this method involves deposition of a precursor material into a micro- or nanoporous template. This template is typically commercially available track-etch polymer filters or anodized alumina, though others have been demonstrated. Depending on both the pore-diameter and the specific chemical interactions between the pore wall and the precursor, the resulting structures may be tubes (hollow) or wires (solid). These structures are referred to as "nano", if one or more of their dimensions are on the nanoscale (< 100 nm). The aspect ratio (length / width), though, is often on the order of 10.

In this embodiment of Li-ion electrodes, a precursor-impregnated polycarbonate template membrane is attached to a section of metal foil. The foil has dual-functionality as it serves as a substrate during synthesis and as a common current collector during electrochemical characterization. The precursor is processed (typically, by aging or heating) into the desired product. Often in the case of battery materials, the template is then removed by plasma etching or dissolution. The result is an electrode that consists of structures that mirror the geometry (length, diameter, and number density) of the pores of the template. These structures will extend from the surface of the current collector like the bristles of a brush. This ensemble is then heated to convert the material into the phase capable of reversibly intercalating Li-ions under the correct conditions (i.e., serving as a Li-ion battery electrode). Each electrode is then electrochemically characterized as a half-cell.

This synthetic method can be contrasted to a general method for commercial Li-ion battery systems.[29] That synthetic process generates particles of active material that range in diameter from approximately 2 to 20 microns. These discrete particles are then mixed into slurry with polymeric binder (PVDf) and a conductive element (carbon) and pressed into a metal mesh current collector. The conductive element is necessary to improve the electronic conductivity of the system, which is very low in this system, because the particles are only point-connected. The polymer is then necessary to simply ensure that the slurry physically contacts the current collector. In our template-synthesis approach, these inactive (not capable of Li-ion intercalation) components are not necessary and are excluded so as not to decrease volumetric and gravimetric energy densities or complicate analysis.[30]

Template-synthesized nanostructured electrodes have demonstrated the ability to deliver more charge per gram (mAh g^{-1}) than a film-control electrode at any given high-rate.[20,24,26,27] This can be attributed to two favorable parameters associated with nanomaterials 1.) decreased solid-state diffusion distance and 2.) decreased effective current density (current / surface area). The high insertion-flux density of Li-ions into/out of the electrode and sluggish solid-state diffusion of Li-ions in the electrode result in concentration polarization of the ions in the electrode material. This limits the rate at which electrode may be discharged and still deliver useful charge. Other labs have come to similar conclusions.[3,31,32]

The intent of this chapter is to demonstrate how nanoscale Li-ion battery design succeeds, where conventional technology fails. We will begin with a discussion of the different templates. Next, we describe both cathodic and anodic materials created by the template-synthesis method. This will be followed by a demonstration of how variations to the general method have solved fundamental questions and improved upon practical limitations of certain Li-ion battery materials. Finally, a plasma-aided method will be discussed that creates a template-replica in a thin-film of a Li-ion anode material.

II. TEMPLATES USED

1. Track-Etch Membranes

Micro- and nanoporous polymeric filtration membranes prepared via the "track-etch" method are available from commercial sources (*e.g.*, GE Osmonics) in a variety of materials and pore geometries.[33] Polycarbonate is perhaps the most common example of a track-etch filter material. Other options of materials include polyester, Teflon®, and polyethersulfone. Electron micrographs of the surface and cross-sections of some of these templates are shown in Figure 1.

The term *track-etch* refers to the pore-production process[34, 35]. The pores of the filters are created by exposing the solid material film to nuclear fission fragments, which leave randomly-dispersed damage-tracks in the film. The high energy (on the order of 2 GeV) of the fragments ensures that the tracks span the entire length of the membrane (typically from 6 to 10 μm). These reactive chain ends at the damage-tracks in the polycarbonate film are then etched with a basic chemical solution, and they become pores. One ion creates one track, which in turn becomes one pore. During production, the pore density is controlled by the duration of time that the polycarbonate film is exposed to the charged particles. Typical

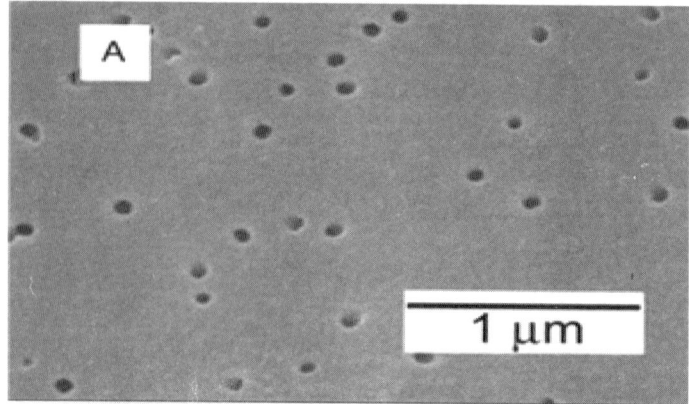

Figure 1. A few examples of porous structures (templates) produced in thin polymeric films using various methods of irradiation and chemical treatment. (A) Surface of polycarbonate; scale bar equals 1 μm.[27] (B) Polypropylene TM (track-membrane) with nearly parallel pores. (C) Polyethylene terephtalate TM with cigar-like pores. B and C are reprinted from Ref. 35, ©2001 with permission from Elsevier.

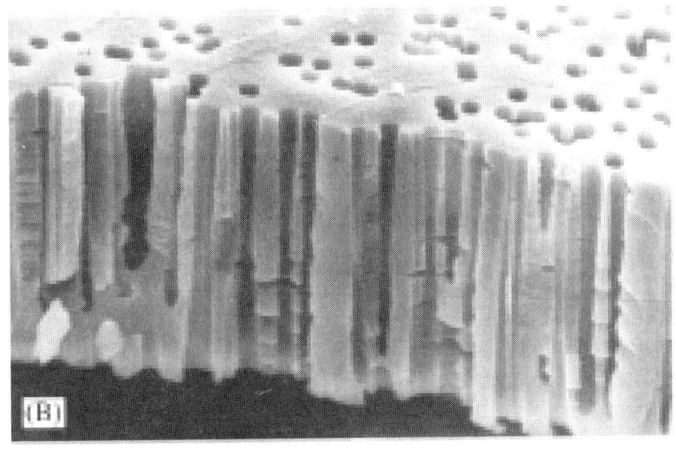

Figure 1. Continuation

track-etch membrane pore-densities are 10^4 to 10^8 pores/cm^2 (Ref. 36).

Varying parameters of this etching solution such as temperature, strength and exposure time dictate the pore diameter. Commercial membranes are available with pore diameters ranging from 10 nm to 20 μm. The microporous material etches to form uniform cylindrical pores, but as the pore diameter is reduced to the smaller nanoscopic dimensions the shape of the pore becomes like a cigar, slightly tapered at the ends. Microscopic investigations of template-synthesized nanostructures prepared within the pores of such nanoporous membranes have shown that the diameter of the pore in the center of the membrane is larger than the diameter at the membrane surface; i.e., cigar-shaped pores are confirmed.[24,37] It has been suggested that this pore geometry arises because the fission fragment that creates the damage track also generates secondary electrons, which contribute to the damage along the track. The number of secondary electrons generated at the faces of the membrane is less than in the central region of the membrane. An alternate suggestion is that the surfactant protective layer adsorbed to the surface of the membrane retards the local etching process.[35] Either suggested mechanism leads to "bottleneck" pores.

In the production of template-synthesized nanostructured battery materials, polymer (specifically polycarbonate) track-etch membranes are the current template of choice. These templates may be easily removed in conditions that do not adversely affect the nanostructures themselves. The wide variety of commercially available pore diameters and densities can generate comparative structures of differing geometries that are key tools for fundamental investigations. The disadvantage that is associated with these types of membranes and their application to electrode materials is their low porosity. These values are typically between 2–10 %. This decreases the ratio of active material to a given footprint area or volume region on the current collector surface.

2. Alumina Membranes

Anodization of aluminum metal in an acidic environment causes the metal to etch in a fashion that leaves a porous structure.[38] These pores are extremely regular, having monodisperse diameters and cylindrical shapes in a hexagonal array. Unlike the track-etch process, this process is systematic and generates an isolated, non-connected pore structure. These membranes are commercially available as filters (Whatman), but only in a very limited pore-diameter.[39] These membranes can be very thick (~60 μm). The pore densities of these alumina filters can be on the order of 10^{11} pores/cm^2, that this about 1000 times the number of pores that are available in the track-etch polycarbonate membranes. This translates to very large porosities (~50%.) However, the only current commercial membrane that has uniform diameter is 200-nm. While smaller effective pore diameters are available, it is a result of "pore-branching" at the end of the 200-nm pore.

Our lab synthesizes "home-grown" alumina templates. The synthetic process is described in detail elsewhere.[40,41] Briefly, a section of high purity Al foil is degreased, electropolished, and then anodized. The anodization is the specific process that forms the pores. Afterwards, the porous structured alumina is removed from the bulk aluminum by a voltage reduction technique. The interfacial oxide layer (barrier-layer) may then be removed by exposure to acidic solution[40] or Ar plasma.[22] The remaining membrane is equivalent on both sides. Electron micrographs of both a surface and cross-sectional view are shown in Figure 2. Control over the synthetic conditions allows us the ability to tailor pore diameters and thicknesses to our specific application. The applications of these alumina templates demonstrated in our lab alone are as diverse as separations for enantiomers,[42] templates for carbon nanotubes,[10] and plasma-etching masks for nanostructured battery materials.[22]

The key advantage of the alumina compared to a track-etch polymer template for energy-storage application is the porosity. This order of magnitude difference in porosity allows much greater volumetric density (g L^{-1}) of battery material. However, alumina filters have much greater mechanical stability and chemical resistivity than polycarbonate. This is an advantage when employed as a filter, but complicates the template dissolution step. Typical methods for alumina dissolution (exposure to a strong acid or base) are too harsh for many battery materials. Recent progress has been made in this field with amorphous MnO_2.[43]

3. Other Templates

The examples of templates discussed to this point have cylindrical-type pore structures. Another type of template that has been employed with battery materials is nanospheres.[32,44] In this method (often called *colloidal crystallization*), spherical particles with diameters of sub-microscale dimensions (typically ~100s of nm) are deposited in a close-packed array. This is commonly accomplished by a solvent evaporation technique. If a solvent evaporates at a slow, controlled rate, it imparts an order to the particles. These particles are typically made of polymers (latex, polysterene, PMMA) or silica. An example of an ordered nanosphere template is shown in Figure 3. Since they are spherical in shape, void-volume exists in the interstitial sites, even when

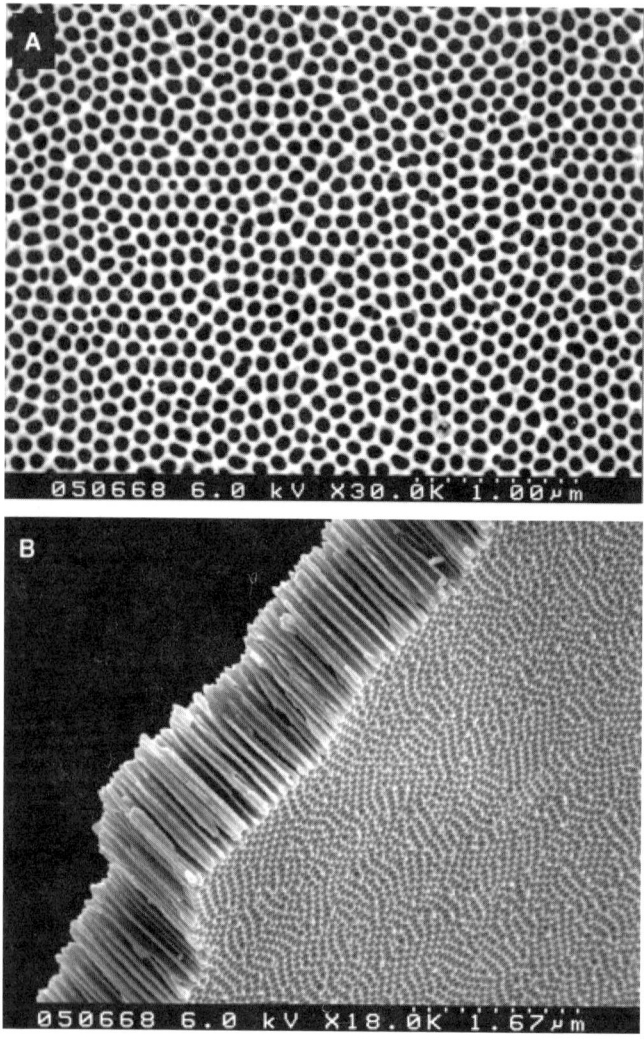

Figure 2. Scaning electron micrographs of a home-grown alumina template. (A) Top view. (B) Side view. Reprinted with permission from Ref. 67. (© 2005), with permission from Wiley-VCH.

close-packed. These interstitial sites become the porous network as the nanosphere array act as a template.

This type of template betters the porosity of the track-etch polymers, yet maintains and ease of removal (dissolution or

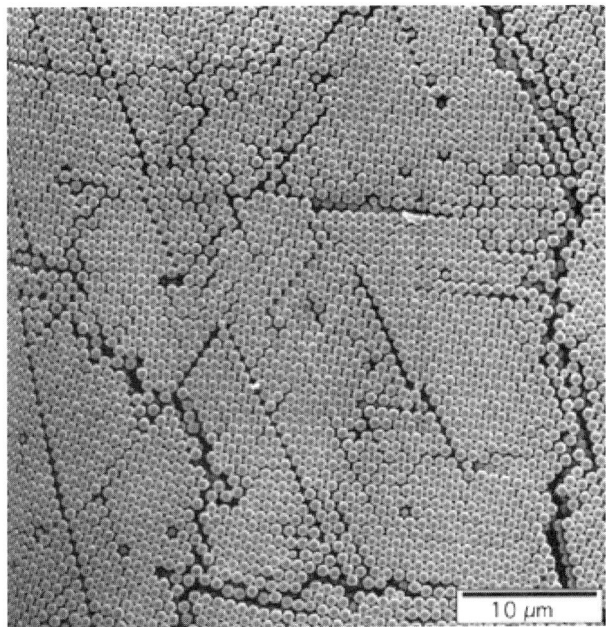

Figure 3. SEM image of PMMA colloidal crystal film fabricated by meniscus deposition of approximately 900 nm PMMA spheres on a Pt-coated glass slide. The film is relatively flat with ordered areas separated by planar and point defects. Reproduced from Ref. 32, (© 2004), by permission of The Royal Society of Chemistry.

calcination.) A distinctive feature of template-synthesized structures that use nanospheres is the connectivity of the resulting system. Research is being conducted to optimize the nanosphere-template synthesis of battery materials and complete their electrochemical characterization.[32, 44]

II. NANOSTRUCTURED CATHODIC ELECTRODE MATERIALS

The *cathode* of a Li-ion battery is the electrode that is operates at the high potential. Typical cathodic active-materials for Li-ion battery systems are transition metal oxides (*e.g.*, Li_xMO_2; where M = Co, Mn, Ni). A popular current research theme is "mixed-metal" materials, where the host structure is doped with another metal to improve the electronic conductivity of the composite[45]. Of course, the quantity of charge capable of being stored is a function of the composition of the host material. This important parameter of stored-charge is thus normalized to unit mass of the host; therefore, it is reported as *specific capacity* (mAh g^{-1}).

As seen in Eq. (1), for each charge/discharge reaction to proceed, both Li-ions and electrons are required. Therefore, both processes (ionic and electronic conduction) are critical (i.e., possibly rate-limiting). Since the solid-state diffusion rate of the Li-ions is an intensive physical property, our strategy becomes to minimize the distance that the ions must traverse. Here we will discuss the fabrication and characterization of template-synthesized cathodic nanostructured V_2O_5. At every rate of discharge the nanostructured electrode was able to deliver more (sometimes dramatically more) specific capacity that the control electrode. This indicates the advantage associated with the template-synthesized nanostructured geometry.

$$V_2O_5 + x\text{Li}^+ + xe^- \underset{\text{charge}}{\overset{\text{discharge}}{\rightleftarrows}} \text{Li}_xV_2O_5 \qquad (1)$$

1. Electrode Fabrication

(*i*) *Nanostructured Electrode*

Template-synthesis makes it possible to routinely create nanostructured electrodes, by restricting particle-growth to the nanoscopic diameter of the pores of the template. The resulting nanostructured electrode has a high surface area to volume ratio of material; therefore, each Li-ion intercalation site resides close to the surface, minimizing the distance that they must solid-state diffuse. For synthesis of nanostructured V_2O_5, a commercially available track-etch polycarbonate filter serves as the template. This membrane has pores that run the entire length of the membrane and are nominally 50 nm in diameter, 6 μm in length and 6 x 10^8 pores/cm^2 in density. A section of this template (~ 1.5 cm^2) is placed on top of a piece of Pt foil. This platinum foil serves as both a substrate for the end-result nanowires and as a common current collector. TIVO (triisopropoxide vanadium) precursor solution is syringed on top of the template membrane. The hydrolysis of this compound in the presence of water vapor

(Eq. 2) is so rapid that this pore-filling step is performed inside of an argon-filled glovebox. This allows the liquid precursor to impregnate the pores of the template, while in the low-viscosity state. This assembly is then transferred to the antechamber which is maintained at a low water content atmosphere.[24] By controlling the moisture content of the atmosphere, we control the rate of reaction of Eq. (2):

$$2 \text{ V(=O)(O-iPr)}_3 \xrightleftharpoons{H_2O \text{ (Water vapor)}} V_2O_5 + [HO-iPr]_3 \quad (2)$$

Vanadium (V) triisopropoxide — Vanadium pentoxide — Isopropanol vapor

The hydrolysis was first done at room temperature for 12 hours in a low O_2 environment, followed by hydrolysis for 2 hours at 80 °C in air. The film of V_2O_5 that covered the upper face of the membrane was then removed by wiping with a damp cotton swab. Oxygen plasma (25 W, 15 Pa, 2 hours) was then used to preferentially etch the organic template.[27] The result of which is nanostructures of the gelled V_2O_5 extending from the platinum foil. To ensure complete conversion to crystalline V_2O_5, the wires were than heated at 400 °C for 10 hours in flowing O_2 gas. A scanning electron micrograph (SEM) of an example of a template-synthesized nanostructured V_2O_5 electrode is shown in Figure 4 .

(ii) Control Electrodes

The control electrode can be either a thin-film[24] or a template-synthesized electrode of microscopic diameter.[27] Thin-film control electrodes are prepared by depositing liquid TIVO onto a Pt foil current collector. This is done as described previously using a micropipette to spread the precursor evenly over the Pt foil surface. Again, hydrolysis proceeds for 12 h in the glove box, after which time the liquid hydrolyzes to a yellow film of TIVO gel. The TIVO gel/Pt composite is then thermally treated in the same way as the nanostructured electrode.[24] This thin-film control electrode is straight-forward to synthesize, but not at a reproducible thickness. This value represents the critical solid-state diffusion distance for the thin-film. However, a template with a *microscopic* pore-diameter would yield a wire of equivalent composition that

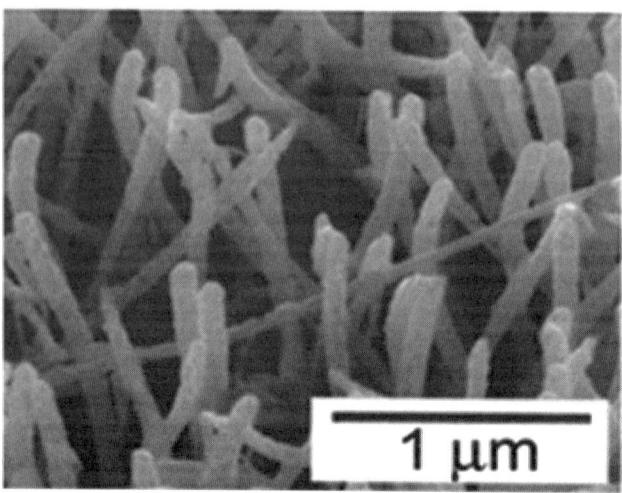

Figure 4. Scanning electron micrograph of a template synthesized V_2O_5 electrodes. Scale bars equal 1 μm. Reproduced from Ref. 27, (© 2005), with permission of Wiley-VCH.

varies only by diameter. This embodiment of a control electrode may alternatively be used. Electrodes of this material have been template-synthesized with diameters as large as 0.8 μm.[27]

2. Structural Investigations

V_2O_5 morphology and crystallinity were investigated with a FE-SEM scanning electron microscope and a X-ray diffractometer (Cu Kα). X-ray diffraction patterns can be compared to accepted literature values to confirm the crystalline-character of the nanostructured electrode (Figure 5). In our latest work, the mass of the electrode material was determined by simple differential weighing of the Pt foil and the Pt/V_2O_5 electrode.[27] Previously, we calculated mass of V_2O_5 post-electrochemical characterization by dissolving the active material in 2 M H_2SO_4 and analyzing the resulting solution for V-ion content via ICP-AES.[24] However, if the mass is determined prior to electrochemical characterization, then applied discharge currents may be normalized to the mass. This allows for direct comparison of electrodes by applying identical C-rates of discharge, as opposed to an arbitrary value of current.

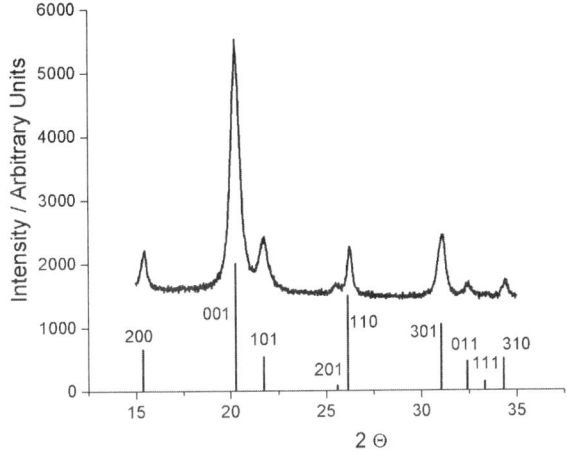

Figure 5. X-ray diffraction pattern of a V_2O_5 electrode created via this method. This compares well to accepted values (JCPDS 41-1426.) The 2Θ range is restricted to eliminate interface from the Pt substrate.

3. Electrochemical Characterization

Chronopotentiometry and cyclic voltammetry (CV) were performed under argon using a three-electrode cell. Li ribbon counter and reference electrodes were used. The electrolyte was 1 M $LiClO_4$ in a 7:3 (v/v) mixture of diethyl carbonate and ethylene carbonate.

(i) Cyclic Voltammetry

We first investigated the Li-ion intercalation properties of V_2O_5 with CV. In this experiment, the charge-discharge reactions are driven as a potential is swept at a constant rate while the current response monitored. A typical CV of a nanostructured V_2O_5 electrode is shown in Figure 6. This figure demonstrates the reversibility of the intercalation reaction. Slow scan-rates (on the order of 0.1 mV s^{-1}) are required to probe the solid-state process. At these conditions, the diffusion of the Li-ions is finite and limited by the radius of the nanostructure.[19] This is observed in the CV by the parameter of ΔE_{pk}, the potential-difference of the peak currents during the anodic and cathodic scan. In the finite diffusion regime, the ideal value is $\Delta E_{pk} = 0$ mV, which is opposed to the semi-

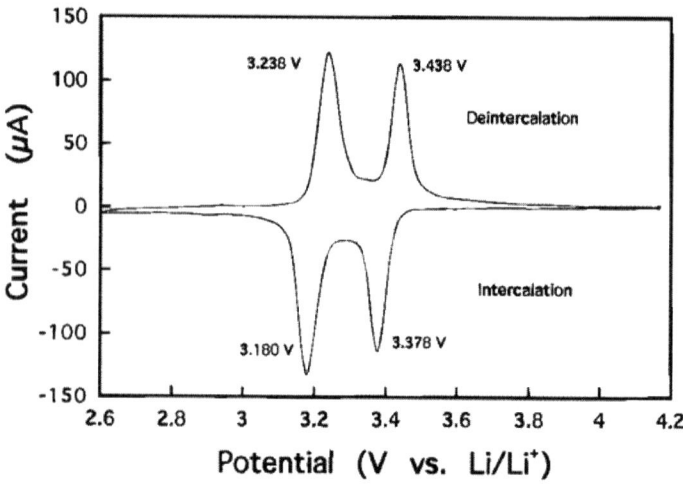

Figure 6. Cyclic voltammogram showing the intercalation and deintercalation of Li$^+$ in the nanostructured V$_2$O$_5$ electrode. Scan rate = 0.1 mV s^{-1}. Reprinted with permission from Ref. 13, (© 1997), American Chemical Society.

infinite diffusion regime's ideal for a one electron process of ΔE_{pk} = 60 mV.[46] These low ΔE_{pk} values indicate that finite diffusion of Li$^+$ is occurring within the nanostructures at the lowest scan rates; i.e., all of the material is intercalated and deintercalated with Li$^+$ during the voltammetric scan. Larger-diameter structures and/or faster scan rates will result in increases in this ΔE_{pk} parameter and diffusional tailing in the more positive peak.[19]

Another notable feature of this CV is that each charge process exhibits two waves, which indicates a phase-transition in the host-material. The intercalation wave in the high potential region (centered around 3.4 V) is seen to encompass a slightly smaller area (charge) than does the lower wave centered around 3.2 V. This trend is similar to that seen in the galvanostatic experiments (see below.) In that experiment, the plateaus in the potential-time curve are analogous to peaks in the CV. The high potential region plateau is seen to be shorter, than the lower potential region plateau. This confirms that slightly more charge from Li-ion intercalation occurs at the lower potential; therefore, the phase-transition of Li$_x$V$_2$O$_5$ occurs at slightly greater than x = 0.5.

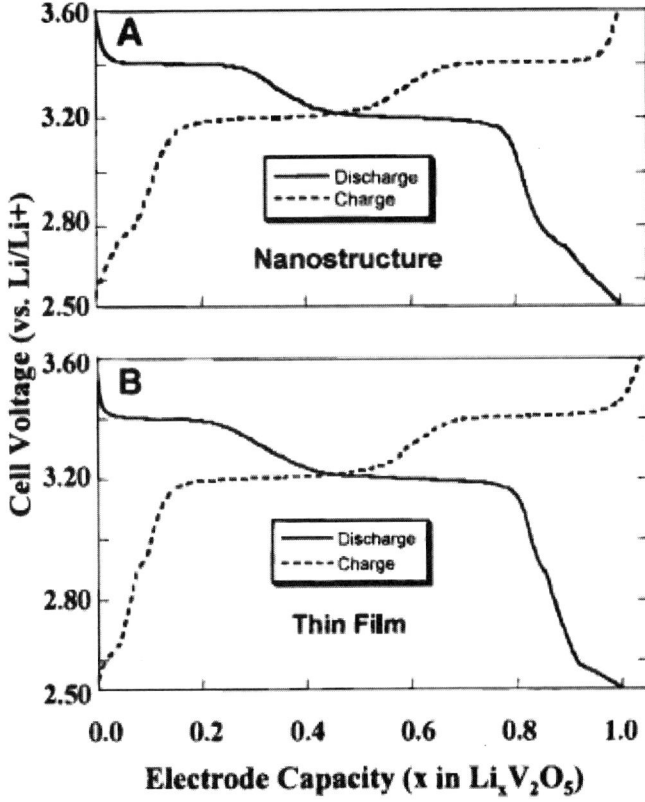

Figure 7. Galvanostatic discharge and charge curves for V_2O_5 electrodes at C/20. A.) Nanostructured electrode. B.) Thin-film electrode. Reproduced from Ref. 24, (© 1999), by permission of The Electrochemical Society, Inc.

(ii) Rate Capabilities

The ability of each of the V_2O_5 electrodes (both nanostructured and thin-film) to deliver charge when discharged at high currents is investigated. We quantify the electrode's *rate capabilities* by measuring the amount of charge that an electrode can store/deliver at increasing rates of discharge. For this series of experiments,[24] each electrode is discharged at increasing rates until the low-potential cutoff (2.5 V) is reached. After each discharge experiment, the system is charged by sweeping the potential to the charge cutoff (3.8 V) at 0.1 mV s^{-1}. Therefore, the electrodes were subjected to the concentration polarization only during the discharge process. If an electrode is discharged at a slow enough rate, then the Li-ions are able to (de)intercalate completely even from the electrode particles of the control. Figure 7 clearly demonstrates this as both a nanostructured V_2O_5 electrode (d ~ 57 nm) and a film (d ~ 250 nm) both obtain near theoretical capacity at the low rate of C/20.

However, the more demanding systems require the higher energy and power densities that are associated with higher rates of discharge (> 5C). Our hypothesis is that the electrode comprised of the smallest particles will deliver the most charge per gram at every rate. This is indeed confirmed by Figure 8, where the nanostructured electrode is shown to deliver greater than 20% of its theoretical capacity at the amazing rate of almost 1200 C.[24] That is 4 times the capacity of the film electrode at that rate. This demonstrates the inherent advantage of using nanomaterials in the design of Li-ion intercalation systems. The small solid-state diffusion distance and high surface area allow nanostructured electrodes to deliver a far superior portion of their theoretical specific capacity when discharged at these high-rates than the film control electrodes.

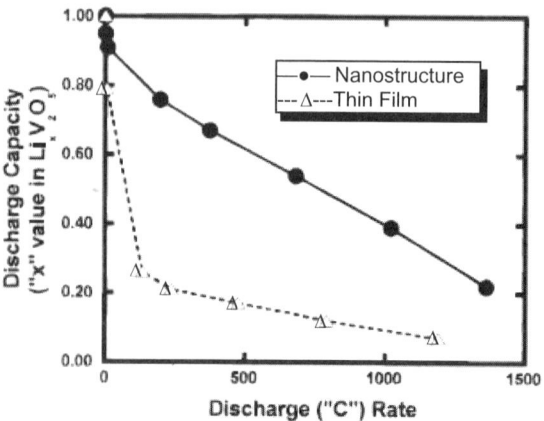

Figure 8. Discharge capacity of typical nanostructured and thin-film electrodes as a function of galvanostatic discharge rate. Reproduced from Ref. 24, (© 1999), by permission of The Electrochemical Society, Inc.

III. NANOSTRUCTURED ANODIC ELECTRODES

In early designs of lithium batteries, the anode was lithium metal foil. This was discontinued, because of safety concerns and electrical shorts associated with dendrientic growth as Li-ions were driven back to the lithium metal. Today in commercial systems, most anodic electrodes are based on carbonaceous materials. For a material to be considered as an anode in Li-ion batteries, it must operate at a sufficiently low reduction potential and reversibly store a high quantity of charge.

Sn-based anodes can theoretically store twice the number Li-ions than can graphite.[47] The Li^+ electrochemistry of the SnO_2-based electrodes is fundamentally interesting because it entails first the irreversible conversion of the tin oxide to metallic tin (Eq. 3) and then the reversible alloying/dealloying of the Sn with Li (Eq. 4):

$$4\,Li^+ + 4\,e^- + SnO_2 \longrightarrow 2\,Li_2O + Sn \tag{3}$$

$$x\,Li^+ + x\,e^- + Sn \rightleftarrows Li_xSn \quad 0 < x < 4.4 \tag{4}$$

It is this alloying/dealloying process that gives this material its charge storage capacity. In theory, as many as 4.4 Li atoms can be stored per atom of Sn, which would give this anode a maximum theoretical charge storage capacity of 781 mAh g^{-1} compared with 372 mAh g^{-1} for graphite. In addition, this charge is stored at high energies, within 1 V of the reduction potential for Li^+ to Li metal.

This material seems to be a suitable companion with template-synthesis fabrication for two reasons. First, the SnO_2 can be synthesized via a sol-gel route.[20] Our group is familiar with this deposition strategy. Also, the major detraction of this material is that it undergoes a dramatic volume change (up to 300%) upon formation of the Li-Sn alloy (Eq. 4), which results in internal damage and loss of capacity during cycling.[48] The void-volume of the end-result template-synthesized nanostructured electrode (*e.g.*, the volume once occupied by the template) may provide the ability for this material to enlarge without causing major structural damage.

Here we describe our investigations of Sn-based anodes[18,20,21,26] as an alternative to the carbonaceous (*e.g.*, graphite) electrodes. For clarity during discussion, it is important to emphasize the distinction that the low-potential anode is charged in the lithiated-state. Therefore, $x = 1$ represents the electrode being charged, and $x = 0$ represents the electrode being discharged.

1. Electrode Fabrication

(i) Nanostructured Electrodes

Nanostructured electrodes are prepared via a solution-based method.[20] In this case, a polycarbonate filter is used as the template; this membrane had a nominal pore diameter of 50 nm and a pore density of 6×10^8 pores per cm^2. This template is immersed into a tin oxide-based sol. This sol is prepared by dissolving $SnCl_2 \cdot 2H_2O$ in a solvent mixture composed of ethanol and hydrochloric acid to yield a 3 M Sn (II) solution. This solution is aged for 24 h to yield a very fine white precipitate. Water is then added, and over a period of 24 h the precipitate suspends to yield a transparent sol. A piece of the template membrane is immersed for 24 h into the sol and then placed on a Pt-foil current collector. At this point, the sol has impregnated the pores of the polycarbonate template.

Because the membrane is at this point wet, good adhesion with the underlying Pt surface is obtained. The membrane surface is wiped of excess solution using a laboratory tissue, and the solvent is allowed to evaporate slowly in air at 80 °C. The polycarbonate template membrane is then etched away in oxygen plasma. This left an ensemble of tin oxide precursor-gel nanostructures protruding from the Pt foil surface like the bristles of a brush.[18,20] Heat-treatment at 440 °C for 2 h in air converts the precursor gel into crystalline SnO_2 (cassiterite, as proven by X-ray diffraction.)[49] From the diameter, density, and length of the nanowires it is shown that these electrodes have 10 cm^2 of SnO_2 area per cm^2 of substrate electrode area.[20]

(ii) Control Electrodes

The SnO_2 control electrode[20] was prepared by dipping a Pt current collector (without a template membrane) into a similar tin oxide-based sol. However, the Sn (II) concentration in this sol was decreased to 1 M in order to lower the viscosity so that thin films are obtained. The aging conditions of the sol are identical to that employed to prepare the nanostructured electrodes, and the heat-treatment procedure to yield SnO_2 is also identical.

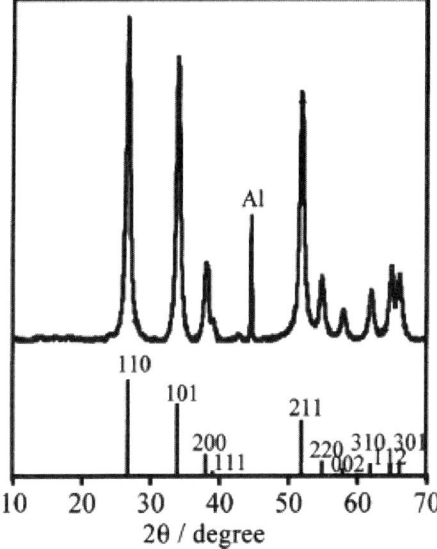

Figure 9. (upper) XRD pattern of SnO_2 powder prepared using the same procedure as the nanostructured electrodes. (lower) Standard pattern for cassiterite, syn-SnO_2. Reproduced from Ref. 20, (© 2001), by permission of The Electrochemical Society, Inc.

2. Structural Investigations

A powder X-ray diffractometer was used to study the crystal structure of SnO_2 powder prepared using the same sol-gel procedure used to prepare the nanostructured SnO_2 (Figure 9) The SEM image (Figure 10A) was obtained by attaching the Pt foil covered with the protruding SnO_2 nanostructures to an SEM sample stub with a piece of copper-foil tape. The transmission electron microscopic (TEM) image (Figure 10B) was obtained by removing the nanostructured SnO_2 from the surface of the Pt substrate by scraping with a razor blade. The liberated nanostructures were dispersed in methanol, and a drop of this dispersion was applied to a TEM grid.

The amounts of the nanostructured SnO_2 and the thin film of SnO_2 were obtained via inductively coupled plasma (ICP) atomic emission analysis.[20] This was accomplished by dissolving the SnO_2 in 1 M HCl and determining the Sn^{2+} concentration of the resulting solution. The mass of SnO_2 was calculated from the concentration of Sn^{2+} in the solution.

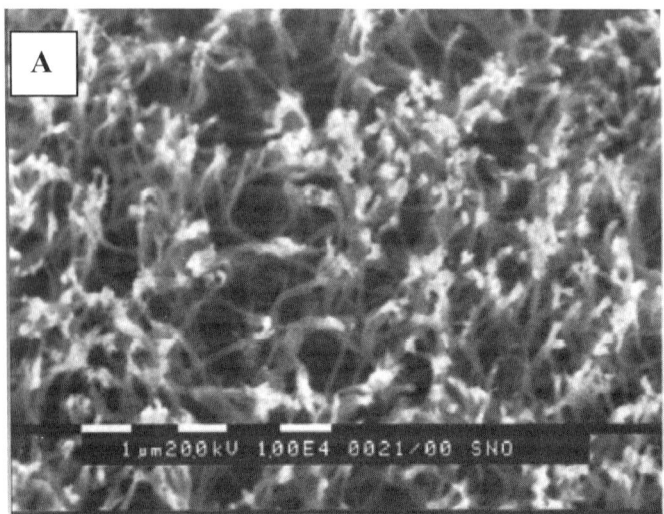

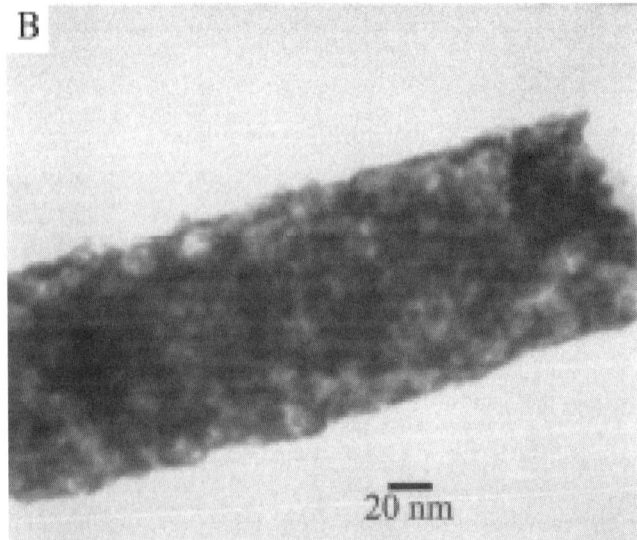

Figure 10. (A) Scanning electron micrograph of the nanostructured SnO_2 electrode. The nanowires are protruding from an underlying Pt current-collector surface. (B) Transmission electron micrograph of a single SnO_2 nanowire before electrochemical cycling. Reproduced from Ref. 18, (© 2000), by permission of The Electrochemical Society, Inc.

3. Electrochemical Investigations

The instrumental setup for the electrochemical characterization of the Sn-based anode is similar to that described previously for the V_2O_5 cathode and is detailed elsewhere.[18,20] SnO_2, however, must be treated uniquely, because of its electrochemical conversion and alloying processes. The first scan was started at the open-circuit potential (2.5 V), and the scan was reversed at a lower (negative) limit of 0.15 V. The return (positive-going) scan was then reversed at a potential of 0.95 V, and the potential was again scanned negatively to 0.15 V. The final positive-going return scan was terminated at 0.95 V. As is typically observed over this potential window, two primary reduction waves are seen on the first negative-going scan, one centered at 0.9 V and the second at 0.25 V. The reduction wave at 0.9 V corresponds to the irreversible conversion of SnO_2 to Sn and Li_2O (Eq. 3). The wave at 0.25 V corresponds to the subsequent alloying of the Sn with Li, and the oxidation wave centered at 0.5 V is associated with the dealloying process. These waves provide the reversible capacity for the Sn anode (Eq. 4).

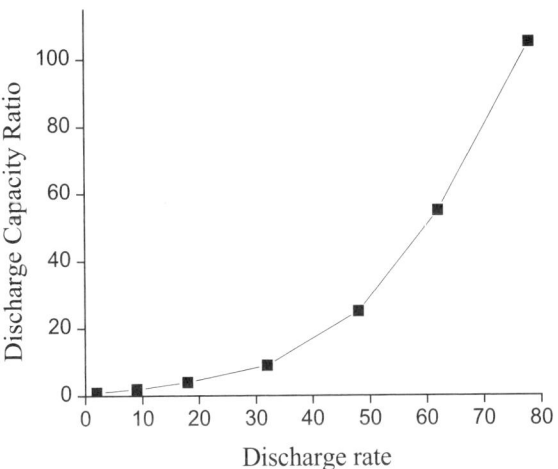

Figure 11. Rate capabilities of Sn-based electrodes. Discharge capacity ratio versus discharge C-rate for the Sn-based nanostructured electrode as compared with the thin-film control electrode. Reproduced from Ref. 26, (© 2002), by permission of MRS Bulletin.

As stated prior, a disadvantage of this material is its structural damage (and subsequent loss of capacity) upon cycling. As such, the ability of this nanostructured Sn-based electrode to withstand the rigors of high rates of discharge and cycle-life was investigated by galvanostatically driving the alloying/dealloying process. The results of these rate capability experiments are shown in Figure 11. Here, we plot the specific capacity delivered of both the nanostructured electrode and the control electrode at various increasing rates of discharge. At the highest C-rate, the nanostructured electrode delivers a capacity that is more than two orders of magnitude higher than the control electrode. Due to the post-characterization analysis of the mass of the active material, it is difficult to apply identical rates of discharge to both systems.

In addition to improved rate capabilities, the nanostructured Sn anode does not suffer from the poor cyclability observed for conventional Sn-based electrodes. This is because the absolute volume change in the nanowires is small, and because the brush-like configuration accommodates the volume expansion around each nanowire. This improved cycle life is illustrated by the data in Figure 12, which show that the nanostructured electrode can be driven through 1400 charge/discharge cycles without loss of capacity. In fact, the capacity increases over the life of the electrode. This occurs because as the expansion of the wires

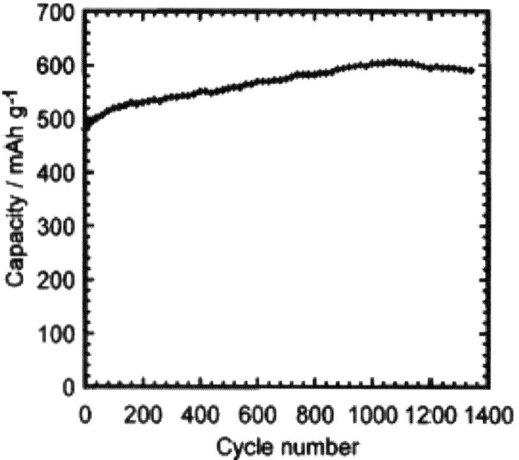

Figure 12. Discharge capacity versus cycle number for the Sn-based nanostructured electrode. Reproduced from Ref. 20, (© 2001), by permission of The Electrochemical Society, Inc.

increases, so does the surface area of the wires. This increase serves to decrease the effective current density to which the electrode is subjected.[18]

The template-synthesized nanostructured Sn-based anode is shown to improve upon the two disadvantages seen with this material. First, by decreasing the solid-state diffusion distance of the Li-ion, the high-rate charge-storage characteristics are improved. Also, the void-volume (once occupied by the sacrificial template) of this nanostructured design accommodates the volume expansion that is detrimental to the Sn-based film's structure upon cycling.

V. NANOELECTRODE APPLICATIONS

1. Low-Temperature Performance

It is well documented that Li-ion batteries show poor low-temperature performance.[50-54] As the temperature of the system is decreased, so is the ability of the system to deliver capacity upon discharge. Specifically, the amount of charge delivered from the battery at temperatures below 0 °C is substantially lower than the amount of charge delivered at room temperature. This precludes their utilization in a number of defense, space and even terrestrial applications.[55] The literature is in conflict as to the fundamental process responsible for this breakdown. There are reports citing the culprit as the cathode material,[52] anode material[53] and the electrolyte.[50] Template-synthesized nanostructured electrodes are ideal tools for such investigations of the fundamental processes of Li-ion batteries. Recently, we were able to identify the rate-limiting process at low temperature as the decreased solid-state diffusion coefficient of the Li-ion.[27]

(i) Electrode Fabrication

The electrodes for this study were nanostructured V_2O_5. Our lab is familiar with the synthesis of this structure, which is discussed earlier in this chapter. The templates were either commercially available nano- or microporous polycarbonate filters. The specifications of each of the three templates are shown in Table 1. The initial synthetic steps were identical to those described previously; however, since larger pore-diameter templates were used, the filling procedure was repeated to ensure

Table 1
Membrane Parameters Needed to Calculate S_c. Reproduced from Ref. 27, (©2005), with permission of Wiley-VCH.

Pore diameter (nm)	Pore density (cm^{-2})	Membrane thickness (μm)
70	6 x 10^8	6
450	1 x 10^8	10
800	3 x 10^7	9

the pores would be completely filled with electrode material.[27] Also, the top-face of the template was wiped clean with a damp swab to remove any surface layer. Again, oxygen plasma etched away the template leaving V_2O_5 structures that mirror the geometries of their respective templates. This ability to control nanowire diameter, length and number density will prove critical in our analysis. The crystalline phase of V_2O_5 was achieved by heating the electrode in flowing O_2 to 400 °C for 10 hrs.

(ii) Strategy

Three electrodes were synthesized using polycarbonate templates with different pore diameters, length and number densities. For this discussion, they will be referred to by the diameter of the resulting structures: 70-nm, 0.45-μm, and 0.8-μm. The mass of V_2O_5 was measured prior to electrochemical characterization using an Ultra-Micro-Balance SC2 (Satorius).[27] This is advantageous because it allowed the applied current to be normalized to the electrode mass. Therefore, it was applied as C-rate, which allows direct comparison between the electrodes of differing mass.

A three-electrode jacketed cell was used to house the 1 M LiClO$_4$ in ethylene carbonate:diethyl carbonate:dimethyl carbonate (1:1:1 v/v/v) electrolyte. This electrolyte solvent mixture has been demonstrated to maintain sufficient ionic conductivity in this experimental temperature range.[56] A Li ribbon reference and counter electrode were used for the electrochemical characterization. This system galvanostatically drives the intercalation and deintercalation reactions of V_2O_5 (Eq. 1.) The discharge reaction was continued until a potential of 2.8 V was achieved, and the charge reaction was terminated when the potential reached 3.8 V. Over this potential window 1 mol of V_2O_5 is known to reversibly intercalate 1 mol of Li$^+$, (i.e., x = 1 in Eq. 1),[57] corresponding to a maximum specific (per g) charge-storage capacity of 148 mAh g^{-1}. From the potential-versus-time data and known mass, the specific capacity of each electrode is calculated.

This experiment was repeated for increasing rates (0.2C to 40C) and decreasing temperature (25 °C to –20 °C.) At all temperatures, we see the decrease in capacity with discharge rate that is typical of Li-ion battery electrodes. However, as expected, this decrease in capacity becomes much more pronounced at low temperatures. Data of this type were obtained for all three of the V$_2$O$_5$ wire electrodes. This allows us to then establish a parameter deemed the Capacity Ratio (specific capacity of nanowire / specific capacity of microwire). If these ratios ($R_{70/0.8}$ and $R_{70/0.45}$) are greater than unity, then at these particular values of current and temperature, the nanowire electrode shows a capacity advantage relative to the microwire electrode.

(iii) *Electrochemical Results*

Looking first at the ratio $R_{70/0.8}$ (Figure 13A) we see that at low discharge rates, $R_{70/0.8}$ is only slightly greater than unity, which means that at room temperature and low discharge rate, both electrodes are delivering nearly 100% of their maximum specific capacities; hence, at room temperature and low discharge rates, there is little capacity advantage for the 70-nm electrode relative to the 0.8-μm electrode. At higher discharge rates $R_{70/0.8}$ increases above unity, indicating that there is now a capacity advantage for the nanowire electrode; however, at room temperature this advantage is modest.

Turning now to the -20 °C data, we find that at the lowest discharge rate the 70-nm electrode delivers twice the specific capacity of the 0.8-μm electrode. Furthermore, $R_{70/0.8}$ increases dramatically with discharge rate; indeed, at the highest rate studied, the specific capacity delivered by the 70-nm electrode is almost two-orders of magnitude greater than the specific capacity delivered by the 0.8-μm electrode. *These data prove our hypothesis – that an electrode composed of nanoscopic particles shows better (indeed, dramatically better) low-temperature performance than an electrode composed of micron-sized particles.*

We continue this analysis with the analogous ratio $R_{70/0.45}$ (Figure 13B). If our hypothesis is correct, the 70-nm electrode should show better low-temperature performance than the 0.45-μm electrode ($R_{70/0.45} > 1$), but at any value of discharge current, $R_{70/0.45}$

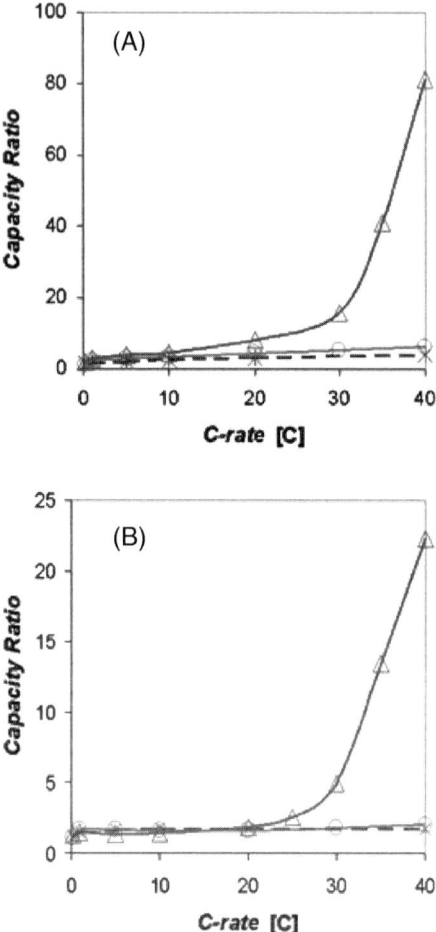

Figure 13. Capacity ratio (see text) versus discharge rate for experiments conducted at 25 °C (X), 0 °C (O), and −20 °C (Δ). (A) $R_{70/0.8}$ and (B) $R_{70/0.45}$. Reproduced from Ref. 27, (© 2005), with permission of Wiley-VCH.

should be less than $R_{70/0.8}$. A comparison of the Δ curves in Figures 13A and 13B shows that this is indeed the case. Let us now further the explanation and analysis by comparing two possible mechanisms responsible for this increased charge-storage ability:

1. decreased diffusion coefficient of the Li-ion or
2. sluggish electron kinetics from larger particles with smaller surface area.

If the solid-state diffusion coefficient is the culprit then low-temperature performance should improve with decreasing particle diameter, and this is exactly what is observed here.

If electrode reaction rate is the culprit, then low-temperature performance should improve with a parameter we designate S_c, the electrode-particle surface area per cm^2 of current collector area.[27] The beauty of the template method is that S_c can be easily calculated from the wire diameter, the pore density of the template membrane, and the wire length (see Table 1). These calculations show that because of the interplay between pore diameter and density in the template membranes, S_c is higher for the 0.45-μm electrode (S_c = 14.1) than for the 70-nm (S_c = 7.9) or the 0.8-μm (S_c = 6.8) electrodes. This indicates that if electrode reaction rate is the culprit, then the 0.45-μm electrode would show the best low-temperature performance, and this is not observed experimentally. Hence, our data show that the temperature dependence of the solid-state diffusion coefficient determines the low-temperature performance of the electrodes studied here.

(*iv*) *Electronic Conductivity*

In commercial Li-ion battery cathodes, it is necessary to add an electronically conductive material, typically carbon particles, to improve electronic conductivity through the electrode. We have shown that this is not necessary with template synthesized electrodes. This is because while the electronic resistance of a single wire (R_f) of the template-synthesized electrode might be high, the electrode is a parallel ensemble of such wires. As such, the total resistance of the electrode (R_t) is given by $R_t = R_f/N$, where N is the number of wires in the electrode.

That electronic conductivity does not dominate the rate capabilities of the electrodes studied here is easy to prove. Assuming the same length, it is easy to show that the electronic resistance of a 70 nm diameter wire is 33 times higher than that of a 0.8 μm diameter wire composed of the same material. If the rate capability was limited by electronic conductivity, the 70-nm electrode would show the worst rate capability and the 0.8-μm electrode would show the best. This is exactly the opposite of what is observed experimentally.

(v) Cycle Life

The order of the series of experiments dictates that each test delivers less charge than the previous (*e.g.*, low to high C-rate of discharge and high to low temperature.) Therefore, it is helpful to examine any adverse effects on the cycle life on the system. After the entire set of experiments was completed, if the system was returned to room temperature (25 °C), the electrodes maintained 98+% of the capacity shown in the original series of tests at that temperature. This is an important control experiment to help validate the data that are the basis for our conclusions.

Using template-synthesized nanostructured Li-ion battery electrodes, we were able to identify the source of the fundamental breakdown that occurs in Li-ion batteries at low temperature as the decrease in diffusion coefficient of Li-ion in the solid-state. This is possible because of the ease at which structures of varying dimensions, but identical composition, are created via template-synthesis.

2. Variations on a Synthetic Theme

(i) Nanocomposite of $LiFePO_4$/Carbon

We have demonstrated the advantages that nanoscale-geometry has on components currently used in Li-ion batteries. Another direct impact of this field is in the development of alternative electrode materials. These proposed materials may be less expensive, more energetic, and more environmentally friendly than the present ones, but are precluded by some inherent limitation. An example was discussed earlier of the ability of nanotechnology to mitigate the detrimental effect of volumetric-expansion associated with proposed high-capacity Sn-based anodes. An alternative cathode material of interest is the olivine-structured $LiFePO_4$ cathode developed by Goodenough and co-workers,[58] which offers several appealing features, such as a high, flat voltage profile and relatively high theoretical specific capacity (168 mAh g^{-1}), combined with low cost and low toxicity.

However, the current designs of cells based on $LiFePO_4$ technology have not shown the ability to deliver high specific capacity at high discharge rates. This limitation is caused by both intrinsically poor electron conductivity and the low rate of Li^+ transport within the micron-sized particles used to prepare the battery electrode. The discharge reaction for $LiFePO_4$ (Eq. 5) entails intercalation of Li^+ (from the contacting electrolyte phase) along with an equivalent number of electrons into the electrode material. A number of approaches have been proposed to improve this material's inherent poor electronic conductivity, including carbon coating,[59] nano-fibril textures,[60] optimized synthesis procedures[61] and foreign metal doping.[62] Though these routes have promise, they have had limited success. For this reason, $LiFePO_4$ is currently not a promising electrode material for high-rate and pulse-power applications.

Here we describe a variation to the previously detailed template-synthesis methods, by pyrolyzing the polycarbonate template.[28,63] The result is a new type of template-prepared nanostructured $LiFePO_4$ electrode—a nanocomposite consisting of monodisperse nanowires of the $LiFePO_4$ electrode material mixed with an electronically conductive carbon matrix. This unique nanocomposite morphology allows these electrodes to deliver high capacity, even when discharged at the extreme rates necessary for many pulse-power applications. This new nanocomposite electrode shows excellent rate capabilities because the nanowire morphology mitigates the problem of slow Li^+-transport in the solid state, and the conductive carbon matrix overcomes the inherently poor electronic conductivity of $LiFePO_4$.

(a) Electrode fabrication

The sol-gel method developed by Croce et al.[64] is employed for the synthesis of the electrode precursor solution. An approximately 1 cm^2 piece of the polycarbonate filter is immersed in a precursor solution of 1 M $LiFePO_4$ in water in 24 hours.[28] This solution is synthesized with ferric nitrate, lithium hydroxide, and phosphoric acid in proportions for a 1:1:1 molar ratio. Ascorbic acid aids the synthesis by forming a complex with the iron, and ammonium hydroxide is used to raise the pH to ~2. The impregnated template was then attached to a Pt current collector and dried in air at 80 °C for 10 minutes. A 10 µL drop of precursor solution was placed on top of the dried filter to increase the amount of active material in the sample. It was dried again under the same conditions. This assembly, template intact, was heated in a reducing atmosphere of flowing Ar/H_2 gas. The temperature was slowly taken over the course of 4 hours from 250 °C to 650 °C and held there for 12 hours. This procedure yields the Fe(II) oxidation state necessary for LiFePO$_4$ and decomposes the template into the carbon necessary for improved conductivity.

(b) *Methods*

The nanocomposite electrode was imaged by SEM. The electrode on Pt foil was prepared for imagining by attaching it to a SEM stub by conductive copper tape. No conductive metal sputtering was required for the composite electrodes, but a thin Au/Pd sputtering was applied to the LiFePO$_4$ (Figure 14C) prior to imaging. The XPS studies were performed and peak positions were all referenced to 70.9 eV for the Pt4f$_{7/2}$ peak (literature value for metallic platinum used as sample support).

For the electrochemical studies we wanted a template with small-diameter pores so that correspondingly small-diameter nanowires of LiFePO$_4$ would be obtained. For this reason we used a template with nominally 50 nm-diameter pores for the electrochemical studies. This template was 6 μm thick and had a pore density of 6×10^8 pores per cm^2 of surface area. To obtain detailed images of the morphology of the nanocomposite structure we found it prudent to use a template with nominally 100 nm-diameter pores. The larger LiFePO$_4$ nanowires obtained from this template are more easily imaged with scanning electron microscopy. This template was also 6 μm thick and had a pore density of 4×10^8 pores per cm^2.

(c) *Imaging*

A lower magnification SEM image of the resulting LiFePO$_4$/carbon nanocomposite electrode is shown in Figure 14A. Because of the relatively low porosity of the template, there is substantial void volume, but in analogy to our prior nanowire electrodes of this type, the LiFePO$_4$ nanowires can be seen crossing through this void space. Higher magnification images (Figure 14B) show that there are carbon particles dispersed through this matrix and that the LiFePO$_4$ nanowires are coated with thin carbon films. To prove that these wires are coated with carbon films, we prepared wires in the same template, but instead of then pyrolyzing the template, we simply removed it quantitatively by burning it away in O$_2$ plasma. Hence, in this sample the wires are not coated with carbon films. An image of these wires is shown in Figure 14C. The wires from the pyrolyzed membrane (Figures 14A and

Nanomaterials in Li-Ion Battery Electrode Design

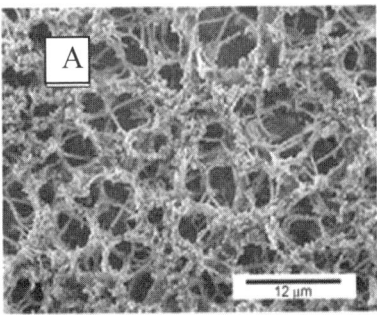

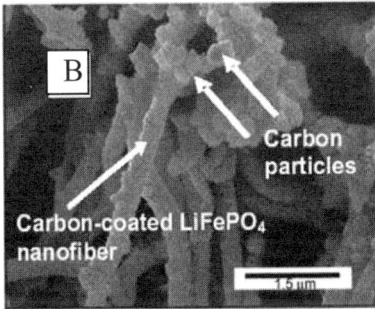

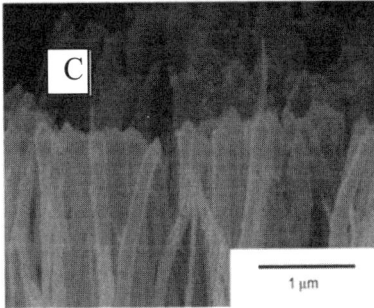

Figure 14. Scanning electron micrographs. (A) Lower magnification image of the nanocomposite LiFePO$_4$/carbon electrode; (B) Higher magnification image of the nanocomposite LiFePO$_4$/carbon electrode. Composite wire diameter is 350 nm; (C) Image of LiFePO$_4$ electrode synthesized by template dissolution method (absent of carbon). LiFePO$_4$ wire diameter is 170 nm.[28]

14B) have a textured surface morphology and have a larger diameter than the wires from the plasma-removed membrane (Figure 14C). Both the larger diameter and the textured surface are due to the carbon coating surrounding the wires from the pyrolyzed membrane.

(*d*) *Carbon analysis*

The presence of carbon in this matrix was confirmed by X-ray diffraction analysis, Raman spectroscopy, and X-ray photoelectron spectroscopy (XPS). The X-ray diffraction data (not shown) exhibit the major peaks of LiFePO$_4$, as well as the d$_{002}$ peak of carbon thin film. The Raman spectra (also not shown) of the nanocomposite LiFePO$_4$/carbon electrode showed the corresponding peaks of the PO$_4$ stretching modes of LiFePO$_4$. Also, bands were able to be assigned to carbon. The presence of carbon was also confirmed by XPS analysis (Figure 15). The high resolution C1s spectrum may be fitted to three peaks with binding energies of 283.2 ± 0.5, 284.7 ± 0.3 and 286.0 ± 0.3 eV. According to Miller et al.,[10] the lowest binding-energy peaks may be assigned to graphitic (283.2 eV) and amorphous (284.7 eV) carbon. The predominance of the 284.7 eV peak indicates that most of the carbon present is amorphous; this is confirmed by the X-ray diffraction data. The peak at the highest binding-energy (286.0 eV) is due to oxygen-containing surface functional groups. Oxygen

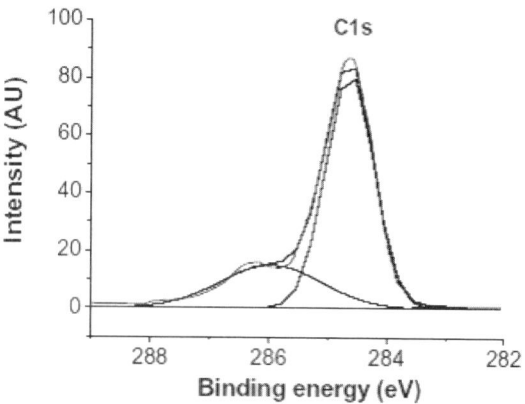

Figure 15. XPS C1s peak for the carbon in the nanocomposite electrode. Relative peak areas show amphorous form is dominant (~ 80%), but graphitic carbon is also present (~ 20%).[28]

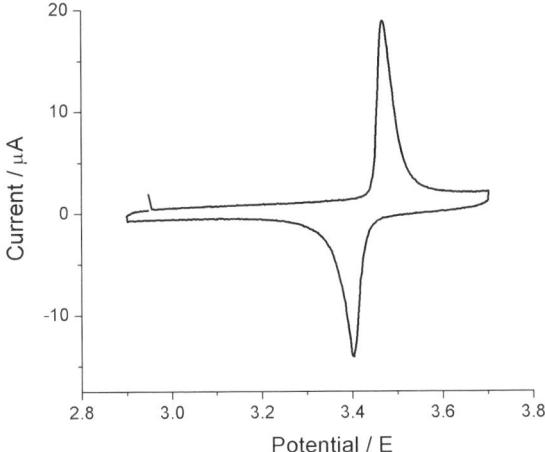

Figure 16. Cyclic voltammogram for the nanocomposite LiFePO$_4$/carbon electrode prepared using a template with 50 nm-diameter pores. Scan rate = 0.1 mV s^{-1}.[28]

functional groups are nearly always observed on carbon surfaces that have been exposed to air.[10]

We estimated the weight percent of carbon in the composite electrode gravimetrically using an Ultra-Micro-Balance SC2 (Satorius). LiFePO$_4$ nanowires were synthesized within the pores of the polycarbonate template on a Pt current collector, and the polymer was pyrolzyed as described previously. The mass of this composite, corresponding to the masses of the Pt current collector, the nanowires and the carbon, was obtained. This composite was then heated in air at 600 °C for 30 minutes to burn off the carbon, and the mass was measured again. The difference between these two masses is the mass of carbon in the composite. Replicate analyses on four identically prepared samples gave a carbon content of 7±4 % in the nanowire/carbon composite.

(*e*) *Electrochemistry*

Cyclic voltammetric and constant current charge/discharge experiments were performed on electrodes prepared in templates with 50 nm-diameter pores in a three-electrode cell similar to those described earlier. Cyclic voltammograms (CVs) for the nanocomposite electrode show reversible waves centered at 3.5 V associated with the reduction and re-oxidation of the LiFePO$_4$

(Figure 16). The difference in peak potentials (ΔE_p) for this nanostructured electrode is 60 mV (Figure 16). This may be contrast to CVs for conventional, non-nanostructured, LiFePO$_4$ electrodes, which at comparable scan rates and in comparable electrolyte solutions show $\Delta E_p > 200$ mV.[64] This clearly shows that our nanostructured LiFePO$_4$ electrodes lack a resistive component that is present in the conventional electrodes. This verifies the major premise of this work – that the conductive carbon matrix overcomes the inherently poor electronic conductivity of LiFePO$_4$. While the mass of the LiFePO$_4$ in the composite was determined gravimetrically (see above), we have found that not all of this LiFePO$_4$ is electroactive. This is because portions of the LiFePO$_4$ in the composite lose ohmic contact with the current collector during pyrolysis.[28] For this reason we determined the electroactive mass from the area under voltammetric waves like those shown in Figure 16. This analysis requires that 100% of the electroactive material is oxidized during the forward scan and reduced again during the reverse scan; i.e., that diffusion in the wires is finite rather than semi-infinite. Because of their small diameters, this is always case for low scan rate voltammograms of our template-synthesized nanostructures.[19] The lack of diffusional tailing in the voltammogram in Figure 16 shows that this is also the case for the nanostructured composite electrodes studied here.

The constant-current discharge curve (lithium insertion, Eq. 5) for the nanocomposite electrode shows the flat voltage plateau centered at 3.5 V, characteristic of LiFePO$_4$. At the lowest discharge rate used (3 C), the specific capacity for the composite is 165 mAh g^{-1}, essentially identical to the maximum theoretical capacity, 168 mAh g^{-1}. While capacity falls off with increasing discharge rate (Figure 17) the electrode retains 36% of its theoretical capacity at discharge rates as high as 65 C.

$$\text{Li}_x\text{FePO}_4 \underset{\text{discharge}}{\overset{\text{charge}}{\rightleftharpoons}} \text{FePO}_4 + x\text{Li}^+ + x\text{e}^-$$

(5)

We have described here a new type of template-prepared nanostructured LiFePO$_4$ electrode.[28] This electrode is a nanocomposite consisting of nanowires of the LiFePO$_4$ electrode material mixed with an electronically conductive carbon matrix. This unique nanocomposite morphology allows these electrodes to

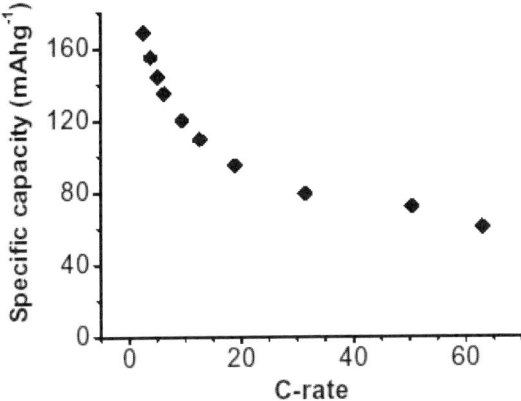

Figure 17. Specific capacity versus C-rate for the nanocomposite LiFePO$_4$/carbon electrode prepared using a template with 50 nm-diameter pores.[28]

deliver high capacity, even when discharged at extreme rates necessary for many pulse-power applications. We are currently working toward developing a commercially viable route for preparing such nanocomposite electrodes.[63]

(ii) Improving Volumetric Capacity

Until this point in the chapter, we have discussed the charge-storage capacity of electrodes on a per-unit-mass basis. Nanostructured electrodes deliver *gravimetric* capacities (mAh g^{-1}) superior to that of the film control electrodes.[24] However, the volume once occupied by the sacrificial template decreases the *volumetric* loading (and thus charge) of the electrode (g L^{-1}, and therefore mAh L^{-1}). This is because the porosity of the template membranes used is low. Low porosity (1.2 % for 50-nm polycarbonate) results in a correspondingly low volume of V$_2$O$_5$ nanowires protruding from the current collector surface. We have developed a two-prong approach to increase the ability of the template-synthesized nanostructured electrodes to deliver higher volumetric capacity.[25] Here, we discuss the results of implementation of these methods.

(*a*) *Strategy*

The first strategy is to chemically etch the polycarbonate template membranes prior to precursor-deposition. The etching process to similar to that which turns latent-tracks into pores (see Section II). Etching was accomplished by immersing the as-received membranes in 6 M NaOH which had been heated to 65 °C. After the immersion period (2 mins for 50-nm and 6 mins for 400-nm), the membrane and the NaOH solution were transferred to deionized water to quench the etch process. The membrane was thoroughly rinsed and then dried under vacuum at room temperature for 1 h on a Teflon sheet. Figure 18 compares micrographs of etched (e50-nm and e400-nm) and as-received (50-nm and 400-nm) membranes.

Another strategy to further improve the volumetric energy density of template-synthesized electrodes is to apply additional TIVO to the surface of the nanostructured parent V_2O_5 electrodes (i.e., increase the volumetric fraction of material).[25] The additional TIVO was applied in two equal increments, and the electrodes

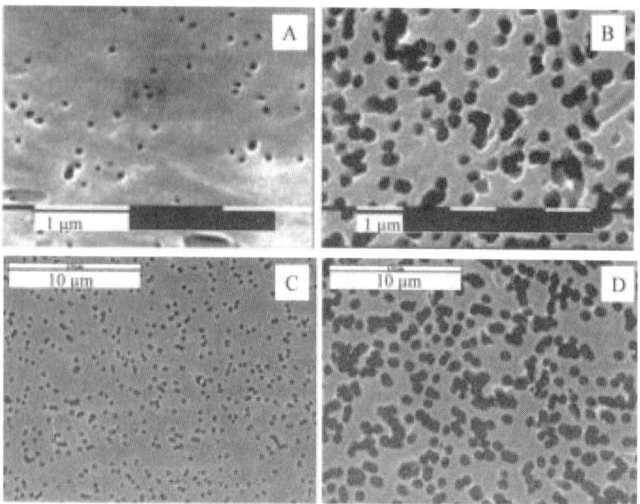

Figure 18. SEMs of as-received 50 nm (A) and 400 nm (C) pore diameter polycarbonate template membranes. Micrographs B and D show the 50 nm and 400 nm membranes, respectively, after etching in 6 M NaOH. Reproduced from Ref. 25, (© 2001), by permission of The Electrochemical Society, Inc.

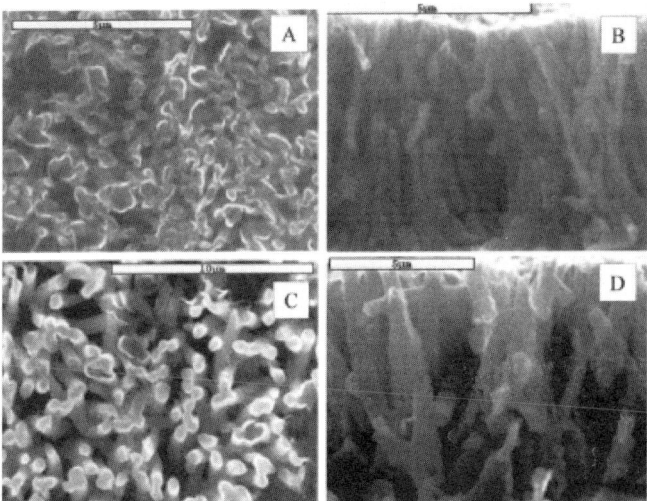

Figure 19. SEMs of template-synthesized electrodes prepared from etched 50 nm (A) and (B) and 400 nm (C and D) polycarbonate template membranes. A and C show the tops of the V_2O_5 wires (normal view) while B and D show the cross sections of these electrodes. Reproduced from Ref. 25, (© 2001), by permission of The Electrochemical Society, Inc.

were again thermally treated as before. Hereafter electrodes prepared by adding TIVO to the e50-nm and e400-nm electrodes are referred to as ''e50-nm + TIVO'' and ''e400-nm + TIVO'' electrodes, respectively.

(*b*) *Electrode morphology*

The SEM images in Figure 19 show normal and cross-sectional views of the e50-nm (4.7 A and B) and e400-nm (4.7 C and D) electrodes. These SEMs show that both electrodes consist of V_2O_5 wires that protrude from the Pt foil current collector surface. The diameter of the wires was determined from such images to be 250 ± 110 nm (e50-nm electrode) and 700 ± 150 nm (e400-nm electrode). These wire diameters are, within the standard deviation, identical to the pore diameters of the etched template membranes. The cross-sectional images (Figures 19B and 19D) show that the lengths of the electrode wires are similar to the length of the respective unetched template membranes.

Figure 20 shows SEMs of the e50-nm + TIVO (Figures 20A and B) and e400-nm + TIVO (Figures. 20C and 20D) electrodes.

Images of the electrode surfaces (Figures. 20A and 20C) show that the wires are encased in a surrounding matrix of V_2O_5 due to the addition of the V_2O_5 precursor to the parent electrodes. The cross-sectional images (Figures 20B and 20D) show the morphologies of these two electrodes differ significantly from each other. Figure 20B shows that the e50-nm + TIVO electrode contains a wire structure similar to that for the e50-nm parent electrode. In contrast, the e400-nm + TIVO electrode has a lower surface area morphology than the other electrodes. The V_2O_5 matrix surrounding the nanowires appears to be porous as shown by the holes on the left side of Figure 20D. The lower surface area of the e400-nm + TIVO electrode would suggest that its rate capabilities should be inferior to those of the other electrodes.

The family of nanostructured electrodes was compared to a thin-film control electrode. This electrode is a continuous film of crystalline V_2O_5 of uniform thickness. SEM cross-sectional images (not shown) were obtained at ten different locations along the thin film edge to determine the average film thickness. A thickness of 2.4 ± 0.2 μm was obtained. As noted above, the electrodes

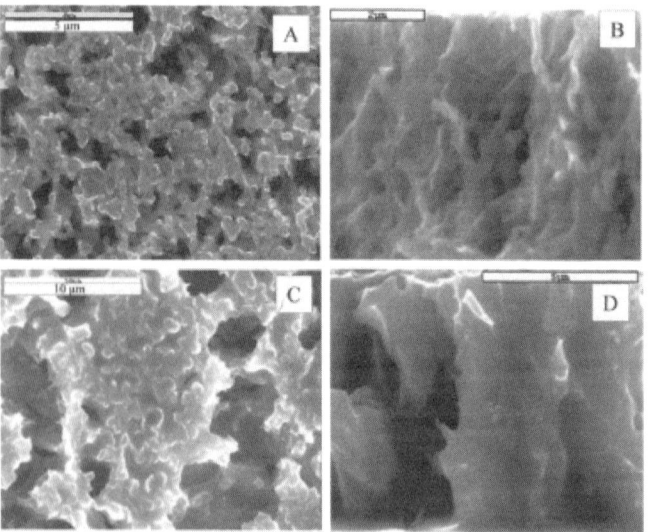

Figure 20. SEMs showing the e50-nm + TIVO (A, B) and e400-nm + TIVO (C, D) electrodes prepared by adding the V_2O_5 precursor to the parent nanowire electrodes. A and C show the tops of the electrodes, B and D show the cross sections of the electrodes. Reproduced from Ref. 25, (© 2001), by permission of The Electrochemical Society, Inc.

Table 2
Electrode Mass Loading, Capacities, Area, and Thickness. Reproduced from Ref. 25, (© 2001), by permission of The Electrochemical Society, Inc.

Electrode description	Geometric mass loading ($\mu g\ cm^{-2}$)	Maximum capacity (μAh)	V_2O_5 geometric area (cm^2)	Electrode thickness (μm)	Volumetric capacity ($\mu Ah\ cm^{-2}\ \mu m^{-1}$)	Geometric capacity ($\mu Ah\ cm^{-2}$)
e50-nm	326	48	1.00	5	9.6 (22%)	48
e50-nm + TIVO	553	57	0.70	5	18 (42%)	82
e400-nm	586	51	0.59	8	11 (26%)	86
e400-nm + TIVO	1081	51	0.32	8	20 (47%)	159
Thin film	700	62	0.60	2.4	43 (100%)	103

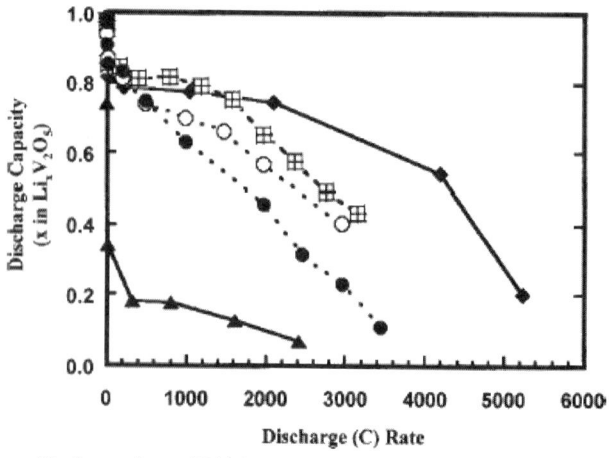

Figure 21. Comparison of Li$^+$ ion storage capacity vs. discharge rate for the template-synthesized and thin-film electrodes. (♦) e50-nm, (⊞) e50-nm + TIVO, (O) e400-nm, (●) e400-nm + TIVO, and (▲) thin film. Reproduced from Ref. 25, (© 2001), by permission of The Electrochemical Society, Inc.

studied here have different morphologies, surface areas, and thicknesses, which are summarized in Table 2.

(c) *Rate capabilities*

Rate capabilities of the various electrodes were evaluated by measuring electrode discharge capacity at increasing discharge (C) rates. The electrode capacities for these experiments were determined using a voltage cutoff of 2.6 V. After each discharge experiment, the electrode was charged at 100 μA from the open-circuit potential to 3.8 V and then equilibrated at this voltage until the current decayed to the first scan value.[25] Probing such high-rates of discharge is then possible, because the effects of concentration polarization are uniform after each charge. In all cases the charge and discharge capacities were equivalent showing that the discharge capacity was 100% faradaic. Complete characterization of the rate capability of each electrode required up to 15 discharge/charge cycles. However, cycle-fading did not complicate the results, because the capacities of the first and last discharge experiments (performed at low discharge rate) were the same within experimental error.[25]

Figure 21 shows plots of discharge capacity as a function of discharge rate for all of the electrodes studied here. (The experimental data are the points; the lines are drawn to aid the eye.) At the lowest discharge rate (between C/24 to C/20) all of the electrodes reach the maximum theoretical capacity for a cutoff voltage of 2.6 V. Figure 21 also shows that at high discharge rates (> 10C), the Li^+ storage capacity of all of the various template-synthesized electrodes is higher than that of the thin-film control electrode. As discussed previously, this is because of the shorter Li^+ diffusion distances and higher surface areas for the template-synthesized electrodes. It is important to point out that we make no assumption about the value of the diffusion coefficient in the nanostructured *vs.* the thin-film control electrodes. The improved rate capability is simply a reflection of the shorter distance Li^+ must diffuse in the V_2O_5 when it is nanostructured. Higher electrode material surface area results in lower Li^+ insertion rate (current) density. Both of these effects delay concentration polarization to higher discharge currents, resulting in higher electrode capacities.

A comparison of the capacities at the highest discharge rates (Figure 22) shows that the e50-nm electrode has the best rate capability. Electron microscopy (Figures 19A and 19B) shows that this electrode contains the smallest diameter wires and therefore the shortest Li^+ diffusion distances. The e50-nm + TIVO and e400-nm electrodes also show good rate capabilities. The cross-sectional SEM images (Figures 20B and 20D) show a wire-like, high surface area morphology for these electrodes which explains their good rate capabilities. The e400-nm + TIVO electrode has the poorest rate capability of the template-synthesized electrodes (but still much better than the thin-film electrode.) SEM images (Figures 20C and 20D) suggest that the Li+ diffusion distances are higher, and the V_2O_5 surface area lower, than for the other template-synthesized electrodes. This accounts for the relatively poor rate capability of this electrode.

Volumetric rate capability is important where electrode volume is a critical issue and high discharge rates are needed. Figure 22A compares the volumetric capacity at various discharge rates for the e50-nm, e50-nm + TIVO, and thin-film electrodes; analogous data for the e400-nm and e400-nm + TIVO electrodes are shown in Figure 22B. The volumetric capacity of the thin-film electrode decreases rapidly to 35% of the maximum value at 160 C and 18% (7.5 µAh cm^{-2} µm^{-1}) at 240 C. The most important point to note is that the 50-nm + TIVO electrode shows volumetric capacities that are two times higher than that of the thin-film electrode at all discharge rates above 200 C. Furthermore, while the e50-nm electrode has the smallest maximum (low-rate) volumetric capacity (Table 2), at discharge rates greater than 1000 C this electrode shows higher volumetric capacity than the thin film electrode.

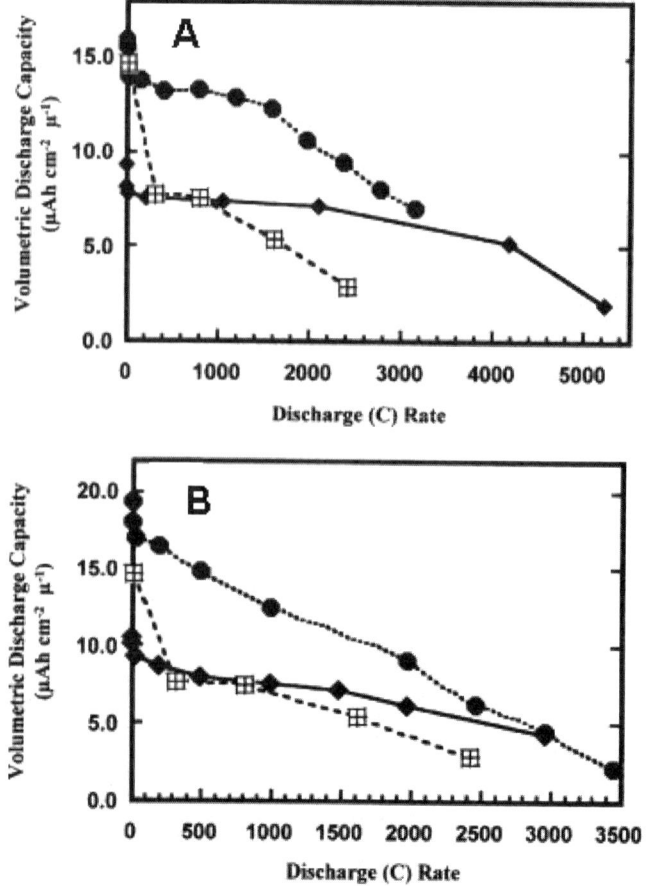

Figure 22. Volumetric capacity *vs.* discharge rate. (A) (♦) e50-nm, (●) e400-nm + TIVO, (⊞) thin-film. (B) (♦) e400-nm, (●) e400-nm + TIVO, and (⊞) thin-film. Reproduced from Ref. 25, (© 2001), by permission of The Electrochemical Society, Inc.

The results for the e400-nm and e400-nm + TIVO electrodes show a similar trend (Figure 22B). The maximum low-rate volumetric capacity of the e400-nm electrode is 26% of the thin film electrode value (Table 2). However, its high rate capability results in a volumetric capacity equivalent to the thin-film at high discharge rates. The e400-nm + TIVO electrode has the highest volumetric capacity of the template-synthesized electrodes and a better rate capability than the thin-film (Figure 21). This ability results in a two times higher volumetric capacity than the thin-film electrode at discharge rates greater than 200 C. These results on volumetric rate capability show, again, the importance of higher surface area and shorter diffusion distances to electrode rate capability. In addition, these experiments show the volumetric rate capability advantage obtained by adding TIVO to the e50-nm and e400-nm electrodes.

An overall objective of our work on template-synthesized nanostructured Li-ion battery electrodes has been to explore the effects of Li^+ diffusion distance within the electrode material and electrode surface area on rate capability. Here, we have demonstrated two new approaches for improving volumetric capacity for template-prepared nanostructured electrodes. The first entails etching the membrane to increase porosity. The second entails adding additional V_2O_5 to a template-prepared nanostructured electrode to increase the quantity of electrode material. A template-synthesized electrode (e50-nm + TIVO) was able to deliver two times greater volumetric capacity ($\mu Ah\ cm^{-2}\ \mu m^{-1}$) than a thin-film control electrode at high discharge rates.

VI. CARBON HONEYCOMB

Until this point, we have detailed the template-synthesis of Li-ion battery electrode materials using polymeric templates. However, the first Section of this chapter describes the benefits of an alumina template compared with a track-etch one. The harsh dissolution conditions necessary to remove alumina templates has limited their use. Here we describe a plasma-assisted method to create a carbon replica of an alumina template membrane. This replica will serve as a nanostructured anode for Li-ion battery design. This Section describes the synthesis of a "honeycomb" carbon framework and its electrochemical characterization.

1. Preparation of Honeycomb Carbon

Our strategy is to create a honeycomb carbon anode by plasma-etching a replica of a porous alumina membrane into a thin carbon film.[22] The first step is the electrochemical synthesis of nanoporous alumina from bulk aluminum. This process, based on Masuda's two-step method,[65] is detailed elsewhere and highlighted in the beginning of this chapter. The resulting membrane has extremely regular well-ordered pores. These pores are approximately 75-nm in diameter and run the entire length of the membrane (~ 1.2 μm.) The membrane has two distinct sides, the solution-facing porous side and the bulk aluminum-facing solid barrier-layer.

The thin-layer of carbon is deposited via chemical vapor deposition (CVD) onto a piece of quartz glass.[22] A CVD reactor houses this quartz-slide target and is heated in argon to very high temperatures (~ 970 °C.) This assembly is held at that temperature while a hydrocarbon gaseous precursor (ethylene) replaces the flowing Ar gas. After two hours, the ethylene is replaced with Ar as the system cools to room temperature. A carbon film of approximately 1 μm thickness covers the quartz slide.

The barrier-layer surface of the alumina membrane was etched in Ar plasma to widen the pores at this surface. The nanoporous alumina membrane was placed, barrier-layer surface up, on top of the CVD carbon film. This assembly was then inserted into the center of the vacuum chamber of a reactive ion etching apparatus (Plasma-Therm 790 series), and Ar plasma was used to etch the barrier-layer surface (Rf power = 200 W, ICP power = 300 W, pressure = 2 mTorr, flow rate = 15 sccm, time = 14 min.) After this time, the conditions were changed as O_2 was introduced to the system (Rf power = 300 W, flow rate Ar = 5 sccm, flow rate O_2 = 10 sccm, time = 2 min.) During this time, the O_2 plasma propagated into the pores of the alumina membrane and etched away the portions of the underlying carbon film beneath the pores. This led to a honeycomb carbon film that is a replica of the pore structure in the alumina membrane mask.

These structures (solution-side and barrier-side of the alumina template, carbon thin-film, and surface of carbon honeycomb) were imaged by SEM. Representative micrographs are shown in Figure 23. X-ray diffraction data (not shown) confirms the presence of carbon and suggests it is disordered in composition (i.e., not graphite.)

Nanomaterials in Li-Ion Battery Electrode Design

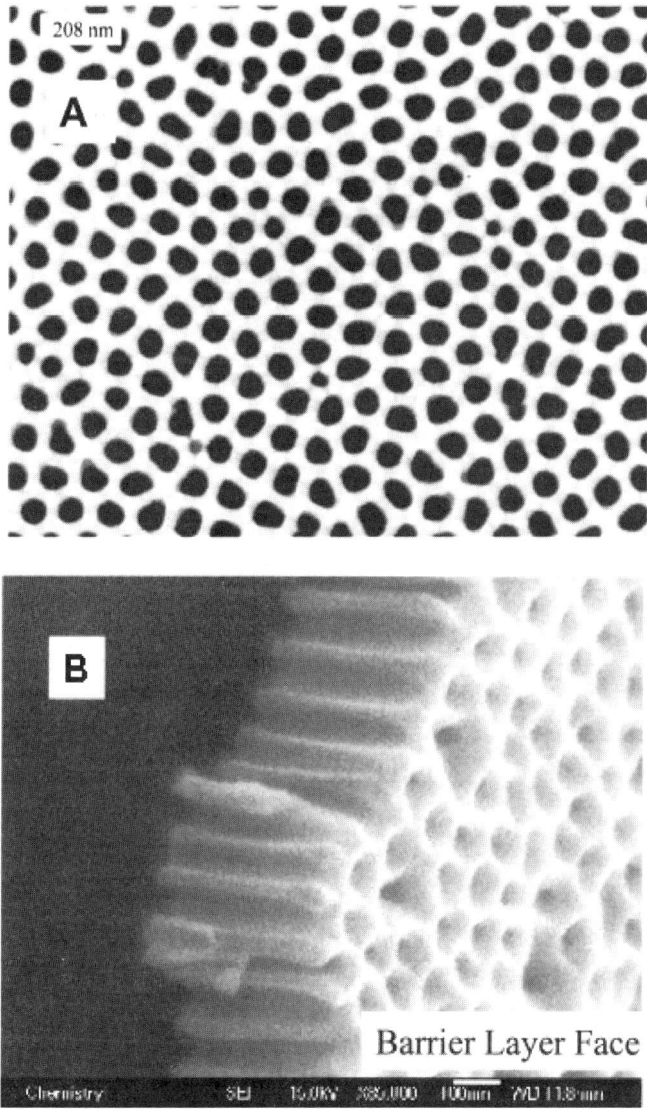

Figure 23. Electron micrographs of the nanopore alumina membrane and the honeycomb carbon. (A) Alumina solution surface; (B) cross section and barrier layer surface after Ar-plasma etching; (C) cross section and surface of as-synthesized (AS) CVD carbon film; and (D) surface of honeycomb (HC).Reproduced from Ref. 22, (© 2003), by permission of The Electrochemical Society, Inc.

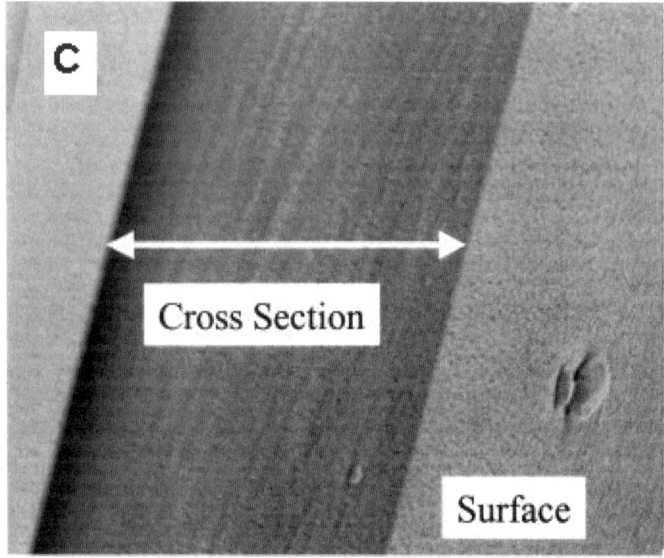

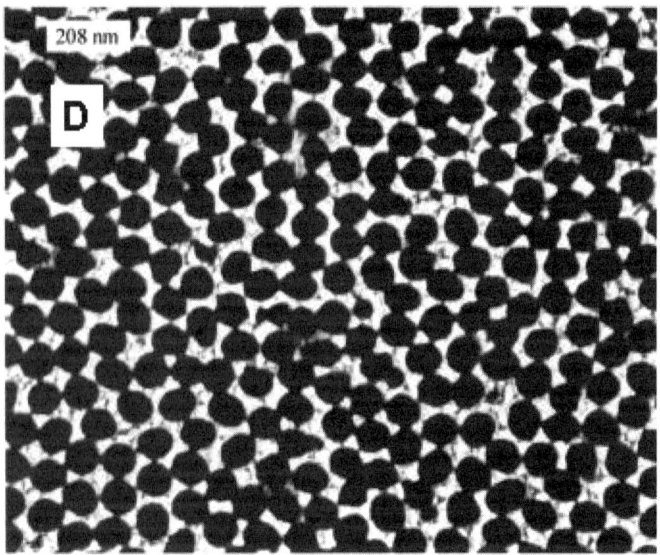

Figure 23. Continuation.

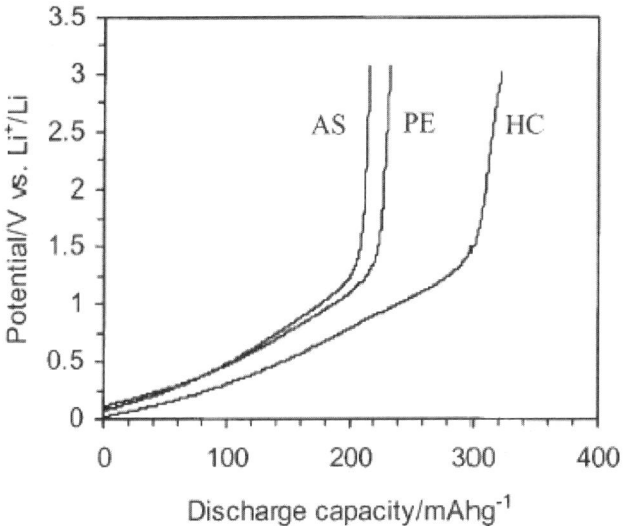

Figure 24. First constant-current discharge curves for the HC, PE, and AS electrodes (0.2 C). Reproduced from Ref. 22, (© 2003), by permission of The Electrochemical Society, Inc.

2. Electrochemical Characterization

Three forms of these structures were electrochemically characterized—as-synthesized thin-film (AS), non-honeycomb plasma-etched film (PE) and honeycomb (HC). The cell is similar to that described in previous experiments. The electrolyte is 1 M $LiClO_4$ in ethylene carbonate: diethyl carbonate (3:7 v/v). The reference and counter electrodes are Li ribbon. The working electrode (carbon) is charged from its open-circuit potential (~ 3 V) to 0 V. This charging reaction intercalates Li-ions as this material acts as an anode. The first-charge irreversible capacity is the largest for HC, because of its increased surface area per gram.[22] The electrodes were charged to the lower potential-limit of 0 V and discharged to the upper potential-limit of 3 V.

The first discharge experiment was performed at a low-rate (0.2 C.) These data for each electrode are shown in Figure 24. The HC electrode delivers almost 90% of the theoretical capacity of graphite (325 of a possible 372 mAh g^{-1}.) There are no potential plateaus in the discharge plot, which confirms the disordered nature of the carbon.[66] The rate capabilities were then investigated.

The specific capacity is calculated from the discharge time, current, and mass of carbon. These data are represented as a Capacity Ratio (Figure 25), similar to the data of the low-temperature experiment discussed earlier. The HC electrode shows a capacity advantage at every rate of discharge. This advantage increases with rate. At the rate of 10 C, the HC electrode delivers 50 times the specific capacity of the thin-film control.

Here we discussed another type of nanostructured carbon Li-ion battery anode. This new type of honeycomb battery material is capable of delivering a higher portion of its theoretical capacity than the control electrode. The porous nature of the electrode allows for penetration of the liquid electrolyte. This ensures that the solid-state diffusion distance of the intercalating Li-ion is minimized. The rate-limiting effect of concentration polarization is delayed to higher rates of discharge for nanomaterials.

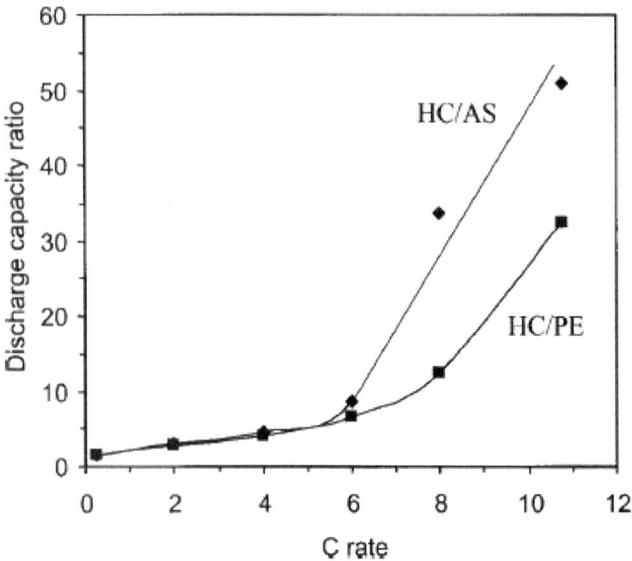

Figure 25. Ratio of the discharge capacities *vs.* discharge C rate for the carbon electrodes. Reproduced from Ref. 22, (© 2003), by permission of The Electrochemical Society, Inc.

VII. CONCLUSIONS

Template synthesis is a general nanofabrication strategy. Our lab has used this method to create nanostructured Li-ion battery cathodes and anodes. Nanomaterials reduce the solid-state diffusion distance of the Li-ion, as the intercalation sites, even at the "core", reside close to the surface. Also, the high-surface area of these materials easily distributes current over a large area (lower current density.) The combination of these two advantageous properties results in the superior (often dramatically uperior) rate capabilities of nanostructured electrodes when compared to microstructured electrodes. In contrast, the concentration polarization of the Li-ion and low surface area per gram of commercial systems preclude current technology from high-rate applications. Nanomaterials are capable delaying these rate- and power-limiting effects to much greater rates of discharge. Therefore, nanostructured electrodes succeed where conventional electrodes fail.

What is the next step for template-synthesized Li-ion battery materials? Our work has used nanomaterials as tools in fundamental research. The next goal is to apply these now-established fundamental principles towards generating a complete nanostructured Li-ion battery device. Another goal is to adapt the template-synthesis method for implementation at the commercial scale for high-rate energy-storage devices. Intercalation of polyvalent ions is also of interest, where the intercalating species carries more than one equivalent of charge. This is an exciting time as the emerging field of nanotechnology attempts to meet the demanding energy-storage challenges of an increasingly portable world.

ACKNOWLEDGEMENTS

This work would not have been possible without the efforts of a number of hard-working and highly motivated graduate students and collaborators, both past and present. They include Charles J. Patrissi, Naichao Li, Guangli Che, Brinda B. Lakshmi, Bruno Scrosati, and Fausto Croce. Financial support from the Department of Energy is also gratefully acknowledged. We also wish to thank The Electron Microscopy Core Laboratory, Biotechnology

Program, and the Major Analytical Instrument Center both at University of Florida, as well as the imaging facilities at Colorado State University.

REFERENCES

[1] T. Nagaura and K. Tozawa, *Prog. Batt. Solar Cells* **9** (1990) 209.
[2] A.H. Tullo, *Chem Eng News* **80** (2002) 25.
[3] R. Moshtev and B. Johnson, *J. Power Sources* **91** (2000) 86.
[4] C.R. Martin, *Science* **266** (1994) 1961.
[5] K. B. Jirage, J. C. Hulteen, and C.R. Martin, *Science* **278** (1997) 655.
[6] C. J. Brumlik and C. R. Martin, *Journal of the American Chemical Society* **113** (1991) 3174.
[7] V. P. Menon and C. R. Martin, *Analytical Chemistry* **67** (1995) 1920.
[8] M. Wirtz and C.R. Martin, *Advanced Materials* **15** (2003) 455.
[9] G. Che, B. B. Lakshmi, C .R. Martin, E. R. Fisher, and R. S. Ruoff, *Chemistry of Materials* **10** (1998) 260.
[10] S. A. Miller, V. Y. Young, and C. R. Martin, *J. Am. Chem. Soc.* **123** (2001) 12335.
[11] G. Che, B.B. Lakshmi, E.R. Fisher, and C.R. Martin, *Nature* **393** (1998) 346.
[12] B. B. Lakshmi, P. K. Dorhout, and C. R. Martin, *Chemistry of Materials* **9** (1997) 857.
[13] B. B. Lakshmi, C. J. Patrissi, and C. R. Martin, *Chem. Mater.* **9** (1997) 2544.
[14] C. R. Martin, *Accounts of Chemical Research* **28** (1995) 61.
[15] C. R. Martin, V. P. Menon, and R. V. Parthasarathy, *Polymer Preprints (American Chemical Society, Division of Polymer Chemistry)* **35** (1994) 229.
[16] G. Che, K. B. Jirage, E. R. Fisher, and C. R. Martin, *J. Electrochem. Soc.* **144** (1997) 4296.
[17] S. Kuwabata, T. Idzu, C. R. Martin, and H. Yoneyama, *J. Electrochem. Soc.* **145** (1998) 2707.
[18] N. Li, C. R. Martin, and B. Scrosati, *Electrochem. Solid St.* **3** (2000) 316.
[19] N. Li, C. J. Patrissi, G. Che, and C. R. Martin, *J. Electrochem. Soc.* **147** (2000) 2044.
[20] N. Li and C. R. Martin, *J. Electrochem. Soc.* **148** (2001) A164.
[21] N. Li, C. R. Martin, and B. Scrosati, *J. Power Sources* **97-98** (2001) 240.
[22] N. Li, D. T. Mitchell, K. -P. Lee, and C. R. Martin, *J. Electrochem. Soc.* **150** (2003) A979.
[23] M. Nishizawa, Mukai, K., Kuwabata, S., Martin, C. R., Yoneyama, H., *J. Electrochem. Soc.* **144** (1997) 1923
[24] C. J. Patrissi and C. R. Martin, *J. Electrochem. Soc.* **146** (1999) 3176.
[25] C. J. Patrissi and C. R. Martin, *J. Electrochem. Soc.* **148** (2001) A1247.
[26] C. R. Sides, N. Li, C. J. Patrissi, B. Scrosati, and C. R. Martin, *MRS Bullet.* **27** (2002) 604.
[27] C. R. Sides and C. R. Martin, *Adv. Mater.* **17** (2005), *in press*.
[28] C. R. Sides, F. Croce, V. Young, C. R. Martin, and B. Scrosati, *Electrochem. Solid St. Letters* **8** (2005) A484.
[29] B. A. Johnson and R. E. White, *J. Power Sources* **70** (1998) 48.
[30] L. E. Fransson, T. Edstrom, K. Gustafsson T., and Thomas, J.G., *J. Power Sources* **101** (2001) 1.

[31] C. H. Lu and Lin S. -W., *J. Power Sources* **97-98** (2001) 458.
[32] J. C. Lytle, H. Yan, N. S. Ergang, W. H. Smyrl, and A. Stein, *J. Mater. Chem.* **14** (2004) 1616.
[33] GE Osmotics, Inc., *Product Guide*, http://www.osmolabstore.com.
[34] R. L. Fleisher, P. B. Price, and R. M. Walker, *Nuclear Tracks in Solids*, University of California Press, Berkely, 1975.
[35] P. Apel, *Radiat. Meas.* **34** (2001) 559.
[36] Poretics, *Product Guide* (1995).
[37] C. R. Martin, M. Nishizawa, K. Jirage, and M. Kang, *J. Phys. Chem. B* **105** (2001) 1925.
[38] A. Despic and V.P. Parkhutik, in *Modern Aspects of Electrochemistry*, Vol. 20, Ed. by J. O. Bockris, R. E. White, and B. E. Conway, Plenum Press, New York, (1989), Ch. 6, p.401.
[39] http://www.whatman.com/
[40] G. L. Hornyak, C. J. Patrissi, and C. R. Martin, *J. Phys. Chem. B* **101** (1997) 1548.
[41] C. A. Foss, Jr., G. L. Hornyak, J. A. Stockert, and C. R. Martin, *J. Phys. Chem.* **98** (1994) 2963.
[42] S. B. Lee, D. T. Mitchell, L. Trofin, T. K. Nevanen, H. Soederlund, and C. R. Martin, *Science* **296** (2002) 2198.
[43] W. C. West, N. V. Myung, J. F. Whitacre, and B. V. Ratnakumar, *J. Power Sources* **126** (2004) 203.
[44] H. Yan, S. Sokolov, J. C. Lytle, A. Stein, F. Zhang, and W. H. Smyrl, *J. Electrochem. Soc.* **150** (2003) A1102.
[45] M. Thackeray, *Nat. Mater.* **1** (2002) 81.
[46] A. J. Bard and L. R. Faulkner, in *Electrochemical Methods: Fundamentals and Applications*, John Wiley & Sons, New York, NY, (1980),
[47] Y. Idota, T. Kubota, A. Matsufuji, Y. Maekawa, and T. Miyasaka, *Science* **276** (1997) 1395.
[48] I. A. Courtney and J. R. Dahn, *J. Electrochem. Soc.* **144** (1997) 2943.
[49] Joint Commission on Powder Diffraction Standards, *International Center for Diffraction Data*, File 41-1445.
[50] E. J. Plichta, M. Hendrickson, R. Thompson, G. Au, W. K. Behl, M. C. Smart, B. V. Ratnakumar, and S. Surampudi, *J. Power Sources* **94** (2001) 160.
[51] M. Salomon, H. P. Lin, E. Plichta, and M. Hendrickson, in *Advances in Lithium-Ion Batteries*, Ed. by W. Van Klinken and B. Scrosati, Kluwer Academic/Plenum Publishers, New York, 2002, p. 309.
[52] G. Nagasubramanian, *J. of Appl. Electrochem.* **31** (2001) 99.
[53] H. P. Lin, D. Chua, M. Salomon, H. C. Shiao, M. Hendrickson, E. Plichta, and S. Slane, *Electrochem. Solid St.* **4** (2001) A71.
[54] C.K. Huang, J.S. Sakamoto, J. Wolfenstine, and S Surampudi, *J. Electrochem. Soc.* **147** (2000) 2893.
[55] B. V. Ratnakumar, M. C. Smart, C. K. Huang, D. Perrone, S. Surampudi, and S. G. Greenbaum, *Electrochim. Acta* **45** (2000) 1513.
[56] M. C. Smart, B. V. Ratnakumar, and S. Surampudi, *J. Electrochem. Soc.* **146** (1999) 486.
[57] C. Delmas, H. Cognac-Auradou, J. M. Cocciantelli, M. Menetrier, and J. P. Doumerc, *Solid State Ionics* **69** (1994) 257.
[58] A. K. Padhi, K. S. Nanjundaswamy, C. Masquelier, S. Okada, and J. B. Goodenough, *J. Electrochem. Soc.* **144** (1997) 1609.
[59] N. Ravet, Y. Chouinard, J. F. Magnan, S. Besner, M. Gauthier, and M. Armand, *J. Power Sources* **97-98** (2001) 503.

[60] H. Huang, S. -C. Yin, and L. F. Nazar, *Electrochem. Solid St.* **4** (2001) A170.
[61] G. Arnold, J. Garche, R. Hemmer, S. Strobele, C. Vogler, and M. Wohlfahrt-Mehrens, *J. Power Sources* **119-121** (2003) 247.
[62] S. -Y. Chung, J. T. Bloking, and Y. -M. Chiang, *Nat. Mater.* **1** (2002) 123.
[63] C. R. Martin, C. R. Sides, F. Croce, and B. Scrosati, A high rate Electrically conductive material and a method of making the same, U.S. Patent pending (2004).
[64] F. Croce, Epifanio, A. D., Hassoun, J., Deptula, A., Olczac, T., and Scrosati, B., *Electrochem. Solid St.* **5** (2002) A47.
[65] H. Masuda, and M. Satoh, *Jpn J. Appl. Phys. 2* **35** (1996) L126.
[66] J. S. Gnanaraj, M. D. Levi, E. Levi, G. Salitra, D. Aurbach, J. E. Fischer, and A. Claye, *J. Electrochem. Soc.* **148** (2001) A525.
[67] M. Kang, S. Yu, N. Li, and C. R. Martin, *Small* **1** (2005) 69.

4

Direct Methanol Fuel Cells: Fundamentals, Problems and Perspectives

Keith Scott[a] and Ashok K. Shukla[b,c]

[a]*School of Chemical Engineering and Advanced Materials, University of Newcastle upon Tyne, UK,*
[b]*Solid State and Structural Chemistry Unit, Indian Institute of Science, Bangalore (India),*
[c]*Central Electrochemical Research Institute, Karaikudi, Tamil Nadu (India)*

I. INTRODUCTION

Fuel cells are chemoelectric engines that convert the chemical energy of a fuel directly into electricity. The process is an electrochemical reaction akin to a battery, but unlike the battery, fuel cells do not store the chemicals internally and instead use a continuous supply of fuel from an external storage tank. Accordingly, fuel cell systems have the potential to solve the most challenging problems associated with the currently available battery systems, namely their insufficient energy at a given weight (specific energy density) or volume (volumetric energy density).[1] Besides, while the leading battery technologies are reaching the practical limits of their energy storage capabilities, commercial fuel cells are still in their infancy. Furthermore, since fuel cells operate without a thermal cycle, they offer a quantum leap in energy efficiency and virtual elimination of air pollution without the use of emission control devices as in conventional energy conversion.[1-3]

A fuel cell consists of two electrodes, an anode to which the fuel and a cathode to which the oxidant are supplied externally, and the electrolyte which separates the two electrodes and allows the ions to flow across it. There are six generic types of fuel cells in various stages of development, namely

- Phosphoric-acid fuel cells (PAFCs),
- alkaline fuel cells (AFCs),
- polymer-electrolyte-membrane fuel cells (PEMFCs),
- molten-electrolyte fuel cells (MCFCs),
- solid-oxide fuel cells (SOFCs), and
- direct-methanol fuel cells (DMFCs).

The most advanced fuel cells in terms of applications and commercialisation are AFCs used successfully in space programs in mid 1960s and PAFCs used for stationary power plants. These fuel cells were intended to power an electric vehicle but AFCs are prone to carbonate fouling due to the presence of carbon dioxide either in a reformate hydrogen fuel or in atmosphere, and the PAFCs are rather complex, too heavy as well as bulky, to fit inside the engine compartment of a car. The low-operating temperature and rapid start-up characteristics together with its robust solid-state construction give PEMFCs a clear advantage for application in cars.[4] The preferred fuel for PEMFCs is hydrogen and while many strategies for providing hydrogen to PEMFCs are being evaluated, the most acceptable option appears to be to generate hydrogen on-board and on-demand from liquid hydrocarbons or methanol.[5] The technical challenge however lies in modifying large-scale industrial processes like steam-reforming or partial-oxidation reactors to light-weight units that can fit inside the car. An elegant solution to the problems associated with the need for gaseous hydrogen fuel is to operate the PEMFCs directly with a liquid fuel. Substantial efforts are therefore being expended on PEMFCs that run on air plus a mixture of methanol and water.[6-10] A solid-polymer-electrolyte direct methanol fuel cell (SPE-DMFC) would be about as efficient as a conventional reformer-based PEFC unit, in both its construction and operation. The realisation of a commercially viable SPE-DMFC is indeed regarded as the 'holy grail' of fuel cell technologies. This article reviews, the advances made in the performance of SPE-DMFCs since its inception.

II. OPERATING PRINCIPLE OF THE SPE-DMFC

Historically, direct electro-oxidation of methanol in a fuel cell has been a subject of study for more than three decades. The early cell designs utilised aqueous sulphuric acid electrolyte at about 60 °C. About 20 years ago, Shell Research Centre in the UK and Hittachi Research Laboratories in Japan built DMFC stacks of up to 5 kW but their power densities were only 20–30 mW/cm^2 even with platinum loadings as high as 10 mg/cm^2, which corresponds to specific power densities of 2–3 W/g of platinum catalyst.

In a fuel cell employing an acid electrolyte, methanol can be directly oxidised to carbon dioxide at the anode according to the reaction,

$$CH_3OH + H_2O \rightarrow CO_2\uparrow + 6\ H^+ + 6\ e^- \qquad (1)$$

The thermodynamic potential (E_a^o) for reaction (1) calculated from the standard chemical potentials at 25 °C is 0.03 V vs. SHE. At the cathode, oxygen gas combines concomitantly with the protons and electrons to get reduced to water through the reaction,

$$\frac{3}{2} O_2 + 6\ H^+ + 6\ e^- \rightarrow 3\ H_2O \qquad (2)$$

The thermodynamic potential (E_c^o) for reaction (2) is 1.23 V (vs. SHE). Accordingly, the net cell reaction is represented as,

$$CH_3OH + \frac{3}{2}\ O_2 \rightarrow 2\ H_2O + CO_2 \qquad (3)$$

The free energy (ΔG^o) for reaction (3) is –702 kJ mol^{-1} and the enthalpy of reaction (ΔH^o) is –726 kJ mol^{-1}. Accordingly, the standard electromotive force (e.m.f.), $E_{eq}^o = -\Delta G^o/nF = (702 \times 10^3)/(6 \times 96500) = E_c^o - E_a^o = 1.20$ V. The potential efficiency (ε_f) of a DMFC for an operational cell e.m.f (E) of 0.5 V is about 40% and the specific energy (W) is $-\Delta G^o/ 3600 \times M = (702 \times 10^3)/(3600 \times 0.032) \sim 6.1$ kWh/kg.

The main drawback of such cells is the very sluggish anode reaction, which coupled with the inefficient cathode reaction, gives rise to very low overall performance, particularly at low temperatures. The performance of the cells utilising sulphuric acid electrolyte is further receded owing to the high internal resistance of the system. In the 1980's, it was realised that a

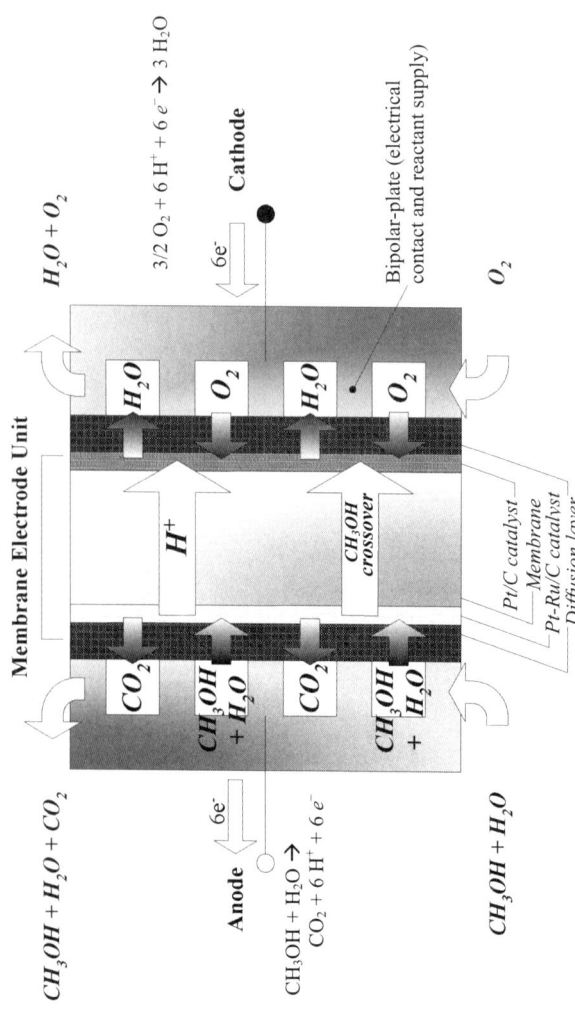

Figure 1. Schematic diagram of the DMFC with proton conducting membrane.

considerable increase in the efficiency might be obtained by using the 'zero gap' cell design principle in which the liquid electrolyte is replaced by a thin proton-conducting polymer sheet such as Nafion® — a perfluoro-sulphonic acid polymer.[11] A DMFC with a Nafion® electrolyte membrane is shown schematically in Figure 1. In a SPE-DMFC, methanol dissolved in water is supplied to its anode, which has the tendency to pass through the membrane electrolyte and hence affect the performance of the cathode.[12-14] Therefore, a fundamental limitation in the practical realisation of such a SPE-DMFC has been the existence of electrochemical losses at both the anode and the cathode arising mainly due to the electrocatalytic restrictions and methanol crossover though the electrolyte membrane associated with the osmotic drag. The typical polarisation curve for a SPE-DMFC along with its constituent electrodes is shown schematically in Figure 2. Although the thermodynamic potential for reaction (1) is 0.03 V (vs. SHE), because of the number of electrons involved, the equilibrium value is not readily realisable, even with the best possible electrocatalysts devised so far. The losses or reduction in voltages from the ideal value are referred to as polarisation or overpotential. Furthermore, because of the high degree of irreversibility of reaction (2), even under open-circuit conditions, the overpotential at the oxygen electrode in a PEFC is about 0.2 V which represents a loss of about 20% from the theoretical efficiency of the PEFCs. The situation is even worse with the SPE-DMFCs where there is an inherent loss of about 0.1 V at the oxygen electrode owing to the crossover of methanol from anode to the cathode.[15] Consequently, the output cell voltage in a SPE-DMFC is much lower than the ideal thermodynamic value and it decreases with increasing current density as shown in Figure 2.

The polarisation curve comprises three distinct regions. Region-I, belongs to the voltage loss at low currents due to the interfacial resistance and is called activation polarisation region. An efficient catalyst would help circumvent the activation polarisation. Region-II is characterised by a linear drop with increasing current and is due to the Ohmic resistance. It is termed as Ohmic polarisation region. A methanol impermeable membrane with high protonic conductivity would help to reduce the Ohmic polarisation.[1] Region-II is followed by a final additional drop at high load currents arising due to the depletion of acceptors at the interface for the transitory species and is termed a concentration

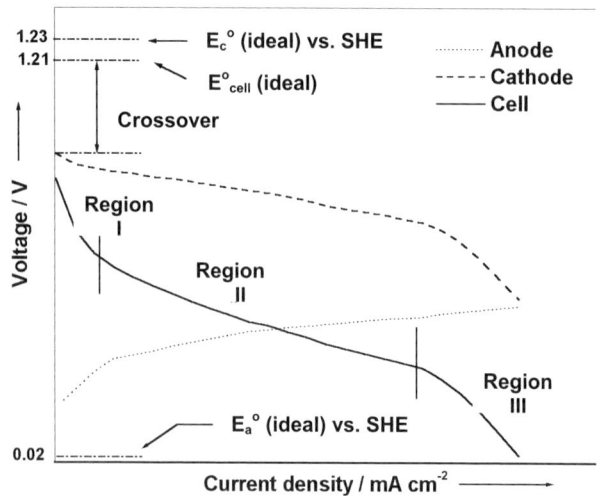

Figure 2. Polarisation curve for a SPE-DMFC along with its constituent electrodes

polarisation region (Region-III).[1] To reduce the concentration polarisation, fuel cell electrodes demand optimally-designed electrode structure and catalyst morphologies. In this context, optimisation modelling of electrode performance and flow fields is an indispensable tool. These aspects are discussed later in Section VII of this article.

III. ELECTRODE REACTION MECHANISMS IN SPE-DMFCS

1. Anodic Oxidation of Methanol

Several schemes have been proffered for the anodic oxidation of methanol. Broadly speaking, the basic mechanism for methanol oxidation can be summarised in two functionalities, namely the electrosorption of methanol on to the substrate followed by addition of oxygen to adsorb carbon-containing intermediates to generate carbon dioxide. In practice, only a few electrode materials are capable of adsorption of methanol. In acidic media, only platinum[16] and platinum-based catalysts[17-19] have been found to show sensible activity, and almost all mechanistic studies have concentrated on these materials. On platinum itself, adsorption of methanol is believed to take place through a sequence of steps described below. The first step is dissociative chemisorption of methanol onto the platinum surface, which involves successive donation of electrons to the catalyst as follows:[16]

$$Pt + CH_3OH \xrightarrow{k_1} Pt\text{-}CH_2OH + H^+ + e^- \quad (4)$$

$$Pt\text{-}CH_2OH \xrightarrow{k_2} Pt_2\text{-}CHOH + H^+ + e^- \quad (5)$$

$$Pt_2\text{-}CHOH \xrightarrow{k_3} Pt_3\text{-}COH + H^+ + e^- \quad (6)$$

where, with the relative values of rate constants $k_1 < k_2 < k_3$ makes Pt_3-COH the major surface species. A surface rearrangement of the oxidation intermediates generates carbon monoxide, linearly or bridge-bonded to Pt-sites according to the reaction,

$$Pt_3\text{-}COH \rightarrow Pt\text{-}CO + 2\ Pt + H^+ + 1\ e^- \quad (7)$$

Water discharge occurs at high anodic overpotentials on Pt with the formation of Pt-OH species at the platinum surface as below:

$$Pt + H_2O \rightarrow Pt\text{-}OH + H^+ + 1\ e^- \quad (8)$$

The ultimate step is the reaction of Pt-OH groups with neighbouring methanolic residues to give carbon dioxide according to the following reaction:

$$Pt\text{-}OH + Pt\text{-}CO \rightarrow 2\ Pt + CO_2 \uparrow + H^+ + 1\ e^- \quad (9)$$

Accordingly, the overall oxidation of methanol to carbon dioxide proceeds through a six-electron donation process.

Platinum alone is not sufficiently active and there is need for a promoter that could effectively provide oxygen in some active form to achieve facile oxidation of the chemisorbed CO on platinum. In the literature, various approaches towards platinum promotion have been attempted. The simplest method is to generate more Pt-O species on the platinum surface by incorporating certain metals with platinum to form alloys such as Pt_3Cr and Pt_3Sn, which then dissolve to leave highly reticulated but active surfaces. A second approach has been the use of surface adatoms produced by underpotential deposition on the platinum surface. A third type of promotion is the use of alloys of platinum with different metals such as Pt-Ru, Pt-Os, Pt-Ir, etc., where the second metal forms a surface oxide in the potential range for methanol oxidation. Among these, Pt-Ru alloy[16-19] has been found to be particularly effective and efforts have been made to enhance the promotion on Pt-Ru based ternary as well as quaternary alloys, namely Pt-Ru-Os and Pt-Ru-Ir, etc.

The fourth type of promotion described in the literature is a combination of Pt with a base-metal oxide such as Nb, Zr, Ta, etc. In addition to electrodeposition or reductive deposition of Pt onto an oxide surface such as Pt-WO_3,[20] attempts have also been expended to study methanol oxidation on perovskite-based oxides with platinum, such as $SrRu_{0.5}Pt_{0.5}O_3$. It also seems that certain amorphous metal alloys, such as Ni-Zr, which form a thick passive oxyhydroxide film, can also facilitate methanol oxidation. Table 1 gives a summary of the species investigated as catalyst promoters that have had a reported positive effect on methanol oxidation.

Among the various methanol oxidation catalysts described above, Pt-Ru and Pt-Sn are the most widely studied catalysts.[16,21,22] The alloying of Sn and Ru with Pt gives rise to electrocatalysts, which strongly promote the oxidation of methanol and related

Table 1
Effect of Catalyst Promoters on Methanol Oxidation

Promotion by	Catalyst promoter	Comment
Alloying and dissolution to produce highly reticulated surfaces	Cr, Fe, Sn	Typically about 100 mV lower potential than Pt.
Surface adatoms	Sn, Bi	
Alloys	Ru, Sn, Mo, Os, Ir, Ti, Re	Ru has the greatest effect. Sn, Mo, Os and Re are substantial promoters.
Metal oxides	Ru	Hydrous Ru oxide as the most active catalyst.
Base metal oxides	W, Nb, Zr, Ta	W oxide is a notable promoter. Small effect for other metal oxides (typically < 100 mV).

methanolic species. On the platinum surface, at low potentials –CO groups are adsorbed while at high potentials chemisorption of –OH groups, takes place during the electro-oxidation of methanol with both the processes succinctly separated. On a Pt-Ru surface, the chemisorption of –OH groups shifts to lower potentials and overlaps with the region where –CO groups are adsorbed on the catalyst as shown in Figure 3. On a Pt-Ru alloy, water discharging occurs on Ru-sites at much lower potentials in relation to pure Pt catalyst as depicted in Figure. 3 according to the reaction given below:

$$Ru + H_2O \rightarrow Ru\text{-}OH + H^+ + 1\ e^- \quad (10)$$

The final step is the reaction of Ru-OH groups with the neighbouring methanolic residues adsorbed on Pt to be oxidized to carbon dioxide according to the reaction:

$$Ru\text{-}OH + Pt\text{-}CO \rightarrow Ru + Pt + CO_2 \uparrow + H^+ + 1\ e^- \quad (11)$$

The following scheme has been envisaged for methanol oxidation on Pt-Sn alloy catalyst:[21, 22]

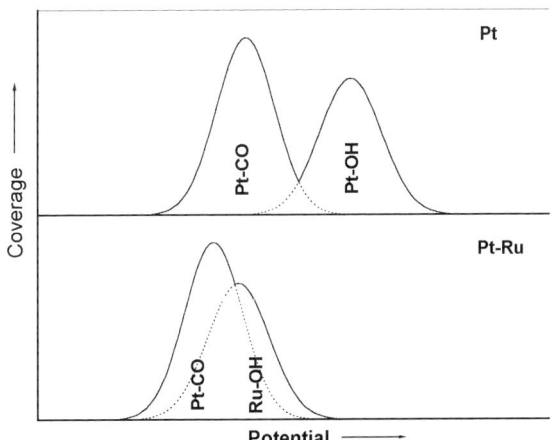

Figure 3. Mechanism of oxidation of methanol on binary catalyst. On a Pt-Ru surface, the chemisorption of –OH groups shifts to lower potentials and overlaps with the region where –CO groups are adsorbed on the catalyst.

$$\text{Sn-Pt} + \text{CH}_3\text{OH} \rightarrow \text{Sn-(Pt-CHO)} + 3\,\text{H}^+ + 3\,e^- \quad (12)$$

$$\text{Sn-(Pt-CHO)} + \text{H}_2\text{O} \rightarrow \text{(Sn-OH)-(Pt-CHO)} + \text{H}^+ + 1\,e^- \quad (13)$$

$$\text{(Sn-OH)-(Pt-CHO)} \rightarrow \text{(Sn-OH)-(Pt-CO)} + \text{H}^+ + 1\,e^- \quad (14)$$

$$\text{(Sn-OH)-(Pt-CO)} \rightarrow \text{Sn-Pt} + \text{CO}_2 \uparrow + \text{H}^+ + 1e^- \quad (15)$$

From the above discussion, it can be surmised that the addition of Ru and Sn to Pt markedly promotes its electrocatalytic activity through the adsorption of oxygenated species on Ru or Sn sites. Obviously, methanol oxidation is more facile on Pt-Ru and Pt-Sn surfaces because the reaction desires the electrocatalyst to be used in a potential regime where labile-bonded oxygen should be present on the surface. In this situation, the supply of active oxygen to the species is of paramount importance, since this, apparently, would facilitate the oxidation of adsorbed methanolic residues to carbon dioxide. It has been documented that with the Pt-Sn alloy catalyst, promotion in methanol oxidation is seen in the low-potential region while Pt-Ru is particularly active in the high-potential region. The obvious pair therefore would have been Sn and Ru but Sn and Ru are not quite miscible, and attempts to form a ternary Pt-Ru-Sn alloy led to expulsion of Ru.

Performance of a catalyst will vary with the nature of the electrolyte due to a change in factors such as ionic conductivity, the degree of adsorption of acid radicals on the catalyst surface and the stability and influence of corrosion. The electrolyte which has received the most intense investigation is sulphuric acid and not surprisingly, since the great interest in PEMCls, is perfluorosulphonic acid in the form of a solid polymer electrolyte. In general, high concentrations of an acid tend to reduce the activity of Pt catalysts, especially above 5 M. Phosphoric acid at high concentrations (> 5M) can deliver superior performance to sulphuric acid at the same concentrations.[23] However, the best performance, at low temperatures (60 °C), is obtained with 3 M sulphuric acid. The oxidation of methanol at high temperature (200 °C) in phosphoric acid (> 96%) on Pt and Pt alloy catalysts (with Sn, Ru, Ti) has given reasonable performance,[24] and has relevance to the use of other PEM systems not based on a perfluorinated sulphonic acid ionomer, e.g., phosphoric acid-doped polybenzimidazole (PBI). Trifluoromethane sulphonic acid (TFMSA) has been examined as an alternative acid for methanol oxidation but no conclusive evidence exists for superior performance and controversially, poisoning of the catalyst is reported from sulphur.[25, 26] It has been reported that the addition of silicotungstic acid ($H_4SiW_{12}O_{38}$) in low concentration (10 mmol dm^{-3}) to sulphuric acid accelerates the reaction rate by up to 100% for methanol oxidation on Pt-Ru catalyst.[27]

Solid-polymer electrolyte DMFCs have been developed to their current status using proton conducting polymer membranes. Two obstacles currently inhibiting the application of DMFCs are the relatively low activity of methanol electro-oxidation catalysts and methanol crossover through the proton-conducting membrane; both of which can be overcome by using hydroxide ion conducting anion-exchange membranes. Alkaline fuel cells have been developed to a significant level; the principal attraction is that the catalysts generally perform better in alkaline than in acid solutions.[28] With an alkaline electrolyte, a lower loading of and, importantly, a wider range of electrocatalyst, (other that Pt), e.g., based on Ni may be used. Aside, direct methanol alkaline fuel cell (DMAFC) does not suffer from poisoning by the chemisorbed methanol fragments observed in acid electrolyte. The principle electrode reactions in a direct methanol alkaline fuel cell (DMAFC) are:

cathode: $\quad \frac{3}{2} O_2 + 3 H_2O + 6 e^- \longrightarrow 6 OH^-$ (16)

anode: $\quad CH_3OH + 6 OH^- \longrightarrow CO_2 + 5 H_2O + 6 e^-$ (17)

In the case of the DMAFC, electro-osmotic water transport is from cathode to anode, which counteracts the diffusion of methanol across the membrane. The operating principle of a DMAFC is shown schematically in Figure. 4.

There is report of methanol oxidation at higher pH using carbonate and bicarbonate electrolytes, and potassium or sodium hydroxide. For the former electrolytes no advantage has be gained in terms of improved catalyst activity over sulphuric acid. In the case of caustic electrolyte, although attractive in terms of achievable current densities, the issue of carbonate fouling of electrolyte is a practical problem of greater significance.

There is a considerable literature on the oxidation of methanol in alkaline electrolytes.[29,30] Methanol oxidation in alkaline electrolyte has been investigated by several research groups, using for example Pt,[31,32] Pd,[33] nickel hydroxide[34] and chemically modified nickel.[35] The oxidation of methanol in alkaline solutions

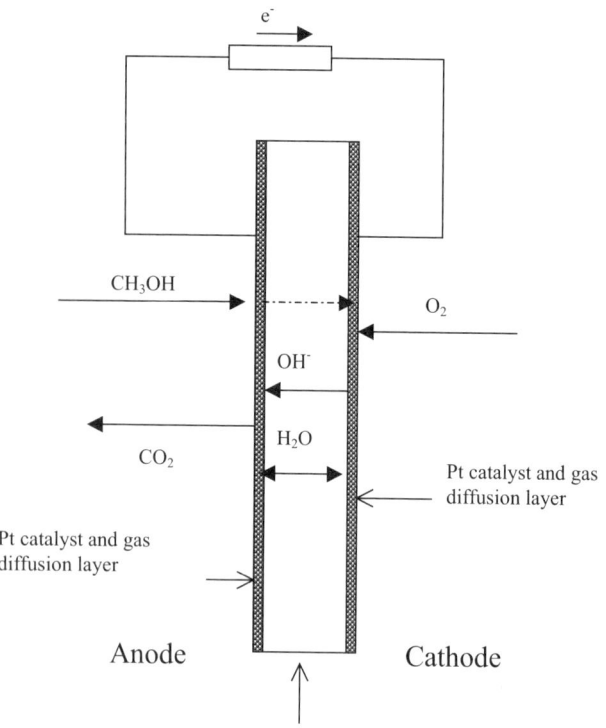

Figure 4. Operating principle of a direct methanol alkaline fuel cell.

is not particularly structure sensitive. The chemi-sorption bonding of (–CHO) on for example Pt is weak, such that further oxidation takes place easily without irreversibly blocking the catalyst sites. A report of methanol oxidation in alkaline solution, on unsupported platinum powder, has shown impressive polarisation characteristics in comparison to that obtained in acid solution.[36] In alkaline solution, current densities for methanol oxidation (at fixed potentials) are at least an order of magnitude higher than in acidic electrolytes around room temperature: at 0.5 V overpotential, acidic ≈ 40 mA cm^{-2} whereas alkaline ≈ 600 mA cm^{-2}. At 50 °C, 1000 mA cm^{-2} is attained at 0.52 V overpotential in alkaline solution. Although these particular data are impressive (estimated power densities: 400–500 mW cm^{-2}) the electrode fabrication method (porous bulk platinum powder, produced by borohydride

Direct Methanol Fuel Cells

reduction, pressed into platinum mesh) is neither practical nor economic for commercial applications. More recently, studies on DMAFCs have been performed with a view of obtaining cheaper, low-temperature polymer electrolyte fuel cell systems.[37]

2. Cathodic Reduction of Oxygen

Oxygen reduction reaction can proceed by two different pathways;[2] namely the direct four electron pathway and peroxide pathway. The direct four-electron pathway in alkaline medium proceeds as,

$$O_2 + 2\ H_2O + 4\ e^- \rightarrow 4\ OH^- \quad (E^o = 0.4 \text{ V vs. SHE}) \quad (18)$$

and in acidic medium, it proceeds as,

$$O_2 + 4\ H^+ + 4e^- \rightarrow 2\ H_2O \quad (E^o = 1.23 \text{ V vs. SHE}) \quad (19)$$

The peroxide pathway in alkaline medium proceeds as,

$$O_2 + H_2O + 2\ e^- \rightarrow HO_2^- + OH^- \quad (E^o = -0.06 \text{ V vs. SHE}) \quad (20)$$

followed by peroxide reduction to OH^- ions,

$$HO_2^- + H_2O + 2\ e^- \rightarrow 3\ OH^- \quad (E^o = 0.87 \text{ V vs. SHE}) \quad (21)$$

or chemical decomposition of peroxide,

$$2\ HO_2^- \rightarrow 2\ OH^- + O_2 \quad (22)$$

In an acidic medium, production of dioxygen through the peroxide pathway is possible as follows:

$$O_2 + 2\ H^+ + 2\ e^- \rightarrow H_2O_2 \quad (E^o = 0.67 \text{ V vs. SHE}) \quad (23)$$

This is followed by either,

$$H_2O_2 + 2\ H^+ + 2\ e^- \rightarrow 2\ H_2O \quad (E^o = 1.77 \text{ V vs. SHE}) \quad (24)$$

or,

$$2\ H_2O_2 \rightarrow 2\ H_2O + O_2 \quad (25)$$

The direct four-electron path does not involve the peroxide species and hence has a higher Faradaic efficiency in relation to the peroxide pathway. However, it has been difficult to find catalysts that could facilitate the direct four-electron pathway for dioxygen reduction. As indicated in Section II, in addition to irreversible losses, there is an additional overpotential observed on the cathode due to methanol crossover in a SPE-DMFC from its anode to the cathode.[14] Therefore, cathodes with high loadings of platinum are usually employed in SPE-DMFCs.[11] However, since platinum has a tendency to be poisoned with methanol, it appears mandatory both to develop methanol impermeable membranes and also methanol tolerant oxygen reduction catalysts for practical realisation of SPE-DMFCs. In recent years, certain Ru-based chalcogenides have shown promise as methanol-tolerant oxygen reduction catalysts.[8,9] However, these materials have much lower intrinsic specific activity for oxygen reduction than platinum.

IV. MATERIALS FOR SPE-DMFCS

From the foregoing, it is clear that for the realization of practical SPE-DMFCs, it is mandatory

1. to increase their operational current-densities making it desirable to develop catalyst materials which would increase the reaction rates both at the methanol anode and the oxygen cathode,
2. to develop methanol-tolerant oxygen-reduction catalysts, and
3. to investigate new electrolyte membranes which would have high proton conductivities and would be temperature resistant and methanol impermeable.

Various developments in these areas are discussed in the following Sections.

1. Catalyst Materials

(i) Anode Catalysts

As perceived earlier in Section III, Pt-Ru happens to be the most commonly used catalyst for the anodic oxidation of methanol in a SPE-DMFC. The reported difference in activities of Pt-Ru catalyst is usually attributable to the method of catalyst production and resulting varying surface compositions. Improvements in catalytic activity can frequently be achieved by dispersing the catalyst on a high surface-area graphitic carbon, such as Vulcan XC-72 and KetzenBlack EC-600D carbon. In catalyst production, the use of $RuCl_3$ with H_2PtCl_6 in aqueous solution produces relatively poor catalysts. Improvement in catalyst dispersion can be achieved by first oxidising the carbon support, chemically or electrochemically, which produces acidic surface oxides capable of complexing with metal cations via an ion-exchange process.[29] The use of $Pt(NH_3)_2NO_2$ with $Ru(NO)(NO_3)_x$ in nitric acid, and $Na_6Pt(SO_3)_4$ with $Na_6Ru(SO_3)_4$ in sulphuric acid produces effective catalysts for methanol oxidation.

The use of nitric acid activates the carbon surface by oxidation and the metal species complex with the surface groups so formed. Similarly, activation of Pt-Ru catalysts by heating in air is reported to impart superior activity than activation in hydrogen. A recent study of $Pt(NH_3)_2(NO_2)_2$-$Ru_3(CO)_{12}$ and $Pt(NH_3)_2(NO_2)_2$-$RuNO(NO_3)_x$ produced catalyst of high activity for methanol oxidation.[38]

It appears that the use of a chloride-free precursor salt for production of Pt-Ru catalyst provides higher catalyst dispersion in relation to chloride-containing salts. In a recent study, it has been reported that the sulfide species, present in the sulfito-complex precursors, help to produce finer catalyst particles. Primarily, impregnation and colloidal procedures have been employed for producing carbon-supported Pt-Ru catalyst. The typical performance of Pt-Ru catalyst from sulphito-complex for methanol oxidation is shown in Figure 5. The use of Pt-Ru carbonyl cluster complexes has been reported for the preparation of carbon nanocomposites.[39]

When producing Pt-Ru catalyst by an impregnation route, the precursor salts are first obtained in their solution form which are subsequently subjected to reduction with an appropriate reducing agent. High surface-area carbon-supported Pt-Ru catalyst has also been prepared by a colloidal dispersion route by Bonnemann et al.[40] Other methods based on thermal decomposition of appropriate high molecular weight Pt-precursors and Pt-carbonyl compounds to produce unsupported high surface area catalysts have also been documented. Beside these methods, there has been interest in co-precipitation, sol-gel and sputtering routes for producing Pt-Ru catalysts.

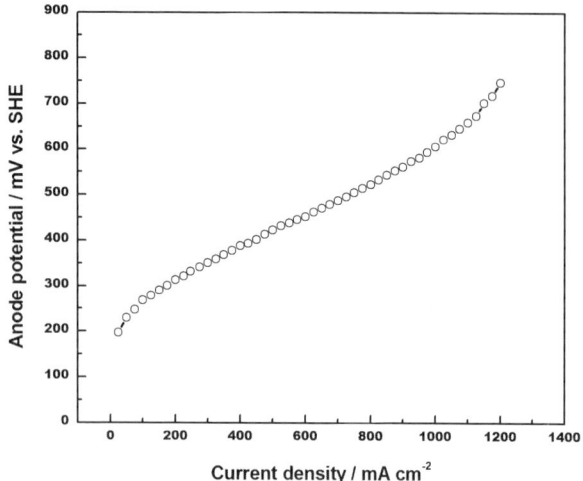

Figure 5. Typical performance of Pt-Ru catalysts produced from sulphito complexes for methanol oxidation.

There is significant interest in the improvement in activity of Pt-Ru catalysts for the DMFC. These improvements are sought in two ways:

1. To introduce ternary or quaternary alloys based on Pt-Ru which have superior activity to the binary catalyst.
2. New synthetic strategies which deliver higher activity for the catalyst.

The development of new routes for the preparation of Pt-Ru alloy nano-composites using the thermal treatment of (η-C_2H_4)(Cl)Pt(μ-Ci)$_2$Ru(Cl)(η^3:η^2-2,7-dimethyloctadienediyl) under appropriate oxidising and reducing conditions using microwave dielectric loss heating has been described by Boxall et al.[41] The method affords Pt-Ru/C nano-composites consisting of Pt-Ru alloy nano-particles highly dispersed on the carbon substrate. In preliminary DMFC tests, these highly active catalysts produced performance significantly better than that achieved with commercial carbon-supported catalyst and comparable with a proprietary unsupported catalyst.

The production of electrocatalysts by electrodeposition, and in particular pulse current and cyclic current electrodeposition, is a potentially attractive route. The production of 15 Å Pt particles onto carbon[42] and also the plating of Pt-Ru, of low loading (< 0.5 mg cm^{-2}), onto teflon bonded carbon supports for the DMFC have been reported.[43]

Pulse electrodeposition has recently been used to produce Pt-based catalysts for methanol oxidation. The catalyst particles, of mean size 15 μm, are formed by electrodeposition from the Pt precursor salts in a Nafion solution previously dried at low temperature.[44] Active electrocatalysts have also been produced by pulsed deposition on carbon particles bonded with Nafion.[45]

The promotion of the activity of Pt_Ru catalysts has been of significant interest. Early work examined the influence of Cr and Ga on Pt-Ru catalyst[46] with only Ga showing any positive effect be it small (approximately 30 mV). Aricó et al.[47] studied the chemical and morphological characteristics of a quaternary Pt-Ru-Sn-W carbon supported catalyst for methanol oxidation and confirmed the promotional effects of Sn, W, Ru (and/or their oxides) together.

Gotz and Wendt[48] demonstrated the use of Mo and W as co-catalyst for Pt-Ru anodes. Catalysts were produced by a colloidal deposition procedure on Vulcan XC-72 carbon or by impregnation of commercial Pt-Ru (Etek) catalyst. Only the ternary catalyst based on W exhibited a noticeable enhancement in methanol oxidation when tested in a DMFC with noble metal loading of 0.4 mg cm^{-2}.

Osmium has been identified as a suitable co-catalyst with Pt-Ru for methanol oxidation by a number of groups. For methanol oxidation electrochemically co-deposited Pt-Os composite electrodes (on glassy carbon) have shown superior oxidation catalysis in comparison to Pt alone. The materials is present as a highly dispersed phase on the substrate surface.[49] Similar performance has been reported for Pt-MoO$_x$ catalysts.[50]

An effective method of catalyst evaluation is combinatorial synthesis. Through this procedure, both ternary catalysts, such as Pt-Ru-Os, and quaternary catalysts, such as Pt-Ru-Os Ir, have been synthesized, with which reportedly, superior methanol oxidation capability than the Pt-Ru catalyst. Unsupported catalyst produced by borohydride reduction of the metal chloride salts, with a composition Pt:Ru:Os of 65:25:10 typically outperformed the Pt-Ru (1:1) catalyst, e.g., at 90 °C, 0.4V, 340 mA cm^{-2}, cf. 260 mA cm^{-2} Pt-Ru.[51]

The addition of Ir to the Pt-Ru-Os catalyst, produced by arc melting or borohydride reduction, is reported to give superior catalysis to Pt-Ru alone.[52] Performance data in a DMFC produced nearly twice the peak power density of that achieved with a Pt-Ru catalyst.[53]

The introduction of transition metal oxides (WO_x, MoO_x, VO_x) with Pt-Ru catalyst has been shown to lead to enhanced catalytic activity for methanol oxidation. The catalyst produced by a modification of the Adams method (hydrogen reduction of a solution of mixed metal oxides) had BET surfaces of 80–120 m^2 g^{-1} and particle sizes of 3–5 nm. The unsupported ternary catalysts of V and Mo (Pt:M ratio 7–8) deposited onto glassy carbon (0.28 mg cm^{-1}) and coated with a thin film of Nafion, showed better activity than Pt-Ru,[54] although there was little difference in the onset potential of methanol oxidation for any of the catalysts, compared to Pt-Ru, i.e., 380 mV (vs. RHE).

Pt/Ru/Ni (5:4:1) electrocatalysts synthesized by reduction with $NaBH_4$ combined with freeze-drying have exhibited superior performance to Pt-Ru catalysts. The onset potential of Pt/Ru/Ni (5:4:1) was lower than that of Pt/Ru (1:1) and the current density obtained with Pt/Ru/Ni (5:4:1) was larger than that for Pt/Ru (1:1). Polarization and power density data in a liquid-feed DMFC with Pt/Ru/Ni (5:4:1) showed a higher catalytic activity than Pt/Ru (1:1).[55] Ternary and quaternary promoters are briefly described in Table 2.

Typical supports used to deposit high surface electrocatalysts for methanol oxidation are carbons such as Vulcan, Ketzen and Black pearls. However, other supports, such as zeolites, carbon

Table 2
Ternary and Quaternary Promoters

Promotion by	Catalyst promoter	Comment
Colloidal deposition on C	W, Sn, Mo	Only W gave significant enhancement.
Adams method	W oxide, Mo oxide, V oxide	All superior. Mo and V oxides substantial effect (100 mV reduction in potential)
Borohydride reduction	Os and Os-Ir	Combinatorial optimisation identified Pt-Ru-Os-Ir as the most active catalyst.
Electrochemical deposition in polyaniline	Mo, W, Co, Sn, Fe, Ni	Mo most active. Sn and Au not active

nano-tubes, conducting polymers and metal mini-mesh, have shown superior activity to these materials.

The electrochemical behaviour of Pt-decorated unsupported Ru catalysts-based anodes has been investigated in direct methanol fuel cells. The preparation procedure enabled surface decoration of the Ru catalysts by Pt nanoparticles. Performance of DMFC anodes with ultra-low Pt loading (0.1 mg cm^{-2}) gave power densities of 150 mW cm^{-2} at 130 °C.[56]

Enhanced activity for Pt(HY) and Pt-Ru(HY) zeolite catalysts has been demonstrated for electrooxidation of methanol in fuel cells The enhanced electrocatalytic activity is explained on the basis of the formation of specific CO clusters in zeolite cages.[57]

A novel system of electro-catalysts based on platinum and organic metal complexes using mixture of platinum tetraammine complex with cobalt quinolyldiamine complex supported on graphite powder have been developed. Although the cobalt complex itself showed little catalytic ability, it enhanced the activity of Pt, several 10-fold, and gave performance comparable to Pt-Ru alloy catalysts.[58]

Preparation of Pt-Ru/graphitic carbon nanofiber nanocomposites as DMFC anode catalysts using microwave processing permits rapid preparation of these nanocomposites and affords metal nanoclusters of nearly uniform size. The Pt-Ru/narrow tubular herringbone GCNF nanocomposite gave DMFC performance comparable to that recorded for an unsupported Pt-Ru colloid.[59]

Multi-walled carbon nanotube-supported Pt (Pt/MWNT) nanocomposites were prepared by reduction of a Pt salt (HCHO reduction) in aqueous solution and in ethylene glycol solution. The Pt/MWNT catalyst prepared by the EG method has a high and homogeneous dispersion of spherical Pt metal particles (size range of 2–5 nm, with a peak at 2.6 nm). The catalysts display significantly higher performance than the Pt/XC-72 catalyst for methanol oxidation; which is attributed to a greater dispersion of the supported Pt particles.[60, 61]

Pt-WO$_3$ supported on carbon nanotubes has also been shown to have promise as possible anodes for direct methanol fuel cell. The carbon nanotube (CNT) are synthesised by template carbonisation of polypyrrole on alumina membrane.[62]

Pt-Ru electrocatalysts have been directly bonded onto a polymer electrolyte membrane by chemical reduction of a mixture

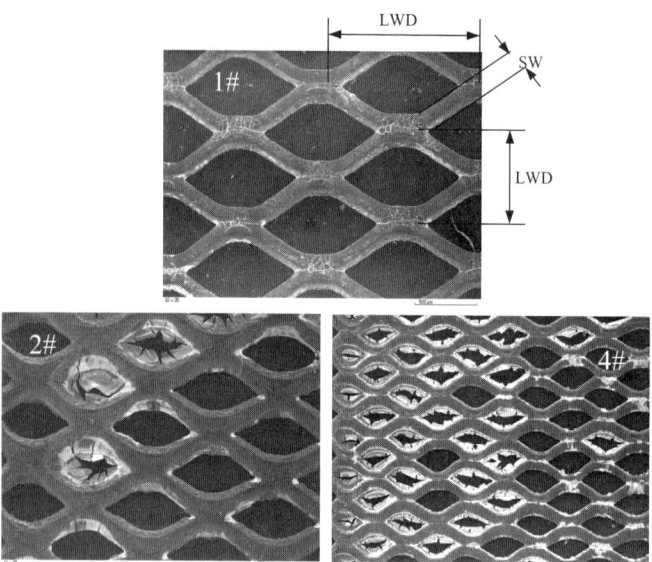

Figure 6. Structure of the minimesh based Pt-Ru anodes. SEM image of PtRu catalysts thermally deposited on Ti mesh, showing the morphologies of different types of meshes

of Pt and Ru complexes impregnated in the membrane. The deposited PtRu particles were embedded in the 3–4 mm region of the membrane surface to form a porous, hydrophilic layer.[63]

The use of metal mini-mesh electrodes for the support of methanol oxidation catalysts has recently been introduced.[64] Figure 6 shows typical structures of the electrode. Typically the mesh is made of titanium and catalysts are deposited onto the mesh by thermal decomposition or electrodeposition. The electrodes have been evaluated for both acid and alkaline PEM cells for methanol oxidation and give performance comparable to or greater than that of state-of-the-art Pt-Ru carbon supported catalysts (Figure 7).

The effective dispersion of the Pt-based catalysts is still a challenge to the production of the most effective methanol oxidation materials. Attempts have been made to produce polymer-based oxidation catalysts. Indeed, catalyst particles dispersed in conducting polymer matrices of poly-aniline,[65,66] poly-pyrrole[66-68] and poly-(3-methyl) thiophene[69] have been examined for methanol oxidation. This procedure, using polymers, which possess high electronic conductivity, ionic conductivity and a

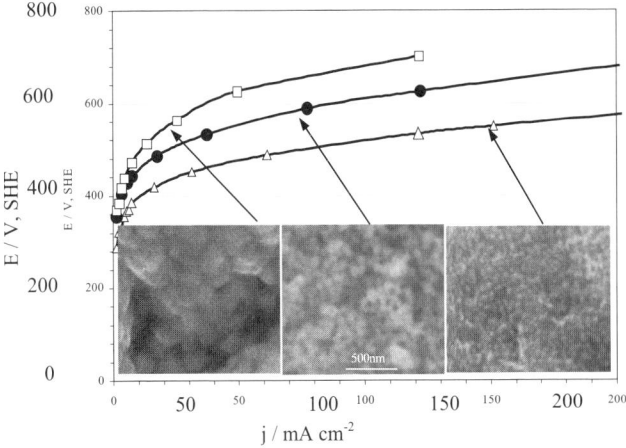

Figure 7. Performance of mini-mesh electrodes in acidic-methanol solution. Galvanostatic polarisation plots in 2 M MeOH + 0.5 M H_2SO_4 (60 °C) on PtRu/Ti and PtRuSn/Ti catalysts thermally formed in argon, showing the correlation of the particle size (from SEM images, insertion) and catalytic activity. Δ: PtRuSn/Ti, 400 °C; ●: PtRuSn/Ti, 500 °C; and ❏ PtRu/Ti, 500 °C.

porous structure, offers a means to support highly dispersed Pt catalysts and maximise catalyst utilisation. Although high-initial methanol oxidation currents were achieved, performance in many cases rapidly deteriorates due either to poisoning of the electrode surface or destruction of the polymer at the high temperature used.

A typical procedure used electropolymerisation to form the conducting polymer films (polypyrrole) onto, for example, a carbon substrate and to electrodeposit Pt into the conducting polymer. Increased effective catalyst electrode area was identified as a contribution to the improved oxidation currents.[68] Recently, Bouzek et al.[70] examined methods for the production of platinum modified polypyrrole films and the influence on methanol oxidation. No catalytic activity was observed with films modified with colloidal platinum particles incorporated during the synthesis or for films synthesised with tetrachloroplatinate complex as a nucleophilic counter ion, which was subsequently reduced. Good activity was observed for catalysts deposited onto films (0.5 μm thick) pre-deposited onto the support and this was better than that achieved with Pt deposited alone on the substrate. Catalyst activity increased with loading in the range of 0.2 to 2.0 mg Pt cm^{-2} for methanol oxidation in sulphuric acid.

Rajeshh and Viswananathan[71] have prepared methanol oxidation catalysts using polymer films of 1,5-dihydroxynaphthalene (1,5-DHN). The polymer films were electropolymerised onto carbon and Pt was electrodeposited from chloroplatinic acid. This structure gave better catalytic methanol oxidation characteristics than Pt alone or Pt co-deposited with the 1, 5-DHN conducting polymer.

Electrocatalytic polyaniline (PANi) electrode has been used for methanol oxidation.[72, 73] The platinum was dispersed onto the PANi-Au coated surface using constant potential and square wave potential cycling from chloroplatinic acid. The polymer electrodes, typically 1.5 μm thick, produced by square wave electrodeposition gave superior performance for methanol and CO oxidation in sulphuric acid.[72-74]

Ternary catalyst of Pt-Ru-X (X = Au, Co, Cu, Fe, Mo, Ni, Sn, W) prepared by electrochemical deposition and dispersion in an electronic conducting polymer, polyaniline (PANI) have been evaluated for methanol oxidation.[75] A particular good catalyst, in comparison to the binary alloy of Pt-Ru, was Pt-Ru-Mo (see Table 3). At potentials up to 500 mV (vs. RHE) an enhancement in current density, by a factor of 10, was achieved by incorporation of approximately 4% Mo in the Pt-Ru (70:26 atomic ratio) catalyst. Fuel cell tests demonstrated that these electrodes outperformed the Pt-Ru anodes, but results were far from optimal and much research is required to produce optimised electrodes that perform better than classic carbon supported materials. Although the catalyst is reported as stable up to potentials of 550 mV (vs. RHE), thermodynamic predictions, from Pourbaix diagrams, suggest that Mo could be oxidised to Mo(III) or to Mo(VI) (MoO_3). Thus, although Pt-Ru may stabilise the Mo species, long-term evaluation of catalyst is required to assess this stability issue. The ternary catalyst with tungsten gave a performance second only to that with

Table 3
Comparison of Ternary PANI Catalysts for Methanol Oxidation (in Methanol in 0.1M $HClO_4$)[75]

Current at 450 mV/mA mg^{-1}	Ternary catalyst component -X
0.6	None
4	Mo
2.1	W
0.15	Sn
2.0	Co

Mo. Other ternary catalysts with good activity for methanol oxidation, i.e., with Co, Ni and Fe lacked stability. Osmium was not considered in this work which would be expected to be stable under potential applied.

Conducting polymers, including chemically prepared poly (3, 4 - ethyldioxythiophene / poly(styrene - 4 - sulphonate) (PEDOT/PSS) and PEDOT/ polyvinylsulphate have been used as high surface area supports for electrocatalysts in oxygen reduction and methanol oxidation. In comparison to carbon supported catalysts, the conducting polymer supported materials showed comparable performance for oxygen reduction but inferior performance for methanol oxidation.[76] Polypyrrole has been used as a support for oxygen reduction with reported loss of stability on use and with potential cycling.[77]

More recently Nafion membranes, modified by in-situ polymerisation of poly(1-methylpyrrole), have been used to reduce methanol crossover in fuel cells.[78] In a related work, highly dispersed platinum and perchlorate doped polypyrrole electrodes have been examined for the electro-oxidation of formic acid.[79] Films were produced by electropolymerisation in presence of $LiClO_4$. Performance of the electrodes was assessed using films containing 0.1 mg cm^{-2} Pt. Oxidation currents increased by a factor of 3 with the perchlorate doped films and by a factor of two with the un-doped films, in comparison to the non-modified catalyst surface.

The catalytic activity of sputter deposited anodes for methanol oxidation has also been reported.[80, 81] Sputter deposited catalysts differ from other catalysts in terms of morphology, surface chemical composition, phase composition, and interfaces with reactant and electrolyte. In fuel cell tests, a sputtered Pt-Ru catalyst with a loading of only 0.03 mg cm^{-2} achieved a power density of 75 mW cm^{-2} at 90 °C, giving power densities per mg of catalyst for methanol oxidation as low as 2.3 W mg^{-1}.

(ii) Oxygen Reduction Catalysts

There are generally four classes of oxygen electrocatalysts:

1. The noble metals, particularly platinum and gold, have been extensively investigated as pure metals,[82] nano-particles[83] and alloys,[82,84-87] and as polycrystalline and single-crystal surfaces.[88] Platinum, in particulate form dispersed on carbon, shows the greatest activity for oxygen reduction in

acid solutions. Among various binary-alloys of transitional-metals with platinised carbon, Pt-Fe alloy catalyst has been reported as potential methanol-resistant oxygen-reduction catalysts.[82]

2. The macrocyclic derivatives of transition-metal compounds[90] e.g., Co and Fe and the ligands, porphyrins, phthalocyanines, tetra-azaannulenes and dimethylglyoxime derivatives.[91-94] However, current densities are generally low, and the dispersion of the catalysts on high-surface-area substrates needs to be ameliorated. This has raised problems of stability and de-metallation with the metallo-complexes.

3. Metallic oxides, particularly of the second and third row transition elements.[95, 96] In alkaline solution, a number of such oxides, including spinel,[97] perovskite,[98] and pyrochlore structures,[99] are particularly active. For oxygen reduction, in acid solutions, the stability of the oxide phase is an issue and the activity declines substantially. There have been reports of oxide catalysis in acid solution, though it has been suggested that, in a number of such cases, *homo*geneous rather than *hetero*geneous mechanisms may play a significant role, particularly when the oxide is acting as a promoter.[100-102]

4. Transition-metal compounds with other non-metallic counter-ions derived from the chalcogenides[103-115] and indeed from other non-metal species. The chalcogenides are frequently highly stable, especially in combination with later transition metals. These new catalysts are particularly noteworthy in that they are active for oxygen reduction but not for methanol oxidation, allowing them to be used in DMFCs even when methanol permeation takes place from anode to cathode.

At present, platinum in particulate form exhibits the highest activity towards oxygen reduction reaction in SPE-DMFCs. But platinum is not tolerant to methanol. In recent years, alternative cathode catalysts to platinum have been researched for SPE-DMFCs, some of which have exhibited tolerance to methanol. Pyrolised transition metal macrocycles have shown such characteristics and particularly Fe and Co tetramethoxyphenyl porphyrin (Fe-TMPP, Co-TMPP) and their derivatives.[116] Binary Fe and Co porphyrin catalysts have been claimed to give better oxygen reduction performance than single metal porphyrin catalyst.[117] Cathode catalysts based on Ru-Mo-Se oxides and other ruthenium-based cluster catalysts containing sulfur or selenium have shown substantial methanol tolerance.[112] These un-optimised catalysts have exhibited good electrochemical response for oxygen reduction, comparable to Pt, which is unaffected by the presence of methanol.

Quite recently, a range of carbon-supported Pt-alloys with Co, Cr and Ni have been evaluated with the aim of improving the tolerance of the catalyst to methanol. Among these, the Pt-Co/C binary catalyst has been identified to be more suitable than Pt/C for SPE-DMFCs.[118]

If either the particle size of platinum electrocatalyst for oxygen reduction is very small or the platinum electrocatalyst is amorphous, the methanol chemisorption energy is lower and the cathode is less susceptible to poisoning.[4] Accordingly, various preparatory routes have been proposed to synthesize platinum and platinum-alloy electrocatalysts with finer particle-sizes.[119] Among these, the most attractive procedure is due to Bonnemann et al.;[120] here platinum dichloride ($PtCl_2$) is suspended in tetrahydrofuran (THF) and treated with tetra-alkyl ammonium hydro-tri-organoborate, which results in a platinum metal colloid solution with a minimal evolution of hydrogen. This colloidal solution is evaporated to dryness under high vacuum, and the resultant waxy residue is mixed with ether. The colloid is then precipitated by addition of ethanol. The grey-black metal colloid powder thus obtained has a particle size between 1–5 nm. Maillard et al.[121] reported that the mass activity of platinum towards oxygen reduction increases continuously with a decrease in particle size from 4.6 to 2.3 nm, whereas mass activity is roughly independent of size in methanol-free electrolyte when the platinum particle size is less than 3.5 nm. The effects of adding a second metal to platinum have also been investigated. Although both Pt-Co/C and Pt-Fe/C have been reported to be methanol-resistant oxygen-reduction catalysts, Pt-Fe/C has been found to exhibit higher activity than Pt-Co/C as a methanol-resistant oxygen-reduction catalyst.

To prepare Pt-Fe/C and Pt-Co/C catalysts, a suitable route starts with platinized carbon prepared by a sulfito-complex route.[122-125] Appropriate amount of Pt/C and ferric nitrate [$Fe(NO_3)_3.9H_2O$] were dispersed in a 1:1 mixture of isopropyl alcohol and Millipore water followed by its ultrasonication for about half-an-hour.[126] The pH of the medium was adjusted to 7

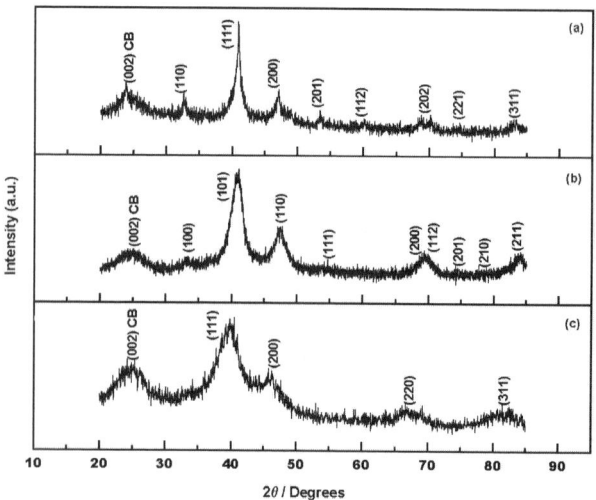

Figure 8. X-ray powder diffraction patterns for (a) Pt-Fe/C, (b) Pt-Co/C and (c) Pt/C catalysts.

with 0.1 M solution of hydrazine and the slurry thus obtained was dried with constant stirring. The resultant mass was alloyed by heating at 750 °C for 1h under flowing hydrogen followed by annealing for 12–15 h in the same atmosphere. Pt-Co/C catalyst was prepared in a similar manner by using $Co(NO_3)_2 \cdot 6H_2O$. The X-ray diffraction pattern (XRD) for Pt-Fe/C, Pt-Co/C, and Pt/C catalysts are shown in Figures 8(a)-(c), respectively. The XRD patterns for Pt-Fe/C and Pt-Co/C exhibit tetragonal structures,[84,127-130] while the XRD pattern of Pt/C shown in Figure 8(c) could be fitted to a face-centered cubic phase.[128]

Iron tetramethoxyphenylporphyrin (FeTMPP) has been reported to be the most active catalyst among transition-metal porphyrins.[93,131] It is prepared by iron insertion into meso-tetramethoxyphenylphorphyrin ($TMPPH_2$). In brief, $TMPPH_2$ is synthesized by reaction of pyrrole with anisaldehyde in propionic acid.[132] The characteristic absorption spectrum of FeTMPP in benzene shows absorption maximum (λ_{max}) in the visible region between 419 and 575 nm. FeTMPP is supported on a high-surface area carbon, such as Vulcan XC-72R and is pyrolyzed at ~ 700 °C in flowing argon.[91,92] In pyrolysed metal porphyrins, the molecular structure of the catalyst is destroyed during the heat treatment, and therefore the metal complex by default is the precursor of the actual active material. It has been proposed that the catalytic site in FeTMPP/C is N_4-Fe, bound to the carbon support.[93,94] This site has been labelled as a low-temperature catalytic site.[133] The other catalytic sites formed at elevated temperatures are not yet fully characterized. However, it is argued that the organic linkages around the iron atom help prevent its oxidation, which is seminal to its catalytic activity.[133] Metal oxides are usually prepared by a solid-state reaction of the component oxides.[134]

The last category of catalyst consists mainly of Ru-Mo-S, RuS, and RuSe.[106,112-114] Among these, RuSe exhibits maximum activity as a selective oxygen-reduction catalyst. RuSe is obtained by reacting a mixture of ruthenium dodecacarbonyl $[Ru_3(CO)_{12}]$ selenium at 140 °C, with xylene in nitrogen under refluxing condition, followed by washing the resultant mass with triethyl ether.[112] These catalysts are deposited onto a Vulcan XC-72R carbon for utilization as oxygen-reduction electrodes. The powder XRD pattern for carbon-supported RuSe (RuSe/C) catalyst is given in Figure 9. This XRD pattern shows all the characteristic peaks due to ruthenium metal and a broad feature at the diffraction angle of $2\theta \approx 25°$ can be attributed to (002) plane of the hexagonal structure of Vulcan XC-72R. The diffraction peak observed at $2\theta \approx 32°$ is due to the (100)-oriented silicon wafer, which was used as the substrate for the catalyst powder. Born et al.[112] reported that

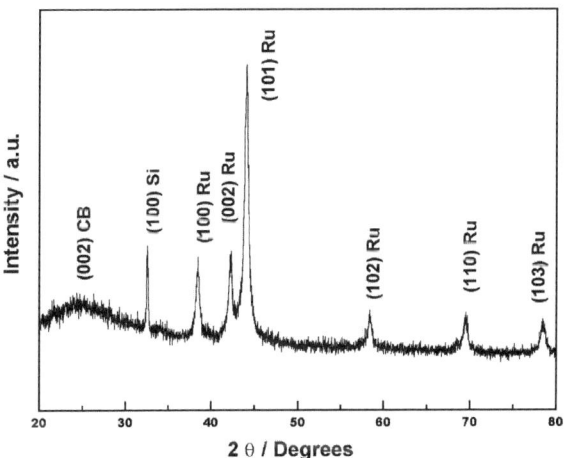

Figure 9. Powder XRD pattern for carbon-supported RuSe (RuSe/C) catalyst.

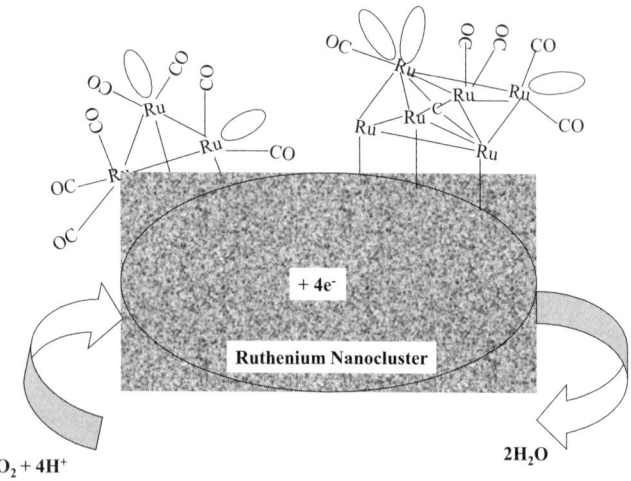

Figure 10. Schematic description of oxygen reduction on a RuSe/C bistructural catalyst.

ruthenium metal, even after refluxing in selenium-containing solution, shows little change in its XRD pattern. However, RuSe exhibits higher catalytic activity in relation to ruthenium metal. According to Born et al., the activity enhancement by selenium is related to an interfacial effect due to the binary structure of the catalyst. Energy dispersive analysis by X-rays (EDAX) of RuSe/C catalyst suggests that the optimum quantity of selenium is ~ 15 at. %.[112] It is noteworthy that the RuSe/C catalyst is different from a metallic-ruthenium surface, and its activity towards oxygen reduction is substantially higher. Because the RuSe/C catalyst is loaded with organic matter made up of carbonyl or carboxylic groups; its high catalytic activity is probably due to an interaction of nanocrystalline ruthenium and carbon ligands. The effect can be twofold:

1. carbon species may stabilize surface ruthenium metal, thus suppressing its oxidation, which otherwise would transform ruthenium particles rapidly into RuO_2, and
2. carbon species may alter the distribution of interfacial electronic states by forming ruthenium complexes.

A schematic description of oxygen reduction on a RuSe/C bistructural catalyst is shown in Figure 10, which depicts catalyst centers comprising of ruthenium clusters with attached carbonyl

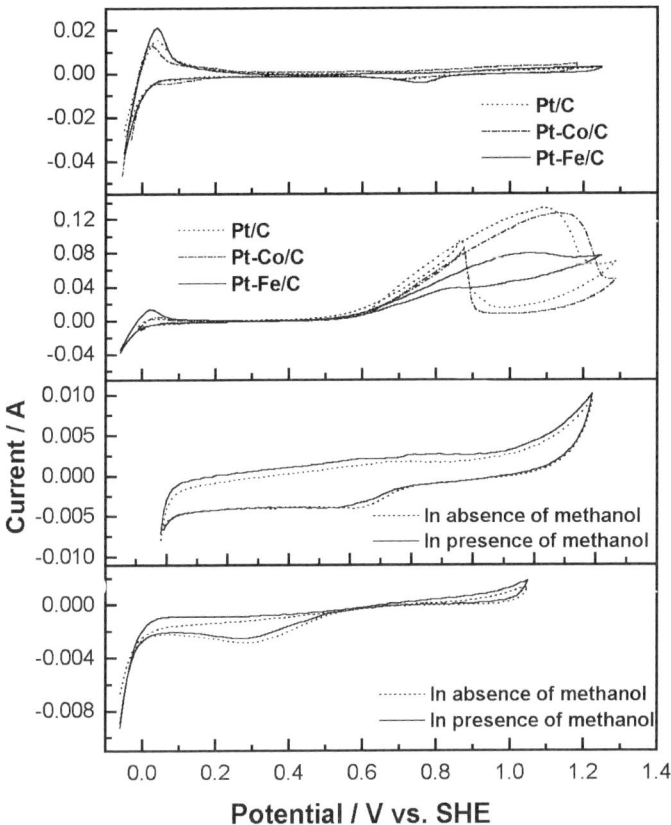

Figure 11. Cyclic voltammogrammes for Pt/C, Pt-Co/C, and Pt-Fe/C catalysts in aqueous sulphuric acid, both with (b) and without (a) methanol.

ligands.[135] Because some of the bonds are dangling (unsaturated), an interaction with oxygen can take place depending on the number of ruthenium sites available in the cluster, which act as electron transfer mediators.

Cyclic voltammogrammes for Pt/C, Pt-Co/C, and Pt-Fe/C catalysts in aqueous sulphuric acid, both with (Figure 11b) and without (Figure 11a) methanol, are shown in Figure 11. The data indicate that of the three catalysts, the methanol oxidation reaction is least favoured on the Pt-Fe/C. Therefore, Pt-Fe/C appears to be a potential selective oxygen-reduction catalyst. Pt-Fe/C has also been reported to be a potential CO-oxidation catalyst.[136] From the XPS,

XAS, and cyclic voltammetric data on Pt/C and Pt-Fe/C catalysts,[137] the higher oxygen-reduction activity of the Pt-Fe/C catalysts in the presence of methanol appears to be primarily due to:
- the higher proportion of active platinum sites in relation to Pt/C, and
- a completely different nearest neighbour environment in the Pt-Fe/C catalyst where, unlike the Pt/C catalyst, the nearest neighbour sites are occupied by Fe, which helps scavenge impurities from the neighbouring active platinum sites.

The cyclic voltammetry data for FeTMPP/C in aqueous sulphuric acid, with and without methanol are shown in Fig. 11(c).[10] The data demonstrate the methanol tolerance of the catalyst towards oxygen reduction reaction. The cyclic voltammetry data for RuSe/C in aqueous sulphuric acid, with and without methanol are given in Fig. 11(d). Here, the data depict the complete absence of methanol oxidation on RuSe/C surface. Accordingly, RuSe/C would be an effective selective catalyst for oxygen-reduction reaction.

(iii) Membrane Materials

One of the key problems impeding the development of SPE-DMFCs, is the properties of the available proton conducting membrane. Owing to the low reactivity of methanol as well as the low conductivity of commercially available proton exchange membranes, namely Nafion^R, at ambient temperatures, it is necessary to have an operating temperature near 100 °C for SPE-DMFCs. Nafion membranes need water inside their skeleton in order to exhibit good proton conductivity at such high temperatures. Although it is quite a difficult requirement for polymer electrolyte fuel cells, it is of little concern in SPE-DMFCs since water along with methanol is constantly circulated during operation. But the problem of methanol crossover associated with the Nafion membranes is detrimental to the performance of an SPE-DMFC since it reduces both the coulombic efficiency of the fuel cell and the cell voltage. Studies by Ren et al.[138] showed a crossover of methanol equivalent to 80 mA/cm^2 at a load current density of 150 mA/cm^2 using a Nafion membrane electrolyte in a liquid-feed SPE-DMFC at 80 °C. They also showed a water transport of 355 mg/cm^2 to the cathode under similar conditions, which causes severe flooding of the cathode. The effects of methanol crossover can be controlled to a certain extent by strictly correlating methanol-feed concentration with the actual demand of the cell. It is therefore mandatory to find electrolyte membranes that would reduce the methanol and water crossover sufficiently. Efforts are therefore been expended to develop methanol impermeable proton exchange membranes either by modifying the available Nafion membranes or by developing altogether new proton exchange membranes.

One early possibility explored was to add different layers of Teflon to the Nafion membrane so as to restrict its permeability to methanol. These membranes showed encouraging results in terms of methanol permeability but the electrical resistance of the membranes increased enormously and they were of no practical relevance in fuel cells. Forthwith membranes comprising metallic blocking layers were proposed. These membranes were composite electrolytes where a film of a methanol impermeable proton conductor, such as metal hybrid, was sandwiched between the Nafion sheets. Such composite electrolyte membranes were found to exhibit lower methanol crossover than Nafion, with the best performance obtained for a Nafion-115/Pt/Pd/Pt/Nafion-115 system. A reduction in methanol crossover was also observed by doping Nafion with Cs^+ ions.

Plasma etching of Nafion and palladium sputtering were also investigated as means to reduce methanol crossover in SPE-DMFCs. Plasma etching increases the roughness of the membrane and decreases the pore size, thereby decreasing its methanol permeability. But plasma etching can potentially remove sulfonic acid groups from the membrane surface and hence reduce its performance. By contrast, palladium sputtering, whilst not affecting proton-transfer rate, has been reported to improve performance in comparison to that with an unmodified membrane. Ion-beam radiation has also been successfully deployed as a technique to increase the roughness of Nafion membranes and enhance cell performance.

Sulphonation of a number of polymers to form ion-conducting membranes has been an active area of research. Sulfonated poly(ether ether ketone) (SPEEK) with different degrees of sulfonation have been prepared and evaluated as proton exchange membrane electrolytes in direct methanol fuel cells. Within a narrow range of sulfonation (ca. 50%), the SPEEK membranes exhibit electrochemical performances comparable to or exceeding that of Nafion at 65 °C, making it an attractive low-cost alternative to Nafion.[139]

Sulfonated phenolphthalein poly(ether sulfone) (SPES-C) membranes with varying sulfonation degrees were synthesized by reaction of PES-C with sulfonic acid as sulfonating agent and solvent. With a 70% degree of sulfonation, the proton conductivity was 3.45×10^{-2} S cm^{-1} at 100 °C, which was close to or superior to that of Nafion®-115 membrane at the same conditions. Methanol permeability of SPES-C was considerable smaller than that of Nafion®-115 membrane. SPES-C membranes appear to be an excellent candidate for use in DMFC applications.[140] Sulfonated naphthalene di-anhydride based polyimide co-polymers for proton-exchange-membrane fuel cells have been reported by Einsla et al.[141]

Among organic-inorganic composite membranes containing zirconium-phosphonates, tin-doped morderites, zeolites or silica, investigated for their methanol permeability, the results on an 80 μm membrane cast from a composite of Nafion ionomer and silica are particularly encouraging. SPE-DMFCs tests conducted with this membrane, and commercial a Pt/Ru anode and a Pt cathode, both with 2 mg/cm^2 Pt loadings, demonstrated peak power densities of 240 and 150 mW/cm^2 with oxygen and air cathodes, respectively. The open-circuit voltage at 145 °C was as high as 0.95 V and methanol crossover as low as 4×10^{-6} mol/cm^2 min, which is equivalent to a methanol crossover current density of 40 mA/cm^2. Recently, a sol-gel route to prepare Nafion/silica composite membranes has been reported by Miyake et al.[142] Nafion-layered silicate nanocomposite membranes for fuel cell applications have also been tested.[143]

The surface properties of inorganic fillers for application in composite membranes-direct methanol fuel cells is discussed by Aricó.[144]

Other Nafion composite materials considered for the DMFC include a Nafion®/montmorillonite nanocomposite membrane formed by direct melt intercalation of perfluorosulfonylfluoride copolymer resin (Nafion® resin) into the montmorillonite and modified montmorillonite (m-MMT). The performance of the MEA using the nanocomposite membrane was higher than that of a commercial Nafion® membrane at high operating temperatures.[145]

Nafion membrane has been impregnated with polypyrrole by in situ polymerization to decrease the crossover of methanol in direct methanol fuel cells (DMFCs). Modified Performance gains result from substantial reductions of the cathode overpotential, while anode overpotentials increase due to the lower conductivities of the modified membranes. Part of the beneficial effect at the cathode appears to be due to lower water crossover from the anode to the cathode.[146]

High proton-conducting Nafion/calcium hydroxyphosphate (CHP) composite membranes have been reported by Park et al.[147] Nafion membranes modified with vinylphosphonic acid (VPA) and 2-acrylamido-2-methyl-1-propanesulfonic acid (AMPSA) polymers have been obtained and characterized. Insertion of VPA into a Nafion membrane increases the power density of DMFCs.[148]

Other composite membranes have also been evaluated for the DMFCs. A hybrid inorganic-organic copolymer was synthesized from 3-glycidoxypropyltrimethoxysilane (GPTS), sulfonated phenyltriethoxysilane (SPS), tetraethoxysilane (TEOS) and H_3PO_4. A proton conductivity of 1.6×10^{-3} S/cm was observed at $100°$ C in a dry atmosphere, which was the highest proton conductivity of a membrane in its anhydrous state ever reported. In an environment with 15% relative humidity (RH), the proton conductivity increased to 3.6×10^{-2} S/cm at 120 °C. The proton conductivity increases with H_3PO_4 contents and relative humidity.[149]

Membranes made from polyvinylalcohol loaded with mordenite, a proton conducting, methanol impermeable zeolite, showed up to twenty times higher selectivity than Nafion.[150] A composite membrane made from a microporous, ceramic and poly(vinylidene) fluoride (PVdF) exhibited a high and stable conductivity, a proton transport mechanism not critically related to the water content and low methanol crossover.[151]

Low cost grafted membranes have been prepared by processes based on electron beam or gamma irradiation, subsequent grafting, cross-linking and sulfonation of a range of polymer films. The polymer films include polyethylene, PVDF and ethylene-tetrafluoroethylene (ETFE). DMFC assemblies based on these membranes show cell resistance and performance values comparable to Nafion®-117. Stable electrochemical performance has been demonstrated during 1 month of cycled operation. Tailoring of grafting and cross-linking properties allows a significant reduction of methanol cross-over while maintaining suitable conductivity and performance levels.[152, 153]

Other radiation grafted poly(ethylene-tetrafluoroethylene), or ETFE, film, which showed higher electric conductivity and lower methanol permeability than Nafion, gave cell performance that was inferior. The analysis of electrode potentials showed larger activation overpotential for both anode and cathodes on the grafted membranes, due to poor bonding of the catalyst layers to the grafted membranes.[154] Polypropylene (PP)-g-sulfonated polystyrene (SPS) composite membranes have been prepared by grafting polystyrene (PS) on microporous polypropylene membranes via plasma-induced polymerization.[155]

Blended polyphosphazene/polyacrylonitrile membranes for direct-methanol fuel cells were prepared from sulfonated poly[bis(3-methylphenoxy)phosphazene] with polyacrylonitrile and then UV cross-linked.[156] The methanol crossover was three times lower than that of Nafion. With a three-membrane composite MEA (a methanol-blocking film sandwiched between two high conductivity membranes), there was a significant decrease in crossover (ten times lower than that of Nafion®-117) with a modest decrease in current-voltage behaviour.[157]

Oxidation of methanol can be promoted substantially by operating the SPE-DMFC at temperatures above 100 °C. The kinetics of oxidation will be accelerated and the influence of poisoning species, such as adsorbed CO species, will be reduced. However, the ionic conductivity of Nafion falls considerably above 100 °C due to loss of water, which is necessary for its conductivity, by evaporation. Composite membranes that exhibit fast-ion proton transport at elevated temperatures are needed for proton-exchange membrane fuel cells operating between 100–120 °C. Several approaches have been pursued to resolve this issue, such as utilizing different proton-conducting ionomer polymers including polyphenylene sulphide sulfonic acid, sulfonated polymides, the perfluorosulfonyl imide form of Nafion, sulfonated polyether ketones, and other novel sulfonated polymeric membranes. Although the influence of temperature and humidity on the proton conductivity of these systems was frequently similar to that for Nafion membrane, they often exhibited inferior performance.

The proton conductivity and methanol permeability of several polymer electrolyte membranes including sulfonated and phosphonated poly[(aryloxy)phosphazenes] was determined at temperatures up to 120 °C. Although the conductivities of the polyphosphazene membranes were either similar to or lower than that of the Nafion®-117membranes, methanol permeability of a sulfonated membrane was about 8 times lower than that of the

Nafion®-117 at room temperature, although the values were comparable at 120 °C. The permeability of a phosphonated phosphazene derivative was about 40 times lower than that of the Nafion®-117 membrane at room temperature and about 9 times lower at 120 °C.[158] Because of its high conductivity and low methanol permeability, the PWA/SPEEK membrane also seems to be an excellent candidate for DMFC applications.[159]

A series of organic/inorganic composite materials based on polyethylene glycol (PEG)/SiO2 have been synthesized through sol-gel processes. Acidic moieties of 4-dodecylbenzene sulfonic acid (DBSA) were doped into the network structure at different levels to provide the hybrid membrane with proton conducting behavior. Some of the hybrid membranes exhibited low methanol permeability without sacrificing their conductivities significantly and were thus felt to be potentially useful in the DMFC.[160]

Another approach is the use of proton-conducting membranes based on acid-impregnated ionomer polymers, such as polystyrene sulfonic acid membranes imbibed with sulphuric acid, Nafion impregnated with 85% phosphoric acid, or non-volatile hetropolyacid impregnated Nafion membranes. A related strategy is to utilize acid-doped non-isomeric polymers, such as phosphoric acid-doped poly-benzimidazole (PBI) or tri-fluoromethane sulfonic-acid doped polyvinylidine-fluoride-hexa-fluoropropylene (PVDF-HFP). Materials, such as polyvinyl Alcohol (PVA), poly-aniline (PANI) and PVDF, showed very low methanol permeability but, however, even when acid impregnated, showed much lower proton conductivity. Of all the materials, only acid doped PBI showed characteristics suitable for the SPE-DMFC.

The use of acid-doped PBI as a polymer electrolyte is attractive in terms of thermal and oxidative stability but in order to attain the necessary mechanical flexibility and proton conductivity, the SPE-DMFC has to be operated at temperatures in excess of 120 °C. In the temperature range of 60–80 °C, despite negligible methanol permeability, the proton conductivity of PBI is too low for it to be a viable option as the main component of the membrane. Film conductivity is explainable mechanistically. Proton conductivity in a Nafion membrane is assigned to the vehicle mechanism by which protons are transported through membrane accompanied by another molecular species such as H_3O^+, $H_5O_2^+$, or $CH_3OH_2^+$. Materials assigned to this mechanism generally show a high degree of methanol crossover and high proton conductivity, which accurately describes the properties of a Nafion membrane. Conversely, PBI exhibits both low proton conductivity and low methanol permeability (2.4% to that of Nafion®-117 membrane), a behaviour described by the Grottus-'jump' mechanism.[161] Theoretically, this mechanism can facilitate high proton conductivity but in practice it is an anomaly. There are potential problems of acid migration, corrosion of cell components, adsorption of anions, and acid volatility with the acid doped or acid impregnated membranes. Other promising alternatives are composite membranes made from blends of acid and base polymers. These membranes are produced by blending acid polymers, such as sulfonated polysulfones (sPSU), sulfonated polyetherketones (sPPEK), sulfonated polyether etherketones (SPEEK), with basic polymers, such as poly-4-vinylpyridine (P4VP), polybenzimidazole (PBI) or a basically-substituted polysulfone (bPSU). In these materials, electrostatic forces from salt formation between acidic and basic groups achieve reversible cross linking of the polymer. Hobson et al. (2002) showed that combination of a thin layer of PBI (~ 20 µm) with Nafion®-117 (~ 50 µm) was more effective than PBI alone.[162] However, despite reducing methanol crossover to the cathode, DMFC performance of the composite membrane is still below that of Nafion®-117, the accepted benchmark. Recently, it is reported that the treatment of Nafion®-117 with dilute PBI/dimethyl acetamide (DMAC) solutions can produce films that exhibit a significantly higher DMFC performance than the parent materials. But, even to-date, the challenge of combining inherent conductivity with a low permeability to methanol remains. The details of the Nafion membrane and its modifications, the arylene main chain polymers such as polybenzimidazole (PBI), polyetherketone (PEEK) and polyethersulfone (PSU) membranes and their modifications, based on the structures of the membrane materials, have been described by Fu et al.[163]

A solution to the problems of methanol crossover lies in the use of mixed-reactant SPE-DMFCs, where unlike the conventional SPE-DMFCs, the fuel and the oxidant are allowed to mix (see Section VI). Such a fuel cell relies on the selectivity of anode and cathode electrocatalysts to separate the electrochemical oxidation of fuel and electrochemical reduction of the oxidant. In such a fuel cell system with no physical separation of fuel and oxidant, there is no longer any need for a gas-tight structure with the stack and hence considerable relaxation of sealing, manifolding and reactant delivery structures is possible.

A relatively new area for direct methanol fuel cells is the use of alkali conducting polymer membranes. The use of such membranes has been demonstrated using, for example, a cross-linked fluorinated polymer containing a quaternary ammonium exchange group (MORGANE^R-ADP).[164]

Radiation-grafted alkaline anion-exchange membranes using polymer films of hexafluoropropylene-co-tetrafluoroethylene, PVDF and FEP for use in low temperature portable DMFCs have been prepared. Vinylbenzyl chloride is radiation grafted onto the polymer film and with subsequent amination and ion-exchange gives the hydroxide ion form membranes, which are reported to be stable at low temperatures.[165, 166]

V. DIRECT METHANOL FUEL CELL PERFORMANCE

Previous research on the DMFC has almost unanimously concluded that perfluorinated polymer electrolytes give significantly better cell performance that the use of other (solution phase) electrolytes. Electrolytes such as triflic acid, perfluoroethane sulphonic acid, perfluoro-octane sulphonic acid and sulphuric acid have are outperformed in terms of methanol oxidation by ionomer membranes such as Nafion.[167] An exception is the use of a basic electrolyte, such as KOH, which however is not a stable electrolyte in the presence of CO_2.

Probably without exception the research and development of the DMFC to reach high performance targets has proceeded with anodes of the binary alloy of Pt and Ru (or Pt-RuO$_x$). The typical ratio of Pt:Ru is 1:1 atomic ratio. The catalyst utilisation in the DMFC is an issue important in its development. It is felt that loadings of platinum of the order of 2 mg cm^{-2} are too high for large power applications, such as transportation, but that for small scale applications such loadings may be acceptable.

In general the performance of the DMFC depends on the MEA construction method, the materials and the operating conditions in the cell. It is well documented that increasing cell temperature above approximately 60 °C causes a significant increase in power performance and that at 90 °C and beyond high power densities above 200 mW cm^{-2}, are achievable. This power performance is typically achieved using pressurised oxygen or air (2–5 bar). The use of oxygen gives the best performance but most terrestrial applications require the use of air. In addition many potential applications of the DMFC will require air under almost ambient conditions of temperature and pressure.

The interest in the DMFCs has blossomed world wide since the early notable efforts at Los Alamos, Jet Propulsion Labs, Siemens, Giner Inc and a number of Universities. Results from these groups include performance of single cells and stacks for transportation applications and mobile equipment.

The feed concentration of methanol is an important consideration in the DMFC. Half-cell studies with high concentrations of methanol up to 8 M have been reported to give superior methanol oxidation performance, than achieved with, for example, 1 M methanol, e.g., 250 mV lower polarisation at 100 mA cm^{-2}.[167]

There is a number of reports on the effect of methanol concentration on the anode performance in an operating DMFC. Reductions in anode potential of up to 200 mV have been claimed at high current densities when a 4.0 M methanol solution is used instead of a 1 M solution.[168] Other research groups have reported that methanol oxidation kinetics are not improved at concentrations above 1 or 2 M (Figure 12). The difference in the observations are due to different methods of catalyst preparation used. However,

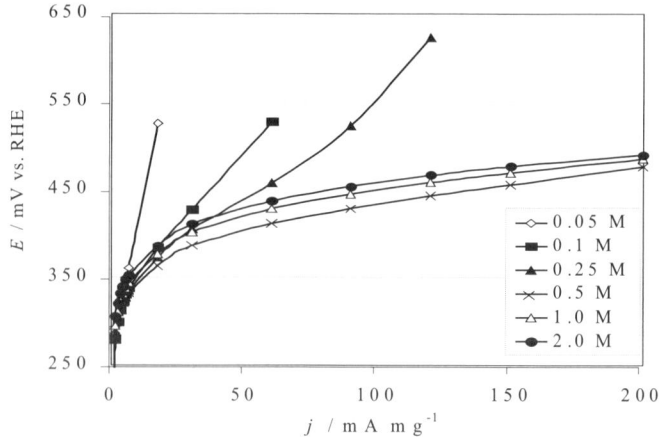

Figure 12. Typical influence of methanol concentration on DMFC anode performance. Galvanostatic plots in x M CH$_3$OH + 0.5 M H$_2$SO$_4$ (60 °C) of the PtRu catalysts formed by a thermal decomposition method on a double layer of titanium mesh. The thermal decomposition was conducted at 400 °C in air for 1 hour; the precursor used was a mixture of 0.2 M PtCl$_3$ and 0.2 M RuCl$_3$ in a 1:1 molar ratio.

regardless of these observations the benefit of reduced polarisation does not translate into higher cell performance due to the problem of methanol crossover. The use of a 4-M solution causes a significant reduction in cathode potential which is greater than the gain in anode potential. Thus typically the open circuit potential was approximately 50 mV lower with the 4-M concentration (c.f. 1-M solution) and there is a greater loss in cell voltage on load. This situation will remain with the DMFC until suitable selective (methanol impermeable) membranes or high performance methanol tolerant cathodes are implemented.

It is interesting to note that the stoichiometric ratio of methanol to water is at a high concentration of methanol (approximately 12 M) which have rarely been considered in experimental cells. At such high methanol concentrations good cell performance has not been achieved at present due to the problem of methanol crossover. In general, minimising the water present in an operating DMFC system is desirable as water is the major product from the cell which brings with it issues of system heat transfer and overall water balance management. However, in practice the DMFC can typically be operated with pumped supply of methanol solution in which the concentration of methanol in the cell is controlled by steady addition of methanol to the feed near inlet. Alternatively, in small-scale, portable devices methanol can be supplied by wicking and/or diffusion.

A factor of potential importance to the DMFC is operation with low concentrations of methanol as a result of high fuel utilisation. Current data suggests that a methanol concentration of 0.5 M was not too far from the optimum for certain applications. Methanol concentrations lower than this can result in significantly lower performance although significant performance has been reported with a methanol concentration of 0.22 M in a vapour-feed system.[170] Representative data for DMFC performance at low methanol concentrations is shown in Table 4.

Early work on the performance of the DMFC at Siemens[171] used relatively high loadings of Pt-Ru catalyst (4–8 mg cm^{-2}) to achieve high power densities (> 200 mW cm^{-2}) at temperatures of 130 °C with pressurised oxygen. More recently, an equivalent power performance has been reported using half the catalyst loading and a lower temperature of 110 °C. However, as air will be the preferred oxidant, at preferably lower pressure, performance evaluation focused on maximising power under those conditions. For example, a DMFC using 1 mg cm^{-2} Pt-Ru anode, with 4 mg cm^{-2} Pt black cathode, operating at 1.5 bar air pressure at 80 °C using 0.5 M methanol, gave power densities of around 90 mW cm^{-2}. Improvements in performance were produced by increasing air flow-rate and temperature.[172]

Table 5 summarises data obtained on single cell performance from a number of research groups over the last 10 years. Hikita et al.[173] and Ren et al.[174] reported high power density performance using MEAs based on Nafion®-112, 115 and 117 membranes. Unsupported Pt-RuO$_x$ catalystk—2.2 mg cm^{-2} (E-Tek)—and Pt black—Johnson Matthey—were used as anode and cathode catalyst respectively. The methanol solution concentration was 1 M and the anode side was pressurised to allow high temperature operation above the normal boiling point. MEA fabrication used a decal transfer of catalyst inks (in Nafion solution) onto the membrane followed by hot pressing. A key to the high power densities (up to 380 mW cm^{-2}) was the production of very thin dense catalysts structures using the unsupported catalyst. In this way, catalysts with relatively poor specific activity (i.e., activities per mg Pt) performed as well as, if not better than, more active supported catalysts due to the higher loadings of Pt in the thin unsupported catalyst layers. This work confirmed that Nafion®-112 gave the better performance at higher current densities (> 150 mA cm^{-2}) with Nafion®-117 superior at the lower current densities. At the lower current densities the thinner membrane suffered from a greater methanol crossover, which caused higher polarisation of the cathode.

Similar power densities and performance to that of Ren[174] have been achieved by Hogarth et al.[175] and Shukla et al.[176] and Aricò[177] using vapour-feed systems operating at high cathode pressures. Typical, high power performance data of the DMFC are shown in Figure 13. The performance of the DMFC generally fell significantly on reduction in air pressure, e.g., 0.15 W cm^{-2} with 2-bar air.

At JPL power densities greater than 200 mW cm^{-2} have been achieved using air at 2.5-bar pressure with catalysts of Pt-Ru (2.5 mg cm^{-2}). It was shown that higher catalyst loadings than 2.5 mg cm^{-2} do not produce a better performance with optimised MEA fabrication methods.[178] Similarly, DMFC development at Los Alamos has realised improved cell performance such that power densities greater than 200 mW cm^{-2} and 260 mW cm^{-2} at 80 °C and 100 °C, respectively, were achieved with 2 bar pressure

Table 4
DMFC Performance at Low Concentration

Reference	Electrolyte	Anode catalyst	Cathode catalyst	Loading mg cm^{-2} A[a]	Loading mg cm^{-2} C[a]	Temp. °C	Anode feed M	Cathode feed and pressure	Power density mW cm^{-2}	Performance at X mA cm^{-2}/V	Performance at 0.5 V mA cm^{-2}
Dohl[169]	N112	Pt-Ru/C	Pt/C	8	43		0.5	O_2	80	100/0.55	135
Scott[268]	N117	Pt-Ru/C	Pt/C	2	2	80	0.25	Air, 2bar	40	90/0.4	
Surampudi[167]	N117	Pt-Ru/C		0.5		60	0.5		70	100/0.5	
Jung[192]	N112,115	Pt-Ru/C	P/C	3	3	120	0.5	O_2,		100/0.6	135
							0.1			75	130
Cruickshank[170]	N117	Pt-Ru/C	Pt/C	2	2	100	0.22 vapour	O_2, 1bar	120	200/0.43	100

[a] A = anode; C = cathode

Table 5
Summary of Performance Attainment of the DMFC (PEM Nafion 117 Pt-Ru Anode, Pt Cathode)

Source and reference	Loading mg cm^{-2} A[a]	Loading mg cm^{-2} C[a]	Temperature °C	Anode feed and concentration M	Cathode feed and pressure bar	Power density mW cm^{-2}	Performance at 400 mA cm^{-2}/V	Performance at 0.5 V/mA cm^{-2}
Shukla[176]	2.5		98	Vapor/2	O$_2$, 5	350	0.5	
Shukla[176]	2.5		98	Vapor/2	Air, 5	220	0.4	
Ren[174]	2.2	2.3	130 / 110	Liquid/1	O$_2$, 3 / Air, 3	380 / 250	0.57 / 0.52	670
Ren[174]	2.2	2.3	130 / 110	Liquid/1	O$_2$, 3 / Air, 3		0.47 / 0.39	450
Surampudi[167]	4		95	Liquid/1	O$_2$, 2.4 / Air, 2.4		0.47 / 0.38	240
Hogarth[175]	2	0.5	97	Vapor/2	O$_2$, 5 / Air	350 / 220	0.5 / 0.4	400 / 120
Arico[177]	2		145	Vapor/1.0	Air, 5.5	240		350
Jung[192]	3		120	Liquid/2.5	O$_2$	260	0.5	
Arico[200]	1	1	95–100	2	O$_2$, 5 / Air, 5	160 / 110		
Hikita[173]	0.65	1.0	80	Liquid/9 vol %	Air	75	200: 0.3	
Arico[200]	1	1	95–100	Vapor/2	O$_2$, 5 / Air	160 / 110	0.4	

Table 5. Continuation

Source/Reference	Loading mg cm^{-2} A^a	Loading mg cm^{-2} C^a	Temperature °C	Anode feed M	Cathode feed and pressure bar	Power density mW cm^{-2}	Performance at 400 mA cm^{-2}/V	Performance at 0.5 V/mA cm^{-2}
Baldauf[195]	1	4	95 80	Liquid/0.5	Air, 1.5	95 80		100
Narayanan[178]	2.5		90	Liquid/1	Air, 2.5	215	0.46	300
Ren[179]	2.1		80	Liquid/1	Air, 2	>200	0.45	300
Ren[179]	1.2		100	Liquid/1	Air, 2	246	0.45	250
Dohle[169]	2.2	4	130	Liquid/2	O$_2$, 3	~170	0.4	230
Arico[183]	2	2	130	Liquid/2	O$_2$, 3 Air, 3	450 280	0.55	~600
Cruickshank[170]	2	2	100 80	Vapor/ 1.7 1.0M	O$_2$, 1 O$_2$, 2	190 180	0.42 0.47	280 350

^{a} A = anode; C = cathode

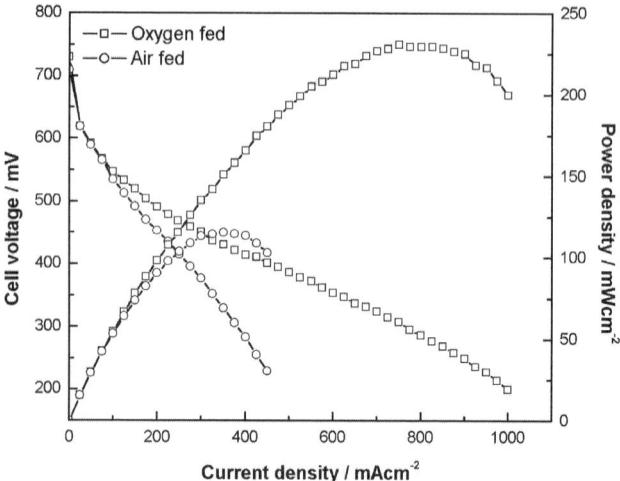

Figure 13. Typical high power performance of the DMFC.

air.[179] Furthermore, an equally good performance has been achieved using Pt-Ru carbon supported catalyst with loading of 1.2 mg cm^{-2}. Overall, the total platinum use in the cells was quoted as 5 g per kW of power. Ren et al.[180] reported that a 50 cm^2 area cell achieved a power density of greater than 80 mW cm^{-2} and typically a current density of 150 mA cm^{-2} at 0.4 V, operating at 60 °C and using air at 0.75 bar and at only 2.5 times the stoichiometric requirement.

The Forschungszentrum, Jülich (Germany), is one of the several institutions which have a DMFC stack development programme targeted towards a 0.5 kW system.[181,182] Reported power performance of cells at 130 °C was approximately 170 mW cm^{-2}, using 2.2 mg Pt-Ru cm^{-2} anode and 4 mg Pt cm^{-2} cathode with a 1-M methanol feed. This performance was typical of that achieved by many groups using standard Pt-Ru anodes. In general, with the new developments in electrode fabrication technology high currents are now achievable which start to approach those of PEM cells.[182] It has been reported[183] that the use of an inter-digitated flow field rather than a serpentine flow field can improve power performance through a claimed enhancement in mass transport. Although the inter-digitated flow field would increase methanol crossover, as indicated both by a lower open circuit voltage (0.71 vs. 0.8 V) and by poorer performance at low

Table 6
Effect of Temperature on DMFC Power Performance[183]

Temperature, °C	Peak Power/ mW cm^{-2}
130	450
120	375
110	300
100	240
90	170

current densities, compared to that with a serpentine flow field, the benefit was seen at the higher current densities and in the peak power. Power densities of 450 mW cm^{-2} and 290 mW cm^{-2} for cells operating on oxygen and air respectively at 130 °C were reported.[183]

The fact that such high power densities, discussed above, are produced now in the DMFC should not disguise the fact that relatively high temperatures are used with relatively high catalyst loadings. Operating at lower temperatures can cause a significant reduction in power performance as illustrated in Table 6.

The majority of DMFC use air feed cathodes, which is the only real practical choice for the majority of systems. Power densities above 200 mW cm^{-2} have been achieved with operating temperatures below 100 °C and with modest air pressures. However, the issue of air stoichiometry has still to be realistically addressed as most systems report data at high excess air.

Wilkinson and Steck[184] reported some data on the influence of air stoichiometry relevant to the operation of the DMFC, for a Pt-Ru/C anode and Pt cathode Nafion®-117 based MEA operating at 115 °C (Table 7). With a 1-M methanol solution, the performance of the cell at 200 mA cm^{-2} remained constant at 0.5 V at air stoichiometries of 2 and above. At a stoichiometry of 1.4 the voltage fell to 0.35 V and rapidly approached zero at lower values. The effect of air stoichiometry using higher methanol concentrations was more severe (Table 7) due to the influence of

Table 7
Effect of Air Stoichiometry on DMFC Performance[184]

Air stoichiometry	Cell voltage at 200 mA cm^{-2}, V		
	1-M methanol	2-M methanol	4-M methanol
6	0.5	0.49	0.4
3	0.5	0.48	0.3
2	0.49	0.42	0.1
1.5	0.46	0.19	
1.3	0.28		

methanol crossover; for example with 4-M methanol a cell voltage of only 0.1 V at twice the stoichiometric air supply. Thus, the problem of high methanol concentrations causing more severe cathode polarisation, due to increased methanol crossover, will be compounded when low stoichiometries of air are used. In large cell stacks, the issue of a variation in performance with position along the flow direction of air will also be a factor in design.

Low temperature performance of the DMFC is particularly relevant to small scale and portable applications, where applications, energy density considerations are of major importance. The effective energy density of methanol in a DMFC operating at 0.5 V with 90% fuel efficiency is 2.25 kWh kg^{-1}. In comparison, for a hydrogen PEMFC operating at 0.7 V to achieve the same energy density requires a storage capability of > 11% by weight, for example, as a metal hydride. This is a demanding requirement for the PEMFC, which does not have the advantage of using a liquid fuel.

The power performance requirement for a refillable DMFC for cellular phones are estimated at approximately 1W.[185] It has been estimated that a fuel cell power source of 50-cm^3 volume and weighing 50 g, would offer a talk time of 10 hours. This would be achieved with 10 cm^3 of methanol at 20% efficiency.

Ren et al.[174] and Surampudi[167] reported DMFC performance of approximately 15 to 35 mA cm^{-2} at 0.5 V, at low temperatures of 30 and 50 °C, respectively. Valdez et al.[186] reported data for a 5-cell stack operating at temperatures of 20 to 60 °C. At 20 °C the power performance was quite low, e.g., 1.5 V at 60 mA cm^{-2}. Increasing the temperature to 40 °C approximately doubled the power density and the stack delivered 2 V at 50 mA cm^{-2}. Subsequently, both laboratories have reported there efforts at producing small scale stacks operating at close to ambient temperature.[179, 186]

The JPL have developed a flat pack monopolar fuel cell which can deliver power densities of > 8 mW cm^{-2}, which relates to single cell performance of 0.3 V at 25 mA cm^{-2}. Gottesfeld[187] reported an air breathing stack design, consisting of two cells connected in series, with a capability of 10 mW cm^{-2}. This design is being developed in collaboration with Motorola.

Energy Related Devices Inc. also reported the development of MicroFuel Cell^{TM} technology for portable devices.[188] The DMFC fabrication used sputtering to produce the electrodes on

Table 8
Influence of Methanol Concentration on Static Feed DMFC Performance

Methanol concentration, wt%	Peak power, mW cm^{-2}	Current density at 0.2 V, mA cm^{-2}	Current density at 0.3 V, /mA cm^{-2}
1.5	~3	12	10
3	~6	26	19
6	9	47	30

microporous supports, enabling fuel cell arrays of only 1 mm to be produced.

The Battery and Fuel Cell Research Centre, Seoul, have developed a 15 cell, monopolar stack with electrodes 6 cm^2 in area.[189] The cells used Johnson Matthey catalyst; 8-mg cm^{-2} Pt-Ru for anode and 8 mg cm^{-2} Pt for cathode. A single cell performance of 9 mW cm^{-2} has been demonstrated under static fuel and air feed conditions. This required a 6 wt % (approximately 2 M) methanol solution enabling current densities above 50 mA cm^{-2} to be achieved, at approximately 0.2 V. The use of lower methanol concentrations limited achievable current densities as shown in Table 8.

For hydrogen PEMFCs the loadings of catalysts have been drastically reduced in recent years due to developments in catalyst electrode structure, cell design and engineering. Performance levels of over 500 mW cm^{-2} are routinely reported with catalyst loadings of around 0.2 mg cm^{-2}. If, for the DMFC, significant power output at such low loadings were achieved then this would make the technology extremely competitive. As an arbitrary target for the DMFC, performance with loadings below 1 mg cm^{-2} are discussed and summarised in Table 9.

At JPL/Giner Inc. performance of 0.5 V at 300 mA cm^{-2} at Pt loading of 0.5 mg cm^{-2} has been reported.[167, 173] and cell tests with 0.65 mg cm^{-2} Pt in the anode catalyst achieved power densities of 75 mW cm^{-2} at 90 °C. The above mentioned catalysts were all carbon supported.

Liu et al.[190] reported the influence of anode catalyst loading on DMFC performance using both supported and unsupported Pt-Ru catalysts. The data suggested that, with the supported catalysts, there was an upper limit to performance at a loading of between 0.5–0.76 mg cm^{-2}, whereas with the unsupported catalysts improvements in performance could still be made by using loadings greater than 2 mg cm^{-2}, e.g., 6. Notably the performance

Table 9
Typical DMFC Performance with Low Catalyst Loadings

Source/reference	Loading, mg cm^{-2} A[a]	Loading, mg cm^{-2} C[a]	Temperature, °C	Anode feed	Cathode feed and pressure, bar	Power density, mW cm^{-2}	Performance at X mA cm^{-2}, V
Hikita[173]	0.65	1.0	80	Liquid, 9 vol%	Air	75	200: 0.3
Surampudi[167]	0.5	6					300: 0.5
Liu[190]	0.76	6	90	0.5 M	O_2	~110	365: 0.3
Liu[190]	0.26	6	90	0.5 M	O_2	~60	200: 0.3
Gotz[48]	0.4	0.4	95	1.0 M	O_2		60: 0.3
Wiltham[81]	0.03	12	90	1.0 M	O_2, 1.5	75	80: 0.3

[a] A = anode; C = cathode.

of the 0.76-mg cm^{-2} supported catalyst was nearly identical to that of the 2-mg cm^{-2} unsupported catalyst. What is also clear from the data of Table 9 is that good power density performance was only achieved when the cathode catalyst had a significantly high loading so as to minimise cathode polarisation due to methanol crossover.

The operation of the DMFC can proceed with the feed either in the form of liquid or vapour. In fact it is feasible to operate with a liquid feed system and, through the internal heat generated in the cell, to transform the cell to a partial (or total) vapour system at the point fuel exhaust. The vapour feed system offers the attraction of better oxidation kinetics through higher temperatures and better gas phase mass transport. However, unless high fuel conversions are achieved in the cell, the system will be mechanically more complex through requirements to separate methanol (and water) from the carbon dioxide exhaust. Vapour phase operation also places additional heat transfer requirements on the system, e.g., to vapourise the aqueous fuel mixture. As a consequence, most cell stack development programmes have progressed with liquid feed systems (see Figure 14), which are mechanically simpler in terms of cooling and system thermal management. In liquid feed systems the exhaust from the anode is a two-phase mixture which requires condensation, or some other means of separation to remove methanol vapour from the carbon dioxide gas. An alternative method is to use a membrane gas separator.

1. DMFC Stack Performance

Small scale up of the DMFC was reported by Shukla et al.[191] and Jung et al.[192] A 3-cell stack of 150 cm^2 area at 90 °C operating with 2.5 M methanol and air as the oxidant produced a current of 58 A at 1 Volt, with a maximum power of 58 W (power density of 150 mW cm^{-2}).

Buttin et al.[193] have developed a 150-W cell stack based on the use of stainless steel (AISI 316) bipolar plates. The plates are surface treated (not plated with Au or Pt) to reduce the effect of corrosion. The cell delivered a power density of 140 mW cm^{-2} compared to 180 mW cm^{-2} for a small scale cell. The difference in performance for the two scales of cell was due to differences in individual cell behaviour associated with flow distribution and internal resistances. The reported variation in individual cell performance in cell stacks has also been observed by others.[194]

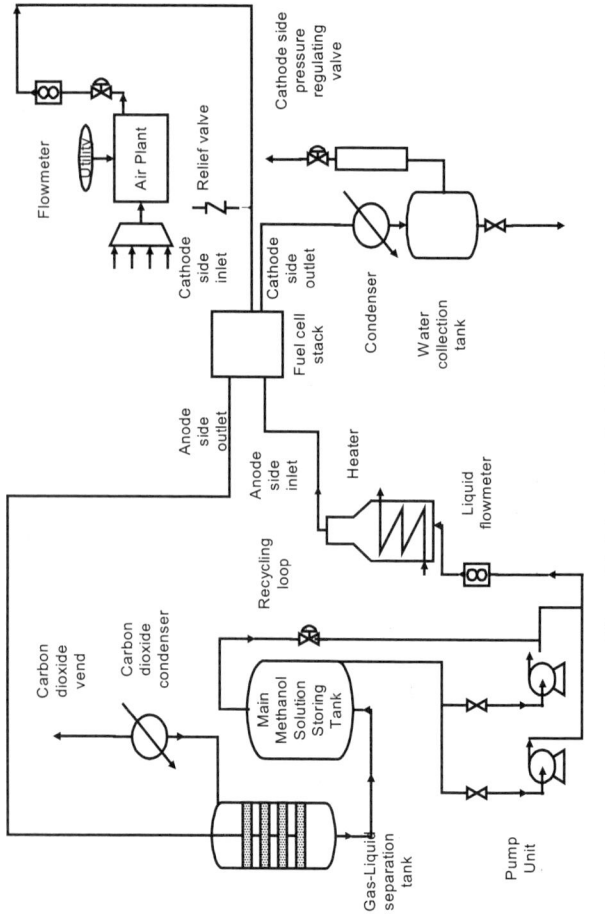

Figure 14. Schematic of a DMFC system.

Siemens, as part of a programme to develop a 1-kW DMFC stack, reported data for a 3-cell assembly with electrode areas of 550 cm^2. Operating on air (1.5 bar) the cell gave 1.4 V at 100 mA cm^{-2} and delivered a power of 87 W at 89 A.[195]

The group at Newcastle have designed and developed a 20-cell stack with a nominal output of 0.5 kW. The stack uses machined graphite bipolar plates and each electrode has an approximate active area of 250 cm^{-2}. Tests for a 3-cell stack have reported a power of 70 W operating at 70 °C and with low-pressure air.

JPL have reported performance data for a 5-cell stack with electrodes of 25 cm^2 area.[186] This stack gave a voltage of 2.2V at 100 mA cm^{-2}, at 60 °C using air supplied at 23 times the stoichiometric excess. The research at JPL has also demonstrated thousands of hours of operation of the cell stacks with no reported deterioration in performance. DMFC development is also taking place with an alternative, low cost replacement to Nafion membrane (unspecified) which is reported to exhibit reduced crossover at similar conductivity.

Continued efforts at the Los Alamos laboratories have lead to developments of DMFC stacks suitable for portable applications.[179] A 50-W stack system, based on a narrow pitch cell of 1.8 mm to minimise volume requirements has been tested. For a stack comprising 5 cells, 45-cm^2 area each, a power of 17 W at approximately 10 A, when operated at temperatures below 60 °C and at ambient air pressure was reported. The stack operated with low flow rates of air, of 2.5–3.5 times stoichiometry produced an effective power density of 300 W dm^{-3}. At a temperature of 100 °C the 5-cell stack produced a power of 50 W, i.e., approximately 1.7 V at 600 mA cm^{-2}. The reported fuel utilisation, achieved by careful cell and electrode design, were impressive, e.g., 82% and 95% with 1-M and 0.75-M methanol. Subsequent scale up of the DMFC has provided a stack of 30 cells with a peak power of 50 W at 14 V operating on 0.5 M methanol at 60 °C.[187]

During the last decade, significant advances have been made in the direct methanol single cell development. Maximum power densities of 450 and 300 mW/cm^2 under oxygen and air-feed operation, respectively and 200 mW/cm^2 at a cell potential of 0.5 V have been achieved for cells operating at temperatures close to or above 100 °C under pressurized conditions, with Pt loadings of 1–2 mg/cm^2. The development of DMFC stacks both for transportation and portable applications has gained momentum.

The rated power output of the DMFC stacks varies from a few watts in the case of portable power sources up to a few kW in the case of portable power generators and hybrid battery-fuel cell vehicles. The best results achieved with DMFC stacks for electro-traction are 1-kW/l power density with an overall efficiency of 37% at the design point of 0.5 V/cell.

Small electronic devices require compact and lightweight power supplies and direct methanol fuel cells offer the potential for double the lifetime of lithium ion batteries. Fabrication of a micro-machined direct methanol fuel cell using traditional micromachining techniques and macro-assembly based on silicon have been demonstrated. Gold and aluminium can be deposited as a current collector.[196]

The Korea Institute of Science and Technology (KIST) have developing passive micro-DMFCs with capacities under 5 W that are expected to be used as portable power sources. Research activities are focused on development of membrane-electrode assemblies (MEAs) and design of monopolar stacks operating under passive and air-breathing conditions. The passive cells showed many unique features, much different from the active ones. Single cells, with active area of 6 cm^2, showed a maximum power density of 40 mW/cm^2 with 4-M methanol solution at room temperature. A six-cell stack having a total active area of 27 cm^2 produced a power output of 1 W (37 mW/cm^2).[197]

2. Alternative Catalysts and Membranes in the DMFC

Thin film technologies like plasma polymerization and sputtering are suitable techniques for realizing membrane electrode assemblies only several microns thick that can be deposited on thin substrates (e.g., silicon wafers, porous foils, or others). Plasma polymerized films exhibit a high degree of cross-linkage and are pinhole free even for films of only a few hundred nanometer in thickness. In case of an electrolyte membrane, e.g., tetrafluoroethylene polymeric backbone and vinylphosphonic acid groups, these benefits yield a reduction of membrane resistance and a decreased methanol crossover. Thin film membrane electrode assembly can be fabricated from porous graphite electrodes (fabricated using an acetylene plasma polymerization process) and plasma polymerized electrolyte membranes.[198]

The development of the DMFC has proceeded mainly with the use of the binary catalysts of Pt-Ru, for which highest power densities have been reported. However, other catalysts with Pt are reported to be superior to or comparable with those of Pt-Ru (Table 10). Several groups report tests with ternary and quaternary catalyst which include Pt-Ru-W-Sn carbon supported catalyst[47] Pt-Ru-Os[19], Pt-Ru-W.[48]

Reddington et al.[199] used a combinatorial method to screen a 645-member array of catalysts containing the five elements; Pt, Ru, Os, Ir, Rh. The study identified a quaternary catalyst $Pt_{44}Ru_{41}Os_{10}Ir_5$ which gave a 40% higher current than the binary Pt-Ru catalyst at a potential of 400 mV. Performance of the catalyst[53] in a DMFC (60 °C, 1.0-M methanol) gave much superior performance to that of a commercial Pt-Ru catalyst (Johnson Matthey). The power density at this relatively low temperature was impressive, approximately 120 mW cm^{-2} compared to Pt-Ru at approximately 60 mW cm^{-2}.

The data on alternative membranes in the DMFC indicates that with suitable MEA fabrication procedures there are realistically good alternative membranes to Nafion (Table 11).

Aricò et al.[200] reported the operation of a DMFC at 145 °C using a Nafion/silica composite electrolyte (80 µm thick). With a Pt-Ru carbon supported anode (2 mg cm^{-2}) a peak power density of 240 mW cm^{-2} was achieved at 600 mA cm^{-2} and 0.4 V. It was suggested that even higher power density could be achieved by a reduction in the internal resistance, i.e., the internal resistance free power density was 450 mW cm^{-2}. Peak power densities of 150 and 90 mW cm^{-2} at 145 and 105 °C respectively, were reported, with an air fed cathode, which was comparable with that achieved for Nafion membranes.[201]

Yang et al.[202] reported data for a Nafion membrane modified with zirconium phosphate. The addition of the zirconium phosphate enabled higher temperatures to be explored without problems of water loss from the membrane. In this way, the performance of the DMFC with this composite membrane was approximately 100 mV greater, at a given current density, than that achieved with identical electrodes using Nafion®-117, at 130 °C. Peak power densities of 380 and 260 mW cm^{-2} were achieved on oxygen and air, respectively.

The performance of a nano-porous membrane, containing immobilised acid, has recently been reported. The membrane offers better temperature stability than Nafion coupled with higher conductivity and lower methanol crossover. Good power performance of 85 mW cm^{-2} are reported at 80 °C.[203]

Table 10
Performance of Ternary and Quaternary Catalysts in the DMFC

Source and reference	Loading, mg cm^{-2} A^a	C^a	Temperature °C	Anode feed and concentration	Cathode feed and pressure	Ternary catalyst Anode catalyst	Performance	Binary catalyst Anode catalyst	Performance
Ley[19]	6	6	90	Vapour, 0.5 M	O$_2$, 0.75 bar	Pt:Ru:Os. 65:25:10	340 mA cm^{-2} at 0.4V	Pt:Ru 50:50	260 mA cm^{-2} at 0.4V
Gotz[48]	0.4	0.4	95	1.0 M	O$_2$	Pt:Ru:W	360 mV at 50 mA cm^{-2}	Pt:Ru	330 mV at 50 mA cm^{-2}
Gotz[48]	0.4	0.4	95	1.0 M	O$_2$	Pt:Ru:Mo	330 mV at 50 mA cm^{-2}	Pt:Ru	330 mV at 50 mA cm^{-2}
Chan[53]	4	4	60	1.0 M	Air	Pt:Ru:Os:Ir	340 m cm^{-2} at 0.3 V	Pt:Ru	160 mA cm^{-2} at 0.3 V

^{a}A = anode; C = cathode

Table 11
Effect of PEM on DMFC Performance (Cathode Pt-C, Anode Pt-Ru)

Reference	Electrolyte	Temperature, °C	Anode feed conc.	Cathode feed and pressure	Power density, mW cm^{-2}	Performance.
202	N117/Zr	145	2M	O_2 Air	380 260	0.5V at 400 mA cm^{-2}
201	N117/Silica	145 145 105	2M	O_2 Air Air	240 150 90	
209	NP-acid immobilised	80 60	1M	O_2, ambient	85 75	0.38V at 200 mA cm^{-2}
203	PBI	200	0.5 mole ratio	O_2, ambient	210	0.42 at 400 mA cm^{-2}
205	PEEK/PSU blend	80	1.5M 0.5M	Air, 1.5 bar	50 42	

Acid-doped PBI membranes have attracted significant interest recently due to the reported very low methanol crossover characteristics.[162] PBI membranes are appropriate for higher temperature operation (> 130°C) as at lower temperatures their conductivity is much lower than that of Nafion. Reported performance of MEAs using the PBI membrane, at temperatures of 200 °C are good; 210 mW cm^{-2} at 500 mA cm^{-2}, using commercial Pt-Ru alloy electrocatalyst bonded with PTFE.[204] Notably this performance was achieved with a methanol to water mole ratio of 0.5, which was an order of magnitude greater concentration than typically used in the DMFC.

Kerres et al.[205] reported performance of the DMFC with a range of new acid-base blend membranes. Comparable performance were reported to that achieved with Nafion. In particular, blend membranes made from sulphonated PEEK with PSU(NH$_2$)$_2$ gave a power density of 50 mW cm^{-2} (0.5 V) at 80 °C using 1.5-bar air.

Cross-linked co-polymers with styrene sulphonic acid have been evaluated in the DMFC by Scott et al.[206] using carbon supported Pt-Ru catalysts. The electrical and power performance of a typical membrane was comparable to that with Nafion®-117 (Figure 15). Although performance was not significantly different from that with Nafion® there was an apparent improvement at higher current densities, which was accredited to a superior electrical conductivity of the MEA.

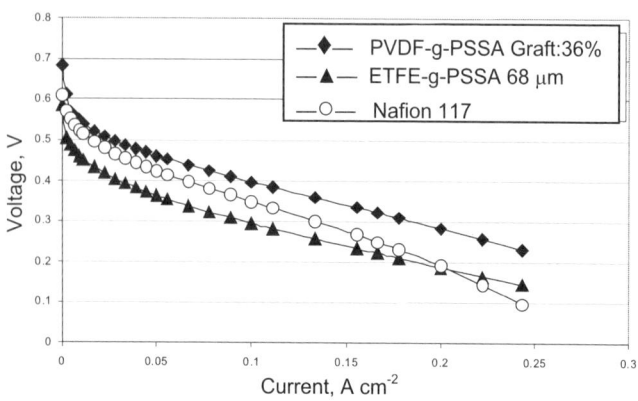

Figure 15. Comparison of performance of the DMFC with Nafion and radiation grafted membranes.

The Fe-TMPP catalyst, supported on carbon, has been used as a methanol tolerant cathode, in a DMFC using a PBI membrane operating at 150 °C.[207] In addition ethanol tolerant catalyst based on Ru-S have been used in the DMFC[208] with concentrations of methanol as high as 6 M. These higher concentrations of methanol result in no deleterious effect to performance, which is comparable to cells using platinum catalysts. Interestingly, the methanol content in the cathode exhaust gas from the DMFC was above 2 M and confirms the high degree of methanol tolerance of these materials acting as oxygen reduction selective cathodes.

3. Alkaline Conducting Membrane and Alternative Oxidants

Solid polymer electrolyte DMFCs have been developed to their current status using proton conducting polymer membranes. Two obstacles, currently inhibiting the application of DMFCs are the relatively low activity of methanol electro-oxidation catalysts and methanol crossover through the proton-conducting membrane[15], both of which can be overcome by research using hydroxide (alkaline) conducting membranes. Alkaline fuel cells have been developed to a significant level. A principle reason is that catalysts generally perform better in alkaline than in acid solutions.[210] With an alkaline electrolyte, a lower loading of and, importantly, a wider range of electrocatalyst (other that Pt) e.g., based on Ni may be used. In the case of methanol, if the fuel cell is operated under alkaline conditions, two major advantages could be gained: it appears that the electro-oxidation of methanol in alkaline solution is structure insensitive,[210] and hence other metals (or oxides of) are as active as platinum towards the oxidation, e.g., Ni.[211] In alkaline conditions, the DMFC does not suffer from poisoning by the chemisorbed methanol fragments observed in acid electrolyte.

The use of a hydroxide form anion exchange membrane will resolve the current problem of methanol crossover from anode to cathode: electro-osmotic water transport from cathode to anode will counteract diffusion of methanol. In the case of the DMAFC, electro-osmotic water transport will be from cathode to anode as has been observed in studies of transport of formaldehyde and ethylene glycol using anion exchange membranes.[212] In this work[212], the cathode generated hydroxide ions during reduction of formaldehyde with no reported adverse effects on performance, even though some oxidation of the organic would occur. Thus OH⁻ ion transfer should potentially preclude transport of methanol from

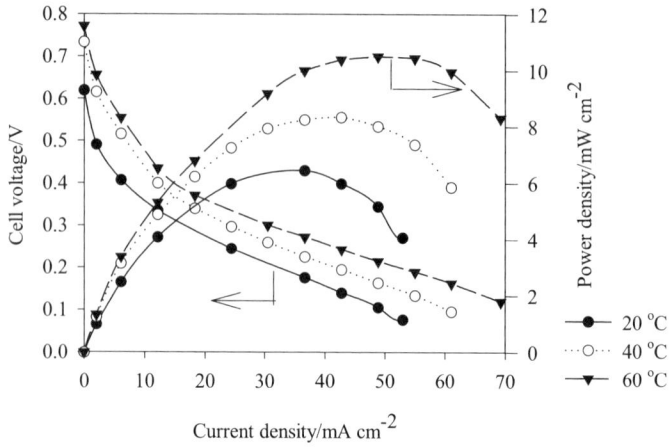

Figure 16. Performance of DMFC with anion exchange membrane. Anode: Pt/C (60 wt% Pt) 2.19 mg cm^{-2} on non-teflonised Toray 90 carbon paper. Cathode: Pt/C (60 wt% Pt) 2.07 mg cm^{-2} with GDL on 20 % Teflonised Toray 90 carbon paper.

anode to cathode in the DMFC, i.e., diffusion of methanol will be counteracted by electro-osmotic transport. The formation of carbon dioxide in the presence of hydroxide, however, raises issues associated with potential carbonate/bicarbonate production.

Alkaline methanol fuel cells are potentially very attractive due to favourable electrocatalysis of both oxygen reduction and methanol oxidation at high pH coupled with the potential use of non-noble metal catalysts such as Nickel. The issue of electrolyte carbonation can be resolved by simply recharging the system with a fresh mixture of methanol and KOH to replace the spent electrolyte containing $K_2CO_3/KHCO_3$. Of course, there will be deterioration in performance as carbonation takes place. Provided carbonation does not interfere with the catalyst performance and stability/longevity then alkaline DMFC may find certain niche applications. Initial results of the performance of a direct methanol fuel cell with an anion exchange membrane (Figure 16) have reported power densities of approximately 10 mW cm^{-2} at room temperature.[213]

The use of alternative oxidants to air (or oxygen) in fuel cells has not been considered greatly because of the convenience in the supply and cost of air. However, an attraction could be the realisation of higher power output. In this respect, the use of hydrogen peroxide is one option that offers the potential for increased power. Such a concept of a methanol-peroxide fuel cell has been reported by Prater et al.[214]

There are two modes of using hydrogen peroxide as an oxidant in a DMFC. One mode would be to allow hydrogen peroxide to decompose on a catalytic surface and then channel the released oxygen to the DMFC cathode. Another mode for using hydrogen peroxide as the oxidant in a fuel cell would be to feed a solution of hydrogen peroxide and water directly to the cathode compartment of the cell.

When hydrogen peroxide is introduced to the cathode compartment of a fuel cell, it is decomposed into oxygen and water at the electrode backing and at the catalyst/electrode-backing interface according to the equation,

$$3 H_2O_2 \rightarrow 3 H_2O + 3/2 O_2 \tag{26}$$

The total cell reaction can be determined by adding the reactants and products of Eqs. (2) and (3),

$$CH_3OH + 3 H_2O_2 \rightarrow CO_2 + 5 H_2O \tag{27}$$

Consumption of hydrogen peroxide is also likely to occur according to the following reaction,

$$3 H_2O_2 + 6 H^+ + 6 e^- \rightarrow 6 H_2O \tag{28}$$

Operating a DMFC with hydrogen peroxide as oxidant can extend the operational environment of DMFCs to locations where the free convection of air is limited, such as in underwater applications.

VI. CONVENTIONAL VS. MIXED-REACTANT SPE-DMFCS

The mixed-reactant fuel cell (MRFC) is a concept, in which a mixture of fuel and gaseous oxygen (or air) flows simultaneously over both sides of a porous anode-electrolyte-cathode structure or through a strip-cell with an anode-electrolyte-cathode configuration. These structures can be single cells or parallel stacks of cells and may be in a planar, tubular or any other geometry. Selectivity in the electrocatalysts for MRFCs is mandatory to minimize mixed-potential at the electrodes, which otherwise would reduce the available cell voltage and compromise the fuel efficiency. The MRFC offers a cost-effective solution in fuel cell design, since there is no need for gas-tight structure within the stack and, as a consequence, considerable reduction of sealing, manifolding and reactants delivery structure is possible. In recent years, significant advances have been made in MRFCs using methanol as a fuel.

One of the first reported mixed-reactant fuel cells used a liquid reactant mixture of methanol and hydrogen peroxide in an alkaline solution.[215] The tests were conducted on a stack of 40 cells using bipolar electrodes with Pt anode and Ag cathode. The stack delivered power at 40 A and 15 V and although was not particularly efficient, it clearly demonstrated the concept of a MR-DMFC. The performance was not as good as that from a separated fuel and oxidant fuel cell as indicated in open cell potentials of 0.41 V for MR-DMFC and 0.81 V for standard configurations. Another reason for possible inferior performance of the MR-DMFC is due to current leakage or bypass losses in the cell stack.

Barton et al.[6] demonstrated the MR-DMFC concept using a strip electrode configuration which was fed with a two phase mixture of methanol solution and air. The cell used Pt-Ru/C anode catalysts and tetramethoxyphenyl porphyrin or RuSeMo/C cathode catalysts. Both of the cathode catalysts exhibited good methanol tolerance for oxygen reduction. The mixed reactant mode cell gave slightly better performance than the standard separated-fuel and oxidant system.

The history of mixed-reactant fuel cells goes back to the 1950's in which tests were performed on mixed hydrogen and air systems. The cells were based on an alkaline electrolyte and used "selective" cathode catalysts of C, Au or Ag, materials that are not good hydrogen oxidation catalysts.[216, 217] The cells were operated in two modes, one in which the fuel and oxygen mixture were first fed to the cathode and then to the anode and the other in which the reactant mixtures were both fed simultaneously to the anode. Although open-circuit potentials of 1 V were obtained, on-load power performance was not impressive, 0.4 V at 4 mA cm^{-2}.

The use of mixed-reactant fuel cells based on hydrogen and oxygen using high temperature and thin film alumina electrolytes was suggested in 1965 with a strip cell design.[218] In the mid 1970's, the cell was experimentally tested using dilute hydrogen and air mixtures and achieved a peak power of 0.32 µW cm^{-2} at a voltage of 0.39 V.[219] More recent reports of high temperature solid oxide fuel cells operating on mixed reactants have been made by

Dyer et al.[220] and Hibino and Iwahara,[221] and the subject was reviewed in 1995.[222] A solid oxide fuel cell (SOFC) with a mixed methane and air feed has been reported,[223] and a low operating temperature solid oxide fuel cell using a hydrocarbon-air mixture and a samaria-doped ceria electrolyte has been reported.[224] The use of a strip cell design for mixed reactant SOFC systems has also been reported.[225,226] In the solid oxide systems, the power densities generated by the mixed reactant cells have been quite respectable. Hibino et al. for example achieved 166 mW cm^{-2} using $BaCe_{0.8}Y_{0.2}O_{3-\alpha}$ electrolyte at 950°C with a Pt anode and Au cathode.[227] In a later study using ethane and air at 500 °C, and a $Ce_{0.8}Sm_{0.2}O_{1.9}$ electrolyte for example, a peak power density of 400 mW cm^{-2} at 0.5 V has been demonstrated.[224, 228]

Although there has been some significant effort in the development of mixed reactant fuel cells with hydrogen and hydrocarbon fuels and some very respectable power performances were achieved, there is inevitably concern over the use of an explosive mixture within a cell where the potential for electrical spark ignition is real. Thus mixed reactant cells will probably need to operate outside the explosive ranges of fuel and air mixtures and use dilute fuel mixtures. One such example is the mixed-reactant direct methanol fuel cell (MR-DMFC), which operates well on dilute fuel compositions in the presence of water.

In a MR-DMFC, aqueous methanol fuel and oxidant oxygen gas (or air) are mixed together before feeding to the fuel cell (Figure 17).[6,229] In mixed-reactant fuel cell, there is no need for gas-tight structure within the MR-DMFC stack providing relaxation for sealing of reactants/products delivery structure.

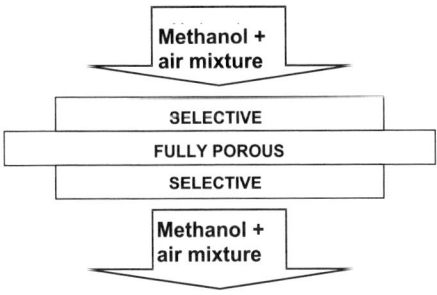

Figure 17. A MR-DMFC. Aqueous methanol fuel and oxidant oxygen gas (or air) are mixed together before feeding to the fuel cell.

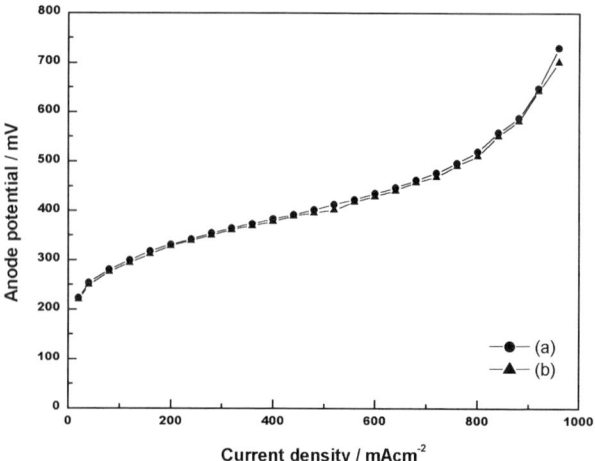

Figure 18. Galvanostatic polarization data for the selective reactants and mixed-reactants anode.

The galvanostatic polarization data for the selective reactants and mixed-reactants anode tests are given in Figure 18(a) and (b), respectively.[9] There was no significant difference between the polarization data of the mixed-feed anode with methanol plus air and mixed-feed anode with methanol plus nitrogen, which is in accord with the findings of Barton et al.[6] This data suggests that there was no parasitic oxidation of methanol with oxygen in the air in the MR-DMFC. It is noteworthy that a DMFC operating at 1-A load would require 7.06×10^{-8} dm^3/s of liquid methanol at the anode, and will result in a CO_2 exhaust of 3.87×10^{-5} dm^3/s (if present as gas) at its anode. This represents about 550-fold volume increase in the anode compartment of the cell during its operation and suggests that CO_2 removal from the anode should improve the cell performance.

In MR-DMFCs, cathode selectivity is paramount and is accomplished by using oxygen-reduction catalysts, which in addition to being tolerant to methanol, do not oxidize it (Refs. 91, 92, 103-114, 230-234). The performance characteristics of certain MR-DMFCs, employing various methanol-resistant oxygen-reduction catalysts at the cathode and a Pt-Ru/C catalyst at the anode, are shown in Figures 19 and 20.[9] The performance curves at 90 °C for the MR-DMFCs employing 1 mg/cm^2 of

Direct Methanol Fuel Cells

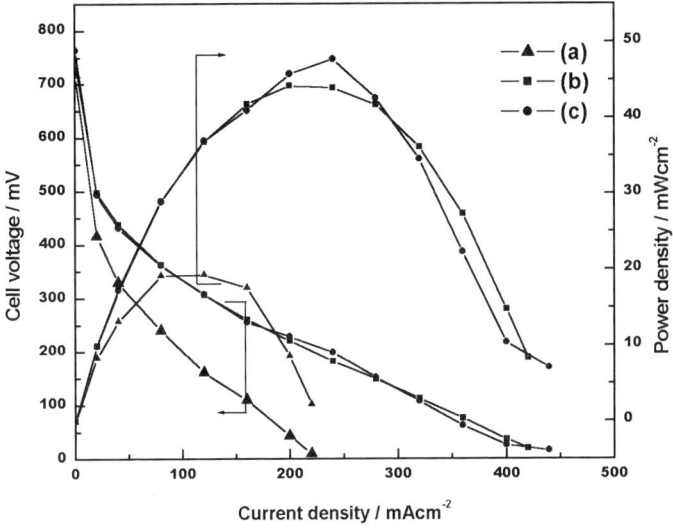

Figure 19. Performance curves at 90 °C for the mixed-reactants DMFCs employing varying amounts of RuSe/C at the cathode: (a) 1, (b) 2.5, and (c) 2 mg cm^{-2}.

FeTMPP/C, CoTMPP/C, FeCoTMPP/C, and RuSe/C at the cathode are shown in Figure 19. Among these, the best performance with a maximum power output of ~ 30 mW/cm^2, is observed for the MR-DMFC employing 1mg/cm^2 of RuSe/C catalyst at the cathode. The performance curves at 90 °C for the mixed-reactants DMFCs employing various amounts of RuSe/C at the cathode are shown in Figure 20. It is found that the maximum output power of approximately 50 mW/cm^2 is obtained for the MR-DMFC with the RuSe/C loading of 2.5 mg/cm^2, while operating the MR-DMFC with methanl plus oxygen. A maximum output power of ~ 20 mW/cm^2 is obtained when operating the cell with methanol plus air.

The cathode polarization curves for oxygen-reduction using Pt/C, Pt-Co/C and Pt-Fe/C catalysts bonded to the DMFC MEAs are shown in Figure 21(a). These data show superior performance for the Pt/C cathode. But as the methanol is passed over the anode, a lower cell performance was found for the cell employing the Pt/C catalyst in relation to both Pt-Co/C and Pt-Fe/C cathodes (Figure 21c).[10] This behaviour clearly reflects the poisoning of the

Pt/C cathode by methanol crossover from anode to cathode. The anode performance was similar for all the cells tested, as shown in Figure 20(b). The Pt-Co/C catalyst exhibits a performance better than that of Pt/C catalyst but is inferior to the Pt-Fe/C catalyst.[10] These data suggests that Pt-Fe/C will be an effective selective oxygen-reduction catalyst and conform to the cyclic voltammetry data on Pt/C, Pt-Co/C, and Pt-Fe/C presented in Figure 11.

Recently, researchers from IFC and UT-Austin have demonstrated that the performance of a MR-DMFC with selective electrodes could exceed that of the conventional DMFCs when the fuel and the oxidant are supplied at identical rates to the anode and the cathode, respectively.[6] They also conducted a design study in which the dimensions of a series of mixed-reactants, strip cell DMFCs were optimized (Figure 22). In the single cell tests, a two-phase reactant mixture of 1 M methanol (3 ml/min) and air (3 dm^3/min) was supplied to both sides of a conventional geometry membrane electrode assembly at 80 °C. The 32-cm^2 MEA consisted of a Nafion-117 membrane coated on one side with a hydrophobic Pt-Ru/C (5mg/cm^2) anode, and on the other side with iron tetra-methoxy phenylporphyrin (FeTMPP), a methanol-tolerant cathode material.

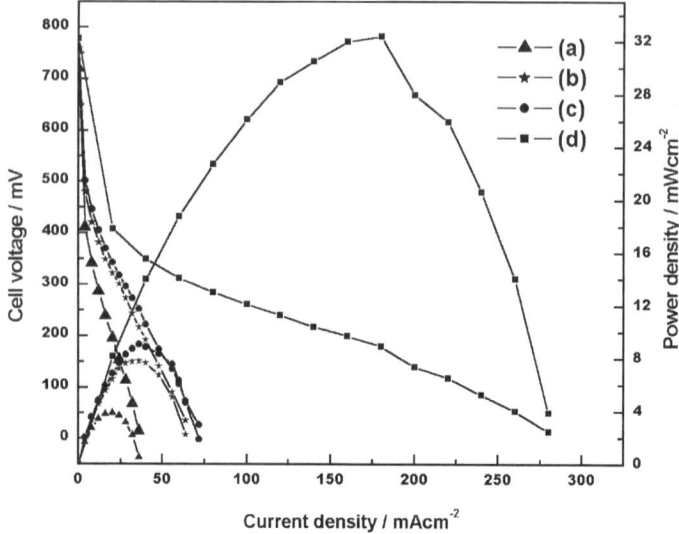

Figure 20. Performance curves at 90 °C for the MR-DMFCs employing 1 mg/cm^2 of (a) FeTMPP/C, (b) CoTMPP/C, (c) FeCoTMPP/C, and (d) RuSe/C at the cathode.

The half-cell experiments demonstrated that, in the system, there was little reaction between oxygen and methanol at the anode and that the main effect of the entrained air or nitrogen in the mixed-feed was to impede mass transport of the feed to the anode at current densities above 100 mA/cm^2. At the cathode, half-cell measurements again showed little difference while operating in mixed-reactant and conventional separated-reactant modes. Consequently, performance of the SPE-DMFCs in mixed-reactant

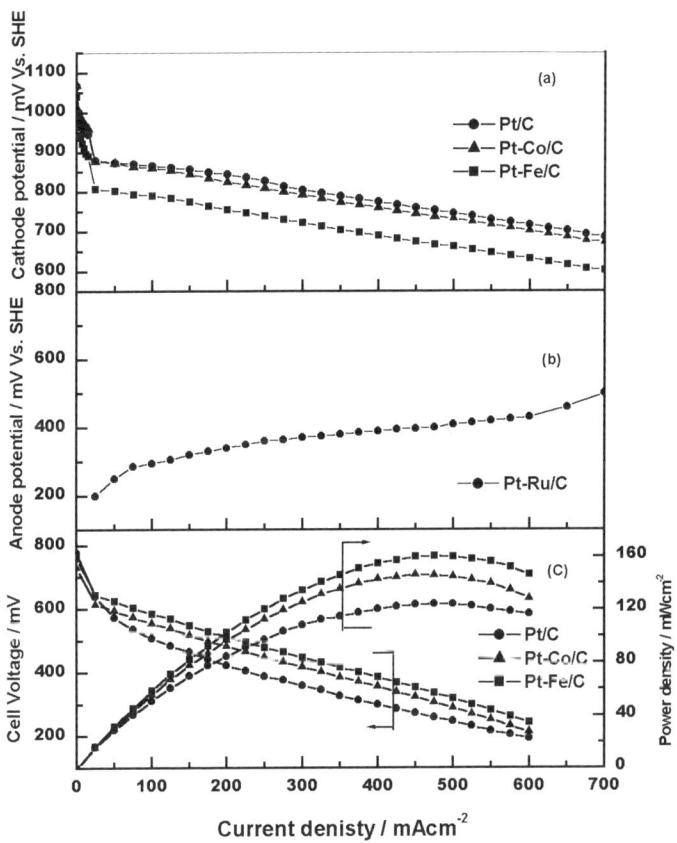

Figure 21. Cathode polarization curves for oxygen-reduction using Pt/C, Pt-Co/C and Pt-Fe/C catalysts bonded to the DMFC MEAs (obtained by oxidizing hydrogen at the anode, which also acts as the reference electrode).

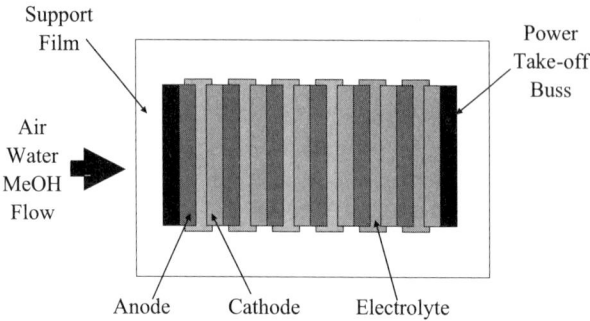

Figure 22. Schematic of mixed-reactants, strip cell DMFCs.

mode was identical to the performance of SPE-DMFCs in conventional mode. In the MR-DMFCs, one may argue that methanol crossover offers a performance advantage, a situation quite opposite to DMFCs. In the latter, methanol leaks constantly from the anode to the cathode causing a lowering in its potential and wastage of fuel through direct chemical oxidation. This has its adverse effect on fuel efficiency in a conventional SPE-DMFC and is one of the primary reasons for various selective cathode materials, such as Ru-Se, being investigated for conventional SPE-DMFCs.

VII. MATHEMATICAL MODELLING OF THE DMFC

Modelling the direct methanol fuel cell can lead to a greater understanding of the cell and its interactions with other components in the system. Because of the similarities with the PEMFC, modelling of the DMFC using polymer electrolyte membranes can follow similar approaches to hydrogen PEM cells; there are of course some crucial differences as will be discussed in this Section.

The DMFC consists of a thin composite structure of anode, cathode and electrolyte. The electrocatalysts in a fuel cell are positioned on either side of the polymer electrolyte, to form the cell assembly. These electrocatalysts are supported on carbon and bonded using, typically Nafion^R, ionomer. In this way a three-dimensional electrode structure is produced in which electronic current movement is through the carbon support and ionic current flow is through the ionomer. The reactants are, in practical operation, fed to the backsides of the electrodes. Flow fields are used to supply and distribute the fuel and the oxidant to the anode and the cathode electrocatalyst respectively. The distribution of flow over the electrodes should ideally be uniform to try to ensure a uniform performance of each electrode across its surface. The flow field allows fluid to flow along the length of the electrode whilst permitting mass transport to the electrocatalyst normal to its surface.

In most practical systems, because single cell potentials are small (< 1V) fuel cells are connected in series to produce useful higher overall voltages (Figure 23). Electrical connection in stacks is usually achieved using bipolar plates, which make electrical connection over the surface of the electrodes. A second function of the bipolar plates is to separate the anode and cathode fluids. These factors introduce several challenges in plate selection and design. For example whilst the flow field enables access of gas to the electrode structure in its open spaces it prevents electrical contact at these points. Electrical contact should be as frequent and as large as possible to mitigate against long current flow path lengths. However large areas of electrical contact could lead to problems of access of reactant gases to regions under the electrical contact.

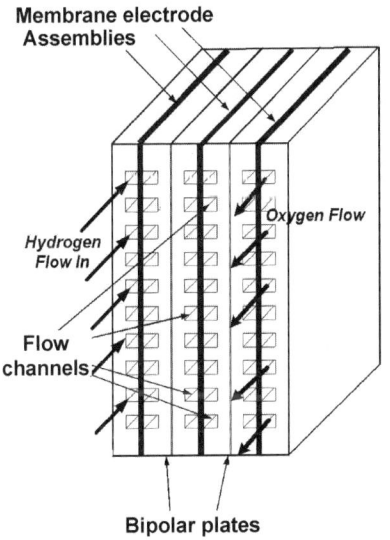

Figure 23. Bipolar connection of fuel cells.

Thus overall the flow field design and the flow therein are a critical factor in fuel cell operation and modelling.

In light of the above requirements and features of the DMFC, modelling of the cell for the purposes of simulation and aiding cell understanding is important. Overall modelling of the DMFC can occur at a series of different levels focusing on one or more different aspects or components in the cell as below:

1. Anode catalysis. The mechanism of methanol oxidation is not known but it is known that simple Butler-Volmer expressions are less than satisfactory.
2. Polymer electrolyte membrane transport and conductivity. As well as facilitating proton transfer from anode to cathode it is known that the crossover of methanol and water is important in determining performance.
3. Cathode catalysis. Oxygen reduction is kinetically a faster process than methanol oxidation, but nevertheless its overpotential behaviour will have a significant influence on overall cell behaviour. In addition, a crucial issue is that of methanol oxidation at the cathode which through a mixed potential has severe repercussions on cell voltage.
4. Fluid transport in porous backing layers. This influences the transport of reactants methanol and oxygen to the catalysts and the transport of products carbon dioxide and water away from the catalysts.
5. Fluid mechanics in the flow channels. This impacts on the local variation of methanol in the cell and is influenced by variations in velocity and pressure as also the associated equilibrium between the methanol and water and carbon dioxide gas. This and many other factors will influence the current distribution over individual cells.
6. Current and potential distribution in the electrocatalyst layers. The electrodes in the DMFC are essentially three dimensional to attempt to provide a high specific area. The current is thus distributed in a direction normal to overall current flow and this distribution should be determined for both the anode and the cathode.
7. Thermal and heat transfer models. The electrochemical inefficient use of methanol leads to significant amounts of heat generation in the cell and thus local variations in temperature and heat transfer rates may influence cell behaviour to a significant extent.

8. Cell stack models. The practical application of multi-cell units requires, in principle, an integration of all the above factors. In addition, due regard must be taken of the variation in temperature in the cells and in the requirements of feeding air and methanol to the many compartments in the cell, i.e., cell manifolds. Such behaviour can see variations in performance from one cell to the other.
9. Dynamics. The application of the DMFC will inevitably impose a varied load demand on the unit and the ability or not of the cell to respond effectively to the load demand will be crucial.

The types of models can vary from simple empirical models to detailed physicochemical models to stochastic models which cover single environments, such as electrodes, to complete fuel cell systems.

1. Methanol Oxidation

There have been various studies of the mechanism of methanol oxidation and several rate/kinetic models are proposed. It is generally thought that the rate determining step is surface reaction between CO_{ads} and OH_{ads}. It has been proposed that methanol and hydroxyl groups are adsorbed on different parts of the surface on carbon supported platinum. Kaurenan and Skou[235] developed a model for methanol oxidation on carbon supported platinum in which the rate of the surface reaction was limiting and that the hydroxyl adsorption was assumed in Nernstian equilibrium and followed a Langmuir adsorption. The rate of adsorption of the intermediate CO was expressed by a Temkin adsorption rate equation. CO and OH adsorption were assumed to occur at different sites of the supported catalyst. Observed experimental limiting currents were accredited to the adsorbed OH-groups reaching saturation coverage. The model overall gave reasonable agreement with experimental observed anode polarisation although aspects of methanol mass transport are not considered.

It has also been suggested that, at low temperatures, methanol adsorption/dehydrogenation may be rate limiting.[236] This assumption has thus been used in a model for the DMFC in which the rate of methanol oxidation is controlled by dissociative chemisorption of methanol on active Pt sites. The mechanism was a slightly simplified version of that proposed first by McNichol.[237] As described in Section III, the oxidation consists of the following four steps:

(R1) dissociative chemisorption of methanol on active platinum catalyst sites,

$$3\ Pt + CH_3OH \Leftrightarrow Pt_3\text{--}COH + 3\ H^+ + 3\ e^- \quad (29)$$

(R2) formation of hydroxylic groups on the active ruthenium catalyst sites,

$$Ru + H_2O \Leftrightarrow Ru\text{--}OH + H^+ + e^- \quad (30)$$

(R3) surface reaction of the absorbed molecules formed in steps (R1) and (R2),

$$Pt_3\text{--}COH + 2\ Ru\text{--}OH \Leftrightarrow Pt\text{--}COOH + H_2O + 2\ Pt + 2\ Ru \quad (31)$$

(R4) surface reaction leading to the release of the reaction product carbon dioxide,

$$Pt\text{--}COOH + Ru\text{--}OH \Leftrightarrow CO_2 + H_2O + Pt + Ru \quad (32)$$

The rates of reaction are given as follows:

$$r_1 = k_1 \exp\left(\frac{\alpha_1 F}{RT}\eta_a\right)\left\{\Theta_{Pt} 3 c^{CL}_{CH_3OH} - \frac{1}{K_1}\exp\left(-\frac{F}{RT}\eta_a\right)\Theta_{Pt_3\text{--}COH}\right\} \quad (33)$$

rate determining step,

$$\begin{aligned}
(2)\ r_2 &= k_2 \exp\left(\frac{\alpha_2 F}{RT}\eta_a\right)\left\{\Theta_{Ru} - \frac{1}{K_2}\exp\left(-\frac{F}{RT}\eta_a\right)\Theta_{Ru\text{--}OH}\right\} \\
(3)\ r_3 &= k_3\left\{\Theta_{Pt_3\text{--}COH}\Theta^2_{Ru\text{--}OH} - \frac{1}{K_3}\Theta_{Pt\text{--}COOH}\Theta^2_{Pt}\Theta^2_{Ru}\right\} \\
(4)\ r_4 &= k_4\left\{\Theta_{Pt\text{--}COOH}\Theta_{Ru\text{--}OH} - \frac{1}{K_4}c^{CL}_{CO_2}\Theta_{Pt}\Theta_{Ru}\right\}
\end{aligned} \quad (34)$$

assumed to be close to equilibrium

The above model generally predicts that methanol oxidation is first order in methanol concentration, which was not confirmed experimentally and is generally contrary to experimental data for methanol oxidation on Pt and Pt-Ru catalysts.

However, despite significant research on methanol oxidation, the mechanism is not fully known and the adsorption of the various reactive intermediates may involve a combination of single site and dual site processes. These species may include those proposed in the mechanism of Freelink[238] for Pt alloys. Nordlund and Lindbergh[239] presented an alternative model in which methanol adsorption was followed by oxidation to adsorbed CO, i.e., step 5 above. After which reaction (8) above occurred as the rate-determining step on the same adsorption sites. They treated the model by applying the stationary state approximation to the surface processes and, from the solution determined, the adsorption of species as a function of electrode potential. The mechanistic model, as an approximation, predicts that oxidation will be a first order process in methanol concentration. They were able to predict a good agreement between experimental half cell polarisation and predictions.

Mayer and Newmann[240,241] have presented modeling and data analysis of transport phenomena in a SPE-DMFC. In contrast to most of the earlier models, which employ a simple Butler-Völmer (BV) relationship for describing the electrode-kinetics for methanol oxidation at the anode, the model due to Mayer and Newmann follows the reaction mechanism proposed by Gasteiger et al.[242]

Scott and Argyropoulos[243] have proposed a one-dimensional potential distribution model of a novel anode (based on Ti mesh) and used a kinetic model, derived from the adsorption model of Nordlund and Lindbergh[239] using the stationary state approximation as follows:

$$j = \frac{6Fk_{10}C_M \, e^{\beta\eta}}{1 + k_{20} \, C_M \, e^{\beta\eta}} \quad (35)$$

This model gave good agreement with experimental data as shown in Figure 24.

However, it is important to state that the mechanistic model alone cannot generally be substantiated for porous high surface area electrodes because of the influences of current distribution, variable geometry and mass transport effects. Thus for modelling, electrode structure and ionic and electronic conduction within, mass transport effects together with a kinetic model of the

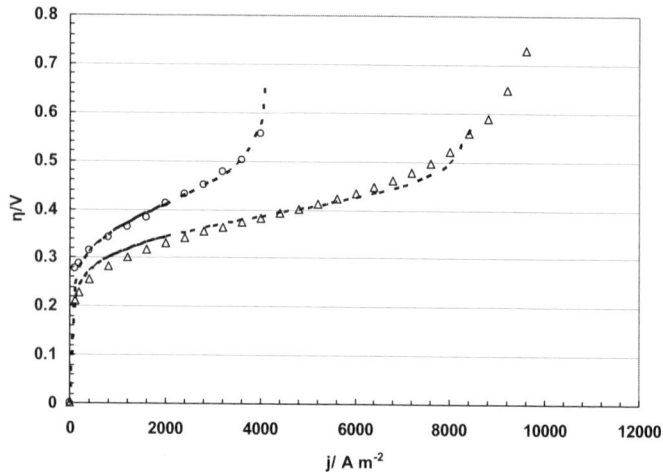

Figure 24. Experimental data for methanol oxidation obtained at 60 and 90 °C and the fit of the data to Eq. (35). $C_M = 1000$ mol m^{-3}. Δ: $k_{10} = 8.64 \times 10^{-10}$ /ms^{-1}, $k_{20} = 5.83 \times 10^{-8}$, and $\beta = 24.91$ (90 °C); ◊: $k_{10} = 8.64 \times 10^{-10}$, $k_{20} = 1.21 \times 10^{-7}$, and $\beta = 21.84$ (60 °C).

oxidation, are required to predict electrode polarisation and thereby cell voltage vs. current density behaviour.

2. Empirical Models for Cell Voltage Behaviour

It is possible to "model" the cell voltage (V) using empirical equations applied in the modelling of hydrogen PEM cells.[244] Such models are useful when the objective of the model is to analyse the system overall, which, not only includes the fuel cell but also the ancillary equipment. The aim of such models may be to explore system interactions and to model material and energy flows, etc. Functional relationships that have been used include:

$$E_{cell} = a - b \log j - cj + dj^e \ln(1 - fj) \qquad (36)$$

$$E_{cell} = a - b \log j - cj + d \exp^{ej} \qquad (37)$$

The equations include several coefficients (a-f), which were fitted (best fit statistically) to the various function components of current density to give an excellent agreement with the full cell polarisation data. However, as with the equivalent PEM cell model, changes in operating conditions, e.g., methanol concentration, produced different values of the empirical coefficients and the predictability of the model, outside a narrow parameter domain, was lost.

However, an alternative approach to empirical modelling has been to relate the empirical coefficients to the chemistry of the system, such as characteristics in the Butler-Völmer kinetic and Nernst equations and to mass transport at the electrode, rather than attempt to obtain the best fit to coefficients.[245] In this way, we obtain an expression for the cell voltage, E_{cell}, as,

$$E_{cell} = E_O^* - b_{cell} \log j - R_e j + C_1 \ln(1 - C_2 j) \qquad (38)$$

where, $E_O^* = E_{Ocell} - \dfrac{RT}{\alpha_c F} \ln\left(\dfrac{p_O^{ref}}{j_{Oc} p_O}\right) - \dfrac{RT}{\alpha_a F}\left(\ln \dfrac{C_{ME}^{ref}}{j_o C_{ME}^N}\right)$,

$b_{cell} = \dfrac{2.303 R T}{F}\left(\dfrac{1}{\alpha_a} + \dfrac{1}{\alpha_c}\right)$, $C_1 = \dfrac{nRT}{\alpha_a F}$, and $C_2 = \dfrac{1}{nFk_{eff} C_{ME}}$.

Using this equation, a good fit to experimental data is produced (Figure 25) using the parameters in Eq. (25), which themselves have a greater physical meaning in relation to the fuel cell electrode and exhibit variations with parameters (temperature, concentration) as predicted from the chemistry of the reaction.

A similar approach has been applied by Nordlund[246] for the anode of the DMFC using the following relationship between current density and potential:

$$j = \dfrac{j_k}{\left(1 + \dfrac{j_k}{j_{\lim}}\right)} \qquad (39)$$

where $j_k = \exp\left[\dfrac{\alpha F(E_a - E_a^*)}{RT}\right]$.

In this equation, j_{lim} was a combined complex power series and hyperbolic function of temperature and methanol concentration and the term E_a^* was a linear function of temperature. A good agreement between experiment and model was thus achieved.

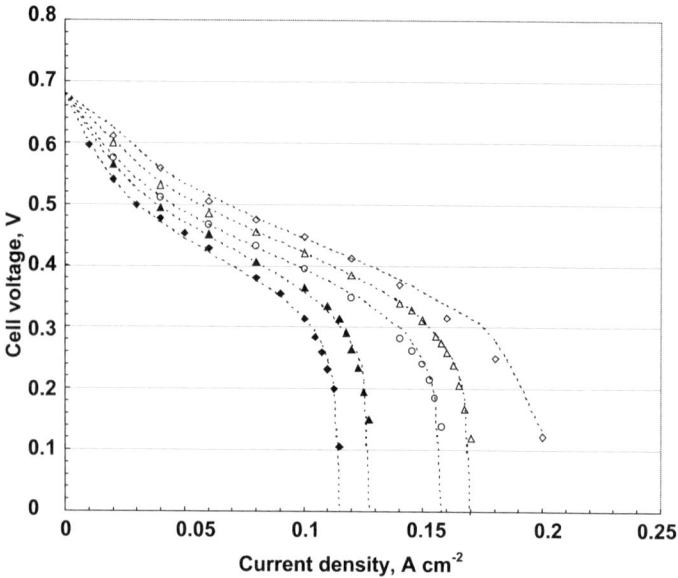

Figure 25. Comparison between experimental data[245] and empirical equation based prediction for a cell operated with 0.5 M-methanol solution supplied at a rate of 1.12 cm^3 min^{-1} with air fed cathodes pressurised at 2 bar. Cell temperatures ♦: 343.15 K, ▲: 348.15 K, o: 353.15 K, ∆: 358.15 K, and ◊: 363.15 K.

An empirical model based on a Canonical Variate Analysis (CVA) state space representation has been developed to predict the voltage response of the DMFC and multi-cell stacks.[247,248] In order to achieve high performance control of a commercial system, it is essential to have a methodology that will accurately predict the stack voltage from a minimum number of sensors and with the smallest time delay after vehicle's start-up. The advantage of CVA-state space modelling is that no a priori knowledge of the system parameters, dynamics, or time-delays is required. The Canonical Variate Analysis approach was able to describe with high accuracy (typically above 90% without model optimisation) the system dynamics (see Figure 26) using only two measurement sources and provided acceptable inferential and one step ahead predictions even when the systems were operated without having reached a steady state.

Direct Methanol Fuel Cells

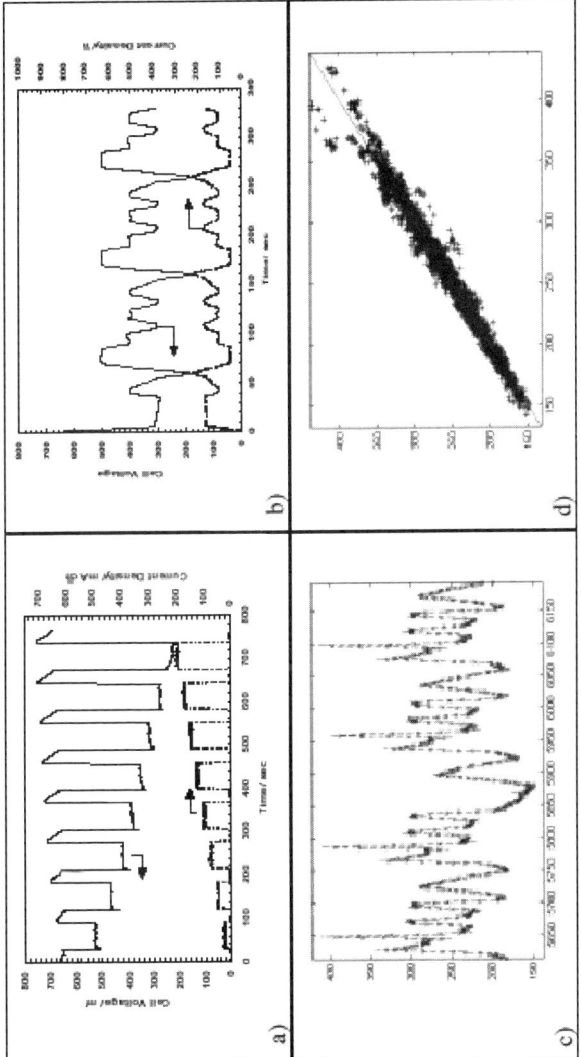

Figure 26. (a) and (b): Dynamic response of small single DMFC cells under variable load conditions. (c) and (d): Comparison of the measured and predicted 3-cell stack voltages with the aid of the CVA mode.

Although empirical and semi-empirical models are useful in modelling in general it is important to establish models that incorporate the physics and chemistry of the system to enable good prediction of behaviour over a broad range of parameters and variables.

3. Membrane Transport

Characteristics of the polymer electrolyte membrane are critical in determining cell behaviour. For the DMFC, the classic material used is the sulphonated perfluoropolymer sulphonic acid membrane, such as Nafion from DuPont, Flemion^R from Asahi Glass Co. and Aciplex from Asahi Chem. Co. Ltd. The structure, properties and applications of the perfluorinated ionomer membranes have recently been reviewed by Wirguin et al.[249] It is well known that methanol readily transports across perfluorosulfonic-acid membranes. In many cases the requirements for the ideal DMFC membrane are thus not only low, or zero, methanol crossover coupled with high ionic conductivity, but also relatively low material cost, ease of MEA fabrication and also, for example in high temperature operation, thermal, chemical and mechanical stability. Consequently, other proton conducting membranes, researched and developed for PEM cells, have been evaluated and, recently, a review on the development of ionomer membranes for fuel cells has appeared.[250]

The transport mechanisms of species such as water, methanol and protons are important in building models that can predict the cell behaviour. Springer et al., charaterised the water transport through polymer membranes in terms of drag and diffusion coefficients.[251] Bernardi and Vergrugge studied proton transport though the ionomer membrane and used a form of the Nernst-Planck equation that includes convection, diffusion and migration.[252] More comprehensive models of the transport processes have since been developed for ionomer membranes, e.g., Um et al.[253] and Gurau et al.[254]

The study of methanol crossover and other characteristics was carried out either as part of the fuel cell performance or as separate membrane characterisation tests. Verbrugge described a simple diffusion model of methanol transport through a PEM, assuming dilute solution theory.[255] Validation of the model with experimental data showed that the diffusion rate of methanol through the membrane was nearly as fast as through water and Verbrugge,[255] from experimental and theoretical results, estimated the effective diffusion coefficient to be 1.15 x10^{-5} cm^2 s^{-1} at 25 °C for Nafion 117. A second model, across a PEM has been used to explain observed experimental data for a vapour feed DMFC.[256]

Kauranen and Skou[257] used an electrochemical method for the measurement of the permeability of methanol in proton exchange membranes with 2-M sulphuric acid supporting-liquid electrolyte. They estimated that the permeability of methanol for the Nafion 117 membrane, at 60 °C, as 4.9 x 10^{-6} cm^2 s^{-1}. The permeability of methanol increased with temperature, according to an Arrhenius law, with an activation energy of approximately 12 kJ mol^{-1}. Others have measured the methanol permeability through Nafion in the absence of electrolyte and obtained comparable values, e.g., 1.72 x 10^{-6} cm^2 s^{-1} at 20 °C.

Ren et al.[258] obtained values of methanol diffusion coefficients in operating DMFCs. In fully swollen membranes, they found that diffusion coefficients varied from approximately 5 x 10^{-6} cm^2 s^{-1} to 38 10^{-6} cm^2 s^{-1} from 30 °C to 130 °C with an activation energy of 4.8 kcal mol^{-1}. The diffusion coefficients were independent of the methanol feed concentration. Narayanan et al.[259] observed an increase of the crossover rate with temperature for Nafion 117 membrane and a decrease in crossover rate with increasing current density due to an increased utilisation of methanol at high current densities.

Cruickshank and Scott[256] studied the permeation rates for water, methanol and water-methanol mixture through Nafion 117. From a simple model of the permeation of methanol in direct methanol fuel cell they estimated the effective diffusion coefficient as approximately 10^{-5} cm^2 s^{-1} for Nafion-117 membranes, at temperatures between 70 °C and 96 °C. The estimated electroosmotic drag coefficient was 0.164 MeOH/H$^+$ for the concentration of methanol used.

Dohle et al.[260] applied the method of Springer to the transport of water and methanol to model a vapour phase DMFC. Kulikovsky[261] and Sundmacher[262] used the Bernardi and Verbrugge model for transport in their DMFC model. The fluid velocity was expressed as a function of electro-potential and pressure using the Schogl's equation, and the flux of species was defined through the Nernst-Planck equation.

4. Effect of Methanol Crossover on Fuel Cell Performance

Several operating parameters that affect methanol crossover are temperature, pressure, methanol concentration, membrane, thickness and current density. The influence of higher methanol concentration, and thus greater rates of methanol crossover, is expected to effect cell voltage in two ways: reduction in the open circuit voltage i.e., increased polarisation at the cathode during practical operation and potentially a modification of the proton conductivity. In fact, several researchers have observed that the open circuit voltage decreases with increasing methanol concentration.[262,263]

This experimental behaviour has been modeled using a mixed potential model in which methanol oxidation and oxygen reduction occurs simultaneously on the cathode. The presence of methanol has a significant effect in shifting the standard potential for oxygen reduction to lower values. It was also found that the cathode electrode performance was significantly lower at higher methanol concentration.[263]

In general, increased methanol crossover is observed at higher temperature, in thinner membranes, at lower cathode side pressure and with higher methanol feed concentrations. Cruickshank and Scott[264] found that pressurising the oxygen side of the DMFC reduced the crossover of methanol, leading to higher cell voltages. Narayanan et al.[265] observed that the effect of pressure on the voltage was more significant at low rates than at high flow rates. Ren et al.[266] measured water permeation rates in operating DMFCs with Nafion 117 membranes and observed that higher pressure reduced the transport at lower current densities. In the DMFC, water drag coefficients varied from 2 to 5 over the temperature range of 20 °C to 120 °C. This water transport would thus have a significant effect on methanol transport in the DMFC.

A simple empirical model of the water transport rates obtained by Ren et al. has been developed in which water transport was by either electro-osmosis or diffusion and electro-osmosis depending upon a critical current density.[267]

In the DMFC, water transport is caused by three different factors, water required for the anode reaction, the electroosmotic water transfer through the membrane with H^+-ions and diffusion across the membrane. Water transfer starts at a finite value at zero current density (diffusion through the membrane) and rises relatively slowly with current density until a "critical current density" (j_{crit}) is reached, from where the water transfer increases linearly with current density. As load is applied to the cell, water accumulates in the cathode pores and the activity of the water effectively increases until the anode and cathode side water concentrations (or activities) are equal, after which the water transfer is by electroosmotic drag. The total water transport (flux) across the membrane is given as:

$$N'_{H_2O} = N^a_{H_2O} + N_{H_2O}$$

$$= \begin{cases} \dfrac{j}{6F} + \dfrac{\lambda_{H_2O} j}{F} + N_{H_2O}\Big|_{j=0} - \dfrac{D^m_{H_2O}}{l_m}\gamma i & if \quad j < j_{crit} \\ \dfrac{j}{6F} + \dfrac{\lambda_{H_2O} j}{F} & if \quad j > j_{crit} \end{cases} \quad (40)$$

where $\gamma = \dfrac{N_{H_2O}\big|_{j=0}}{j_{crit}} \dfrac{l_m}{D^m_{H_2O}}$.

The empirical model coefficients, λ_{H2O}, $N_{H2O}|_{j=0}$ and $j_{crit,}$ identified from data of Ren et al.[266] are given by Scott et al.[267] The total methanol flow at the anode (reaction, diffusion, and electro-osmosis) is expressed as:

$$N'_{ME} = \dfrac{j}{6F} + \dfrac{18}{\rho_{H_2O} F}\lambda_{H_2O} C^a_{ME} i + \dfrac{D^m_{ME}}{l_m} C^a_{ME} \quad (41)$$

In this equation methanol transfer flux in the MEA is brought about by reaction at the anode and transfer across the membrane by a combination of diffusion and electro-osmotic drag with water.

A model for the DMFC, which accounts for transfer of methanol across the membrane and its effect on the anode side methanol concentration and thus anode polarisation has been developed which gives agreement with experimental data.

A model of methanol transfer by diffusion and electro-osmosis, across a PEM was used to explain observed experimental data for a vapour feed DMFC.[264] This model included an empirical relationship for the influence of cathode side pressure on the transfer of methanol.

5. Mass Transport and Gas Evolution

Mass transport is a factor that generally limits the performance of solid polymer electrolyte fuel cells that operate at relatively high current densities. For hydrogen fuel cells, these mass transport limitations are predominantly associated with oxygen transport in the gas diffusion layers and in the catalyst layers. For the DMFC, mass transport limitations are also significant at the anode. For modelling, first it must be decided whether the DMFC is operated as a vapour feed system or as a liquid feed system. Furthermore, the pressure of cell operation is important, as this will have an effect on the extent, if any, of carbon dioxide gas evolution in the cell when operated with a liquid feed. For example, with a pressure of say 5–10 bar, which is not unreasonable, and operating at high solution flow rate, or low methanol utilisation, gas evolution may not occur in the cell, or may be limited to small regions in the flow channel and not occur in the catalyst of porous gas transfer backing layers.

Under approximate ambient pressure conditions, high rates of gas evolution will occur from the cell as has been observed in detailed flow visualisation studies, as shown in Figure 27.[268] Bubble nucleation in the DMFC will occur within the porous regions of the fuel cell when capillary forces and pore sizes permit. Bubble evolution behaviour was significantly affected by operating conditions of flow, pressure, temperature and current density. Direct methanol fuel cell flow beds have also been fabricated with stainless steel mesh flow beds.[269,246]

The evolution of carbon dioxide is known to influence the mass transport of methanol to the anode in the cell. In the liquid fed DMFC, mass transport limitations may arise due to counter current gas, liquid flow in the gas diffusion and catalyst layers. Limiting current densities for methanol oxidation at SPE based electrodes have been reported by Ravikumar and Shukla[263] and Kauranen and Skou.[235,270] The latter attributed this limiting current behaviour to saturation coverage of absorbed OH on the platinum surface.

The DMFC is a complex system based on porous electrocatalytic electrodes. Its operation as a liquid-feed system is complicated by the evolution of carbon dioxide gas from the anode surface of the membrane electrode assembly (MEA). For the vapour-fed DMFC, mass transfer limitations occur at the anode, similar to those for the oxygen cathode, although the cell is fed with methanol/water vapour and produces CO_2, flowing counter current to the fuel. In general the treatment of flow and mass transport in flow fields and porous backing layers will require the coupling of, and solution of, equations of motion or momentum

Direct Methanol Fuel Cells

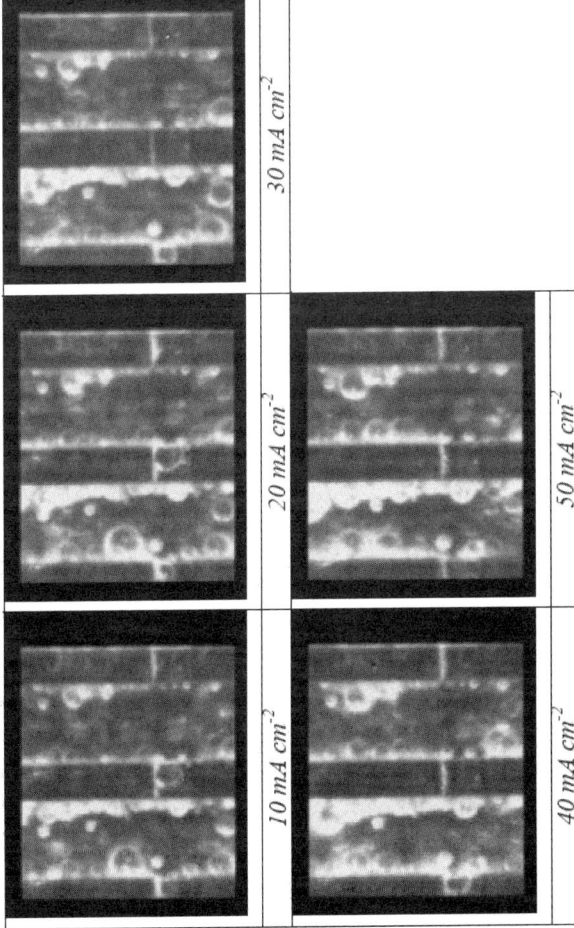

Figure 27. The effect of current density in the gas evolution patterns for the large cell. Anode inlet flow rate: 103 cm3 min-1; cell temperature: 75 °C; active area: 102.0 cm^2.

transfer, equations of continuity, equations of multi-component mass transfer, source equations for material transfer generation or consumption and boundary conditions.[271] This will enable the prediction of the effect of variation in fluid velocity on material transfer to and from the MEA in the cell and thus its effect on local electrode potentials and consequent current distribution.

For a multi-component mixture of gases the diffusion coefficient of a species in the gas mixture depends on the nature, the concentration and fluxes of the other species present. For an isothermal gas in the absence of a pressure gradient, the diffusion flux of a component i is given by the Stefan Maxwell equation as,

$$\nabla x_i = \sum_{j=1}^{n} \frac{RT}{PD_{ij}} \left(x_i N_{j,g} - x_j N_{i,g} \right) \qquad i = 1,2,...,n \qquad (42)$$

where N is the molar flux, n is the number of components, x is the mol fraction, P is the pressure, T is temperature, R is the gas constant and D_{ij}^{eff} is a binary diffusivity for the i-j pair in the porous medium.

In the gas diffusion layer, where gas can flow and diffuse, an effective diffusion coefficient, D_{ij}^{eff}, can be defined which allows for the porous structure of the layer, i.e., its porosity, ε, and tortuosity, using a Bruggeman-type relationship, e.g., $D_{ij}^{eff} = D_{ij}\varepsilon^{1.5}$.

The model considers the case of one-dimensional diffusion in the porous media, normal to the MEA plane. Pressure changes in the flow channel and porous media are assumed to be small. From the relevant equation, the influence of diffusion in a porous backing layer or flow field can be determined and the effective concentration of reactant gas, i.e., methanol and oxygen, at the edge of the catalyst layers calculated to enable estimation of electrode polarisation phenomena.

In the case of a Vapour-feed DMFC, the fuel to the anode is a dilute mixture of methanol in water vapour at a maximum molar ratio of approximately 1:27 methanol to water (2-M methanol solution). Thus, the water mole fraction will not change significantly and is assumed constant. Mass transfer of this system is described by the Stefan Maxwell equation. The assumptions, that the crossover of methanol is small in comparison to water transferred across the membrane and the effective diffusivity for methanol in water and carbon dioxide are equal, gives the mole fraction of methanol in the anode side, x_l as

$$\frac{x_{li} + \frac{B}{A}}{x_{1c} + \frac{B}{A}} = \exp\left[\frac{k_{12}l_{ca}(1+\alpha_W)j}{6F}\right] \qquad (43)$$

where, i and c refer to the catalyst interface and the flow channel interface respectively, $B/A = -(1 + \alpha_w)^{-1}$, α_w is the ratio of total water flux to the molar water flux and $k_{12} = RT/(P D_{12}^{eff})$.

At relatively high current densities the mole fraction of methanol can fall significantly below the feed value and thus illustrates why mass transport limitations are observed in the DMFC.[272] In addition, methanol mass transport limitations may also arise in the catalyst layers of the MEA.

6. DMFC Electrode Modelling

To model the electrodes in the DMFC, a one-dimensional model is typically used which contains the required elements of ionic transport, current flow, kinetics and mass transport.[271] For transport in a fuel cell, we in general must consider several interactive phenomena, which predict both dynamic and steady state behaviour. The governing equations for heat and material transfer in the diffusion and reaction layers, and electronic and ionic conduction are based on:

- material balances of gases which take into account changes in gas voidage simultaneously with gas partial pressure and allow for a change in gas volume associated with a change in temperature;
- material balance of water and methanol vapour which include the influence of condensation or evaporation of water and methanol, depending upon the saturation partial pressure and the content of water in the vapour;
- material balance of liquid water (and methanol) that includes an appropriate description for the mechanism of water transport;
- energy balance of gas and solid phase which predict the local values of temperature;
- gas transport;

- the local values of current density as computed from an appropriate kinetic equation(s), e.g., the Butler-Volmer equation;
- variation in local potential and thus current density due to ionic (proton) conduction;
- electronic conduction in the catalyst and catalyst support solid phase; and
- volumetric current balances.

In the development of a model the following assumptions are frequently adopted: temperature of the gas phase is identical to the solid phase at every position and no heat transport in the gas phase. In a DMFC, oxygen reduction at the cathode leads to water formation which, if the gas phase is fully saturated, is present as liquid. This liquid phase can change the effective porosity of the diffusion layer by part filling the pores in which gas flows. This therefore influences the gas mass transport rate in the porous structure. The extent to which the porous structure is filled with liquid depends upon the governing mechanism for water transport through the structure. In addition the transport of reactant gases and thus local partial pressures are generally affected by both Stefan-Maxwell diffusion, Knudsen diffusion and friction pressure losses. Knudsen diffusion arises when the mean free path of the gas molecules is of a similar magnitude to the pore dimensions, i.e., molecular and pore-wall interactions. Pressure changes are due to friction and can be simply modelled on the basis of laminar flow in a capillary or Poiseulle flow.

The model of the gas flow field or porous diffusion layer cannot be considered in isolation as there is clearly significant interaction with the catalyst reaction layer and also the membrane in the fuel cell. In the case of the catalyst layer the equations include chemical transformations associated with local reactions. The physico-chemical effect of, for example, oxygen consumption and water generation affects the material balances of these two species and the energy balance. The diffusion layer is described by a set of differential equations, for gas and water transport, without the source term for material generation/consumption by electrochemical reaction.

7. Cell Models

A simplified model of the DMFC has been developed by Scott et al.[264] in which the diffusion of reactant methanol vapour and oxygen are modelled in terms of an effective diffusion coefficient and in which ion (proton) transfer is modelled by an effective conductivity in the structure. Solution of the model using Butler-Völmer kinetic equations for methanol oxidation and oxygen reduction gives the current distribution in the electrocatalyst layers, from which can be determined the overall electrode polarisations in the cell. In practice, this is an oversimplified picture of coupled reactant transport and ionic movement in catalyst layers that are covered with thin layers of ionomer (Nafion) and water.

Scott et al.[262,268,269] have developed several relatively straightforward models for single phase and two phase operation in liquid feed cells and focused on the important influence of methanol diffusion which limits performance. Dohle et al.[260] have developed a one dimensional model for the vapour-feed DMFC which included the effect of methanol concentration on the cell performance and methanol crossover.

A one-dimensional model for predicting the cell voltage of the DMFC, which includes mass transfer behaviour associated with the two-phase flow in the MEA diffusion layer has been produced.[262,268] The two-phase flow model is based on capillary pressure theory and momentum balance equations for counter current gas and liquid flow and is used to determine the effective gas fraction in the porous layer. Agreement between the model and the experiment is generally good over the full range of current densities. Clearly the model is applied to the particular type of MEA used in this study and it remains to be seen whether it can be used to predict behaviour of other MEA structures and materials.

Sundmacher et al.[236] studied both the static and the dynamic response of a SPE-DMFC and showed that methanol crossover in the cell can be reduced by pulsed methanol feed. Wang and Wang[273] have developed a two dimensional model of a liquid-feed DMFC which includes diffusion and convection of gas and liquid phases in the backing layers and flow channels. The model allows for anode and cathode kinetics and methanol crossover. Anode kinetics was assumed to be zero order in methanol concentration. The influence of crossover is expressed as a parasitic current density, which affects both cell open circuit potential and cathode kinetics. The model is validated against experimental data and notably predicts the observed limiting current behaviour of the DMFC, which is said to be due to limited supply of oxygen at the cathode affected by methanol crossover.

Divisek et al.[274] have also formulated a two phase (water and gas) DMFC model using species and conservations equations as used by Wang and Wang. In the model the permeability of species is a function of *capillary saturation* that in turn depends on the capillary pressure. Mass transport between gas and liquid is modelled in terms of evaporation (or condensation) rates that are a function of surface area and temperature.[274]

A potential distribution model[275] of a liquid feed DMFC that accounts for two-phase flow and methanol crossover has been developed on the basis of Nernst-Planck equation, Stefan-Maxwell diffusion equations and Butler-Völmer kinetics. The model of the anode and the fuel cell has been shown to give good agreement to experimental data. The model incorporates an approximate analytical integration of the Butler Völmer equation over the catalyst layer to reduce the computational time required to solve the model. The model provides good predictions of anode polarisation behaviour and fuel cell operating performance (Figure 28).

8. Single Phase Flow

It is open to argument whether two-phase flow exists in the backing layers or especially in the catalyst layers. Gas formation will depend upon suitable conditions that facilitate gas bubble nucleation, in particular capillary/pore size and wettability, but also

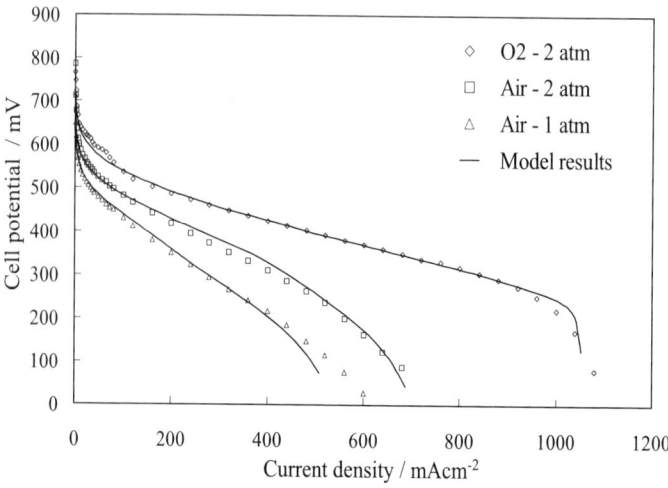

Figure 28. Cell polarization data obtained at 90 °C for 1.5-M aqueous methanol and the model results.

the cell pressure, i.e., carbon dioxide solubility. In fact, high-pressure operation has been suggested as one means of alleviating problems associated with gas evolution in the cell. In the absence of carbon dioxide gas, diffusion mass transport can be described by Stefan Maxwell equations for the three components, namely methanol, water and carbon dioxide.

Baxter et al.[276] have developed a single-phase one-dimensional model for the DMFC based on a simplified structure in which the anode comprises liquid-filled pores bounded by supported catalyst covered by a layer of ionomer. Methanol and water flow, in the ionomer "bond" layer normal to the electrode, are defined in terms of effective diffusion coefficients plus electro-osmotic drag. Carbon dioxide transfer is defined in terms of diffusion alone. Species transfer across the ionomer layer is defined by an effective mass transport coefficient. In solution, species transport is modelled using dilute solution theory with transport by effective diffusion and convection. The ionic transfer is based on the potential gradient in the bond layer with electro-osmotic drag, i.e., function of the gradients of the electrochemical potentials of protons, water and methanol.

Kinetic limitations are shown to be a dominant factor in the performance of the system and at high current density polymer electrolyte conductivity and anode thickness are important factors in determining the overall polarisation behaviour. The model predicts that the variation of methanol in the anode catalyst layer is small and thus could lead to a simpler model than used by Baxter et al.[276] The model is however not validated against experimental data and does not consider mass transport of species in other regions of the electrode assembly.

Nordlund and Lindbergh[239] developed an agglomerate model for the porous DMFC anode using kinetic expressions for methanol oxidation based on the formation of adsorbed methanol, CO and OH species. The transport of methanol in the spherical agglomerate was described by radial (Fick's) diffusion. It was shown that the mass transport does not limit cell performance but that liquid phase mass transport was of importance at lower methanol concentrations.

9. Two- and Three-Dimensional Modelling

Kulikovsky et al.[277] modelled a vapour-feed DMFC in two dimensions and simulated the current distribution and concentration distribution in the various layers of the membrane electrode assembly. The model was based on the Stefan Maxwell and Knudsen diffusion mechanism for gas transport in the backing layers, the continuity equations for the gases, Poisson's equation for potentials of the carbon catalyst support phase and membrane phase and Butler-Völmer kinetics. The influence of hydrodynamics was not considered.

The model showed that there were two-dimensional distributions in oxygen concentration, overpotential and current density. The model identified potential dead zones in the active layers in front of the fuel channel where the reaction rate was small. Current density was also higher at positions of the electrocatalyst layer facing the edges of the current collectors. It has been shown that it is theoretically possible to remove up to 50% of the electrocatalyst from the areas facing the central parts of the flow channels without significant loss in power performance. From an analysis of the problem of current distribution with standard parallel channel flow fields, it was proposed that, to prevent the shielding of catalyst layers by the flow field current collectors, embedded current collectors be positioned inside the backing and catalyst layers.[277] Thus, current collection is normal to the flux of protons into the membrane and results in a uniform current distribution.

Kulikovsky[278] also modelled a liquid-feed DMFC based on methanol transport through the liquid phase and in the hydrophilic pores of the anode-backing layer, but ignored the effect of methanol crossover.

The flow in fuel and oxidant supply channels of fuel cells is usually laminar unless high stoichiometric excess of fuel or oxidant is used. Hence, as an approximation, the flow in porous flow-fields or in porous backing layers can also be considered to be laminar. The influence of hydrodynamics in the flow-fields is to change the local values of flow velocity, which has a direct influence on the mass transport or diffusion flux of species. In general, this variation in velocity occurs in three dimensions, which can result in a three dimensional variation in diffusion. Consequently, in the DMFC, we can expect that there will be multi-dimensional variation in local reactant gas partial pressure and thus local current density. However, this is a difficult aspect to model and at least initially model development has focused on one-dimensional systems.

10. Dynamics and Modelling

There is little information published on the time varying performance of the DMFC except with regard to stability studies. The effect of current pulsing on performance has been reported.[279] More detailed studies of the dynamic voltage response under varied current loads have been also reported for small and large scale cells[280,281] under a range of different operating conditions (see Figure 26). The cell responds rapidly and reversibly to changes in magnitude and rate of change of load. Under dynamic operation, the cell voltage response can be significantly better than that achieved under steady state operation. Open circuit potentials are also increased, by up to 100 mV, by imposing a dynamic loading strategy. In addition, the study reports the dynamic characteristics of a large-scale cell and cell stack, and explains the differences in cell response.

Modelling of the dynamic behaviour of the DMFC has been limited to only a few studies, although, in principle, most steady state models can be readily adopted for dynamic simulation by introducing time derivatives as indicated in Section VII.7. Dynamics are important from the point of assessing cell and system stability to fluctuation in variables as well as control. Sundmacher et al. have extended their steady state models to simulate dynamic operation.[282,285] Through simulation of the pulsing of methanol feed solution concentration, it was shown that an enhanced cell response (increased cell potential) was maintained as shown in Figure 29. This enhancement, confirmed experimentally, was due to the reduction in the impact of methanol crossover on oxygen reduction. The dynamic model was also used to simulate the operation of the DMFC in a vehicular application.

An empirical model based on a Canonical Variate Analysis (CVA) state space representation has also been developed to predict the dynamic voltage response of the DMFC and multi cell stacks.[247,248]

11. Stack Hydraulic and Thermal Models

A model for the liquid feed, direct methanol fuel cell (DMFC), based on the homogeneous two-phase flow theory and mass conservation equation, which describes the hydraulic behaviour of internally manifolded cell stacks has been developed.[283] The model predicts the pressure drop behaviour of an individual DMFC cell and is used to calculate flow distribution through fuel cell stack

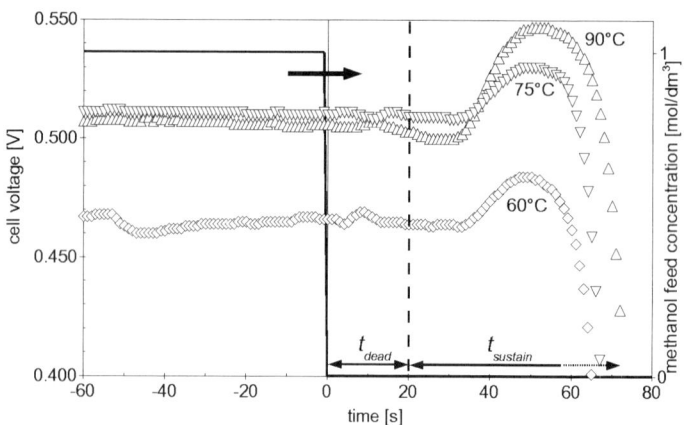

Figure 29. The effect of pulsed methanol-solution flow on DMFC performance. From Shultz (with permission of the author). [285]

internal manifolds. A mathematical model of large scale liquid feed, direct methanol fuel (DMFC) cell stacks which describes the thermal, hydraulic and chemical equilibrium performance has also been presented.[284] The model incorporates two-phase flow, pressure drop behaviour with internal thermal and heat transfer characteristics and chemical equilibria to calculate the flow distribution to individual cells and the temperature distribution throughout the stack. The model provided estimations of the stack behaviour and partly explains the variation of individual cell voltage observed in DMFC cell stacks.

VIII. CONCLUSIONS

Direct methanol fuel cells have reached a high level of development and are now almost universally referred to as the 6th-fuel cell type. In terms of applications they are set to function as power sources for a range of mobile applications. This position has largely been brought about by the convenience of storage of the liquid fuel. It is true however to say that the power performance of the DMFC does not match that of the sister fuel cell using hydrogen as fuel. Even with an order of magnitude greater catalyst loading (cf. to hydrogen PEMFC) the power performance of the DMFC is at best half that of the PEMFC. This makes it unlikely that the DMFC will see applications in transportation except perhaps in niche area.

For an expansion in the applications of the DMFC the search for improved electrocatalysts for the anode and for the cathode where methanol tolerance is preferred must continue. In conjunction with these electrocatalyst developments, further improvements in membrane materials are quite urgently required. One potential way of overcoming some of the power limitations of the DMFC is to operate at much higher temperatures than currently used (ca. 200 °C) to promote much more rapid oxidation and reduction kinetics. This change will require the use of temperature stable membranes.

Of the more recent developments in DMFCs there is great potential in the use of mixed reactant systems, largely because of the great simplicity in fuel cell stack design that will result in much reduced cell stack costs.

LIST OF SYMBOLS

D_{ij}^{eff} effective binary diffusivity coefficient of the i-j gas pair
E_0^o standard state reference potential
C_i concentration of species i
C_{me} concentration of methanol
D_{ij} binary diffusion coefficient between components i and j
E^{cell} cell voltage
F Faraday's constant
j local current density
j_0 kinetic current density
j exchange current density
k_{lim} rate constant
K_i limiting current density
m equilibrium constant of step i
N membrane thickness
N_j number of electrons involved in reaction
P molar flux of the component j
r_i pressure
R reaction rate of step i
T_i universal gas constant
X temperature
    mole fraction of species i

Superscripts and Subscripts

A   anode

crit   critical values
eff   effective (i.e., average) conditions
g   gas
H$_2$O water
me   methanol
m   membrane

Greek Symbols

λ   electroosmotic drag coefficient (number of water molecules carried per proton)
α   charge transfer coefficient
α$_w$ ratiuo of total water flow to molar water flux
β   exponent in Eq. (7)
∇   gradient or spatial derivative
ε   volume fraction of pore (gas) volume
η   overvoltage

REFERENCES

[1] J. Larminie and A. Dicks. *Fuel Cell Systems Explained,* Wiley, New York, 2000.
[2] K. Kordesch and G. Simader, *Fuel Cells and their Applications*, VCH, Weinheim, 1994.
[3] A. Lamm and J. Muller, System design for transport applications, in *Handbook of Fuel Cell: Fundamentals, Technology and Applications,* Vol 4, Ed. by W. Vielstich, A Lamm, and H. A. Gasteiger, J. Wiley and Sons, England, 2003, Chapter 64.
[4] A. S. Aricò, S. Srinivasan, and V. Antonucci, *Fuel Cells* **1** (2001) 1.
[5] V. Klouz, V. Fierro, P. Denton, H. Katz, J. P. Lisse, S. Bouvot-Mauduit, and C. Mirodatos, *J. Power Sources* **105** (2002) 26.
[6] S. C. Barton, T. Patterson, E. Wang, T. F. Fuller, and A. C. West, *J. Power Sources* **96** (2001) 329.
[7] M. A. Priestnall, V. P. Kotzeva, D. J. Fish, and E. M. Nilsson, *J. Power Sources* **106** (2002) 21.
[8] A. K. Shukla, C. L. Jackson, K. Scott and G. Murgia, *J. Power Sources* **111** (2002) 43.
[9] A. K. Shukla , K. Scott, C. L. Jackson, and W. R. Meuleman, *J. Electrochem. Soc.* **150** (2003) A12131.
[10] A. K. Shukla and R. K. Raman, *Annu. Rev. Mater. Res.* **33** (2003) 155.
[11] A. K. Shukla, C. L. Jackson, K. Scott, and R. K. Raman, *Electrochim. Acta* **47** (2002) 3401.
[12] N. Munichandraiah, K. McGrath, G. K. S. Prakash, R. Aniszfeld, and G. A. Olah, *J. Power Sources* **117** (2003) 98.
[13] B. Gurau and E. S. Smotkin, *J. Power Sources* 112 (2002) 339.
[14] A. Heinzel and V. M. Barragán, *J. Power Sources* **84** (1999) 70.
[15] G. Murgia, L. Pisani, A. K. Shukla and K. Scott, unpublished information.
[16] P. S. Kauranen, E. Skou, and J. Munk, *J. Electroanal. Chem.* **404** (1996) 1.
[17] V. S. Bagotzky and Yu. B. Vassilyer, *Electrochim. Acta* **12** (1967) 1323.

18. W. H. Lizcano-Valbuena, V. A. Paganin, and E. R. Gonzalez. *Electrochim. Acta* **47** (2002) 3715.
19. K. L. Ley, R. Liu, C. Pu, Q. Fan, N. Leyarovska, C. Serge, and E. S. Smotkin, *J. Electrochem. Soc.* **144** (1997) 1543.
20. A. K. Shukla, M. K. Ravikumar, A. S. Aricò, G. Candiano, V. Antonucci, N. Giordano, and A. Hamnett, *J. Appl. Electrochem.* **25** (1995) 528.
21. A. S. Aricò, H. Kim, V. Antonucci, A. K. Shukla, M. K. Ravikumar and N. Giordano, *Electrochim. Acta* **39** (1994) 691.
22. A. S. Aricò, V. Antonucci, N. Giordano, A. K. Shukla, M. K. Ravikumar, A. Roy, S. R. Barman and D. D. Sarma, *J. Power Sources* **50** (1994) 295.
23. M. R. Andrew, B. D. McNichol, R. T. Short and J. S. Dury, *J. Appl. Electrochem.* **7** (1977).
24. D. A.Landsman and F. J. Luczak, Report FCR-3463 United Technologies, Power Systems, 1981.
25. V. B. Hughes, B. D. McNicol M. R. Andrew R. B. Jones, and R. T. Short, *J. Appl. Electrochem.* **7** (1977) 161.
26. A. Adams and R. T. Foley, *J. Electrochem. Soc.* **127** (1980) 2646.
27. G. T. Burnstein, C. J. Barnett, A. R. Kucernak, and K. R. Williams. *Catalysis Today* **38** (1997) 425.
28. E. H. Yu, K. Scott, and R. W. Reeve, *Journal of Electroanalytical Chemistry*, **547** (2003) 17.
29. B. D. McNichol, D. A. J. Rand, and K. R. Williams, *J. Power Sources* **83** (1999) 15.
30. S. Wasmus and A. Kuver, *J. Electroanal. Chem.* **461** (1999) 14.
31. A. V. Tripkovic, K. D. Popovic, J. D. Momcilovic, and D. M. Drazic, *Electrochimica Acta.* **44** (1998) 1135.
32. D. M. Drazic, A. V. Tripkovic, K. D. Popovic, and J. D. Lovic, *J. Electroanal. Chem.* **466** (1999) 155.
33. J. Prabhuram, R. Manoharan, and H. N. Vasan, *J. Appl. Electrochem.* **28** (1998) 935.
34. A. Kowal, S. N. Port, and R. J. Nichol, *Catalysis Today* **38** (1997) 483.
35. A. Ciszewski and G. Milczarek, *J. Electroanal. Chem.* **413** (1996) 137.
36. R. Manoharan and J. Prabhuram, *J. Power Sources.* **96** (2001) 220.
37. E. H. Yu and K. Scott, *Journal of Power Sources*, **137** (2004) 248.
38. Y. Takasu, T, Fujiwara, Y. Murakami, K. Sasaki, M. Oguri, T. Asaki and W. Sugimoto, *J Electrochem. Soc.* **147** (2000) 4427.
39. C. W. Hills, M. S. Nashner, A. I. Frenkel, J. R. Shapley, and R. G. Nuzzo, *Langmuir* **15** (1999) 690.
40. H. Bönneman, R. Brinkmann, P. Britz, U. Endruschat, R. Mortel, U. A. Paulus, and G. J. Feldmeyer, *J. New Mat. Electrochem. System*s **3** (2000) 199.
41. D. L. Boxall, G. A. Delugga, e. A. Kenik, W. D. King and C. M. Lukehart, *Chem. Mater.* **13** (2001) 891.
42. H. Choi, H. S. Kim, and T. H. Lee, *J Power Sources* **75** (1998) 230.
43. M. C. Kimble, K. D. Jayne, A. S. Woodman and E. B. Anderson, Development of corrosion-resistant coatings for fuel cell bipolar plates, *American Electroplaters and Surface Finishers Society (AESF), Proceedings AESF SUR/FIN'99*, **6** (1999) 717.
44. M. S. Loffler, H. Natter, R. Hempelmann, Th. Krajewski, and J. Divisek, *J. Phys. Chem. Chem. Phys.* **3** (2001) 333.
45. C. Coutanceau, A. F. Rakotondrainibe, and A. Lima, *J. Appl. Electrochem.* **34** (2004) 61.

46. G. L. Troughton, Ph.D. Thesis Newcastle upon Tyne University, 1993.
47. A. S. Arico, P. Creti, N. Giordana, V. Antonucci, P. L. Antonucci, and A. Chuvilin, *J. Appl Electrochem.* **26** (1996) 959.
48. M. Gotz and H. Wendt, *Electrochimica Acta.* **43** (1998) 3637.
49. Y. Zhu and C. R. Cabrera, *Electrochem. and Solid State Letters* **4** (2001) A45.
50. H. Zhang, Y. Wang, E. Fachini E C. R. Caberra, *Electrochem and Solid State letters* **2** (1999) 437.
51. K. L. Ley, R. Liu, C. Pu, Q. Fan, N. Leyarovska, C. Segre, and E. S. Smotkin, *J. Electrochem. Soc.* **144** (1997) 1543.
52. R. Liu, H. Iddir, Q. Fan, G. Hou, A. Bo, K. Ley, E. Smotkin, Y. Sung, H. Kim, S. Thomas and A. Wieckowski, *J Phys. Chem. B.* **104** (2000) 3518.
53. B. Gurua, R. Viswananthan, R. Liu, T. J. Lanfrenz, K. L. Ley, E. S. Smotkin, E. Reddington, A. Sapienza, B. C. Chan, and T. E. Mallouk, *J. Phys. Chem.* **102** (1998) 9997.
54. K. Lasch, L. Jorissen and J. Garche, *J. Power Sources* **84** (1999) 225.
55. J. H. Choi, K. W. Park, B. K. Kwon and Y. E. Sung, *J. Electrochem. Soc.* **150** (2003) 973.
56. A. S. Arico, V. Baglio, E. Modica, A. Di Blasi and V. Antonucci, *Electrochem. Comm* 2004, **6** 164.
57. P. V. Samant, and J. B. Fernandes, *J. Power Sources* **125** (2004) 172.
58. T. Okada, Y. Suzuki, T. Hirose, and T. Ozawa, *Electrochim. Acta* **49** (2004) 385.
59. E. S. Steigerwalt, G. A. Deluga, and C. M. Lukehart, *J. Nanosci. Nanotechnol.* **3** (2003) 247.
60. W. Z. Li, C. H. Liang, W. J. Zhou, J. S. Qiu, Z. H. Zhou, G. Q. Sun and Q Xin, *J. Phys. Chem. B* **107** (2003), 6292.
61. W. Z. Li, C. H. Liang, W. J. Zhou, J. S. Qiu, H. Q. Li, G. Q. Sun, and Q. Xin, *Carbon* **42** (2004) 436–439.
62. B. Rajesh, V. Karthik, S. Karthikeyan, K. R. Thampi, J. M. Bonard, and B. Viswanathan, *Fuel* **81** (2002).
63. N. Fujiwara, K.Yasuda, T. Ioroi, Z. Siroma, and Y. Miyazaki, *Electrochim. Acta* **47** (2002) 4079.
64. C. Lim, R. G. Allen, K. Scott, and S. Roy. *J. of Applied Electrochemistry*, **34** (2004) 929.
65. F. Ficicioglu and F. Kadirgan, *J. Electroanal. Chem.* **431** (1997) 179.
66. M. F. Strike M. De Rooij H. Kouldelka, M. Ulmann, and J. Augustynskki, *J. Appl. Electrochem.* **22** (1992) 922.
67. M. Hepel. *J. Electrochem. Soc.* **145** (1998) 124.
68. K. H. Xue, C. X, Cai , H.Yang , Y.M. Zhou, S. G. Sun, S. P. Chen, and G. Xu, *J. Power Sources* **75** (1998) 207.
69. S. Swathirajan and Y. M. Mikhail, *J. Electrochem. Soc.* **139** (1992) 2105.
70. K. Bouzek, K. M. Mangold and K. Juttner, *J. Appl. Electrochem.* **31** (2001) 501.
71. B. Rajeshh and B. Viswananthan, *Indian J. Chem.* **39**A (2000) 826.
72. W. T. Napporn, H. Lahorde J. M. Leger, and C Lamy, *J. Electroanal. Chem.* **404** (1996) 153.
73. A. M. Castro Luna, *J Appl. Electrochem.* **30** (2000) 1137.
74. A. Kitani, T. Akashi, K. Sugimoto, and S. Ito, *Synthetic Metals* 121 (2001) 1301.
75. A. Lima, C. Coutanceau, J. M. Leger, and C. Lamy, *J. Appl. Electrochem.* **31** (2001) 379.
76. M. C. Lefebvre, Z. Qi, and P. G. Pickup, *J. Electrochem. Soc.* **46** (1999) 2054.
77. Z. Qi and P. G. Pickup, *Chem. Commun.* (1998) 15.
78. N. Jia, M. C. Lefebvre, J. Halfyard, Z. Qi, and P. G. Pickup, *Electrochem. Solid State Letters* **3** (2000) 529.

[79] I. Bercerik and F. J. Kadirgan, *J. Electrochem. Soc.* **148** (2001) D49.
[80] O. Enea and A. H. McEvoy, *Electrochimica Acta* **36** (1998) 1441.
[81] C. K. Witham, W. Chun, T. I. Valdez, and S. R. Narayanan, *Electrochem. and Solid State Letters*. **3** (2000) 497.
[82] N. M. Marković, T. J. Schmidt, V. Stamenković and P. N. Ross, *Fuel Cells* **1**(2) (2000)105.
[83] L. Geines, R. Faure, and R. Durand, *Electrochim. Acta* 44 (1998) 1317.
[84] T. Toda, H. Igarashi, H. Uchida, and M. Watanabe, *J. Electrochem. Soc.* **146** (1999) 3750.
[85] T. Toda, H. Igarashi, H. Uchida, and M. Watanabe, *J. Electrochem. Soc.* **145** (1998) 4185.
[86] S. Mukerjee, S. Srinivasan, M. P. Soriaga and J. McBreen, *J. Electrochem. Soc.* **142** (1995) 1409.
[87] S. Mukerjee, S. Srinivasan, M. P. Soriaga and J. McBreen, *J. Phys. Chem.* **99** (1995) 4577.
[88] S. Štrbac, R. R. Adžić, 1996. *J. Electroanal. Chem.* **403** (1996)169.
[89] S. Štrbac, R. R. Adžić, *Electrochim. Acta* **39** (1994) 983.
[90] R. Jasinski, *J. Electrochem. Soc.* **112** (1965) 526.
[91] G. Q. Sun, J. T. Wang, and R. F. Savinell, *J. Appl. Electrochem.* **28** (1998) 1087.
[92] S. Gupta, D. Tryk, S. K. Zecevic, W. Aldred, D. Guo, and R. F. Savinell, *J. Appl. Electro chem.* **28** (1998) 673.
[93] S. L. Gojkoić, S. Gupta, and R.F. Savinell, *J. Electroanal. Chem.* **462** (1999) 63.
[94] H. Kalvelage, A. Mecklenburg, U. Kunz, and U. Hoffmann, *Chem. Eng. Technol.* **23** (2000) 803.
[95] J. B. Goodenough, *Prog. Solid State Chem.* **5** (1971) 39.
[96] S. Trassatti (Ed.), *Electrodes of Conducting Metallic Oxides,* Part B, Elsevier, Amsterdam, 1981.
[97] M. R. Tarasevich and B. N. Efremov, in *Electrodes of Conducting Metallic Oxides,* Part A, Ed. by S. Trassatti, Elsevier, Amsterdam, 1981, pp. 221-60.
[98] H. Tamura, H. Yoneyama and Y. Matsumoto, in *Electrodes of Conducting Metallic Oxides,* Part A, Ed. by S Trassatti, Elsevier, Amsterdam, 1981, pp. 261-300.
[99] J. B. Goodenough, A. K. Shukla, C. Paliterio, K. R. Jamieson, A. Hamnett, and R. Manoharan, British patent 842254 (1985).
[100] H.C. Chiu and A.C.C. Tseung, *Electrochem. Solid State Lett.* **2** (1999) 379.
[101] L. Gurban, A. Teze and G. Hervé, *C.R. Acad. Sci.* II C **1** (1998) 397.
[102] O. Savadogo and P. Beck, *J. Electrochem. Soc.* **143** (1996) 3842.
[103] V. Trapp, P. A. Christensen, and A. Hamnett, *Faraday Trans.* **92** (1996) 4311.
[104] C. Fischer, N. Alonso-Vante, S. Fiechter, and H. Tributsch, *J. Appl. Electrochem.* **25** (1995) 1004.
[105] R. W. Reeve, P. A. Christensen, A. Hamnett, S. A. Haydock, and S. C. Roy, *J. Electrochem. Soc.* **145** (1998) 3463.
[106] P. J. Sebastian, F. J. Rodriguez, O. Solorza, and R. Rivera, *J. New Mater. Electrochem. Syst.* **2** (1999) 115.
[107] O. Solorza-Feria, S. Citalán-Cigarroa, R.Rivera-Noriega, and S. M. Fernández-Valverde, *Electrochem. Commun.* **1** (1999) 585.
[108] S. Durón, R. Rivera-Noriega, M. A. Leyva, P. Nkeng, G. Poillerat and O. Solorza-Feria. *J. Solid State Electrochem.* **4** (2000) 70.
[109] S. Durón, R. Rivera-Noriega, G. Poillerat, and O. Solorza-Feria, *J. New Mater. Electrochem. Syst.* **4** (2001) 17.
[110] F. J. Rodriguez and P. J. Sebastian, *Int. J. Hydrogen Energy* **25** (2000) 243.
[111] N. Alanso-Vante, P. Bogdanoff, and H. Tributsch, *J. Catalysis*. **190** (2000) 240.

[112] M. Bron, P. Bogdanoff, S. Fiechter, I. Dorbandt, M. Hilgendorff, H. Schulenburg, and H. Tributsch, *J. Electroanal. Chem.* **500** (2001) 510.
[113] V. Le Rhan and N. Alonso-Vante, *J. New Materials for Electro Chemical System.* **3** (2000) 331.
[114] H. Tributsch, M. Bron, M. Hilgendorff, H. Schulenburg, I. Dorbandt, V. Eyert, P. Bogdanoff, and S. Fiechter. *J. Appl. Electrochem.* **31** (2001) 739.
[115] N. Alonso-Vante, B. Schubert, and H. Tributsch, *Materials Chemistry and Physics* **22** (1989) 281.
[116] B. Bittins-Cattaneo, S. Wasmus, B. Lopez-Mishima, and W. Vielstich, *J. Appl. Electrochem.* **23** (1993) 625.
[117] S. L. Gojkovic, S. Gupta, and R. F. Savinell, *J. Electroanal. Chem.* **462** (1999) 63.
[118] M. Neergat, A. K. Shukla, and K. S. Gandhi, *J Appl. Electrochem.* **31** (2001) 373.
[119] H. G. Petrow and R. J. Allen, U. S. Patent 3992331 (1976).
[120] H. Bönnemann and W. Brijoux, *Angew. Chem. Int. Ed. Engl.* **30** (1991) 1312.
[121] F. Maillard, M. Martin, F. Gloaguen, and J-M. Léger, *Electrochim. Acta* **47** (2002) 3431.
[122] M. K. Ravikumar and A. K. Shukla, J. Electrochem. Soc. **143** (1996) 2601.
[123] H. G. Petrow and R. J. Allen, US Patent 3992331 (1976).
[124] H. G. Petrow and R. J. Allen, US Patent 3992512 (1976).
[125] H. G. Petrow and R. J. Allen, US Patent 4044193 (1975).
[126] Z. Wei, H. Guo, and Z. Tang, *J. Power Sources* **62** (1996) 233.
[127] S. Sun, C. B. Murray, D. Weller, L. Folks, and A. Moser, *Science* **287** (2000) 1989.
[128] A. K. Shukla, M. Neergat, P. Bera, V. Jayaram, and M. S. Hegde, *J. Electroanal. Chem.* **504** (2001) 111.
[129] M. Neergat, A. K. Shukla, and K. S. Gandhi, *J. Appl. Electrochem.* **31** (2001) 373.
[130] S. Mukerjee and S. Srinivasan, *J. Electroanal. Chem.* **357** (1993) 201.
[131] D. Chu and R. Jiang. U. S. Patent 6245707 (2001).
[132] P. Stevens, A. K. Shukla, and A. Hamnett, *I. Chem. E. Symp. Ser.* **112** (1989) 141.
[133] M. Bron, S. Fiechter, M. Hilgendroff, and P. Bogdanoff, *J. Appl. Electrochem.* **32** (2002) 11.
[134] P. Hagenmuller, *Preparative Methods in Solid State Chemistry*, Academic Press, New York, 1972.
[135] P. Bogdanoff, S. Fiechter, H. Tributsch, M. Bron, and I. Dorbandt, German Patent 10035841-A1 (1999).
[136] H. Uchida, H. Ozuka and M. Watanabe, *Electrochim. Acta* **47** (2002) 3629.
[137] A. K. Shukla, R. K. Raman, N. A. Choudhury, K. R. Priolkar, P. R. Sarode, S. Emura, and R. Kumashiro, *J. Electroanal. Chem.* **563** (2004) 181.
[138] X. Ren, T. E. Springer T. A. Zawodzinski, and S. Gottesfeld, *J. Electrochem. Soc.* **147** (2000) 466.
[139] M. Gil, X. L. Ji, X. F. Li, H. Na, J. E. Hampsey, and Y. F. Lu, *J. Membr. Sci.* **234** (2004) 75.
[140] L. Li, L. Xu, and Y. X. Wang, *Acta Polym. Sin.* **Aug. 4th** (2003) 465.
[141] B. R. Einsla, Y. T. Hong, Y. S. Kim, F. Wang, N. Gunduz, and J. E. Mcgrath, *J. Polym. Sci. Pol. Chem.* **42** (2004) 862.
[142] N Miyake, J. S. Wainright and R. F. Savinell, *J. Electrochem. Soc.* **148** (2001) A905.

143. J. M. Thomassin, C. Pagnoulle, D. Bizzari, G. Caldarella, A. Germain, and R. E. Jerome, *Polymers* **46** (2005) 11389.
144. A. S. Arico, V. Baglio, V Di Blasio, A. Di Blasi, E. Modica, P. L. Antonucci mand V. Antonucci, *J. Power Sources* **128** (2004) 113.
145. D. H. Jung, S. Y. Cho, D. H. Peck, D. R. Shin, and J. S. Kim, *J. Power Sources* **118** (2003) 205.
146. E. B. Easton, B. L. Langsdorf, J. A. Hughes, J. Sultan, Z. G. Qi, A. Kaufman, and P. G. Pickup, *J. Electrochem. Soc.* **150** (2003) C735.
147. Y. S. Park, M. Y. Jang, T. Hatae, H. Itoh, and Y. Yamazaki, *Electrochemistry* **72** (2004) 165.
148. Z. Florjanczyk, E. Wielgus-Barry, and Z. Poltarzewski, in *Solid State Ionics* **145** (2001) 119.
149. S. W. Li and M. L. Liu, *Electrochim. Acta*, **48** (2003) 4271.
150. B. Libby, W. H. Smyrl, and E. L. Cussler, *AICHE J.*49 (2003) 991.
151. M.A. Navarra, S. Materazzi, S. Panero, and B. Scrosati, *J. Electrochem. Soc.* **150** (2003) A1528.
152. A. S. Arico, V. Baglio, P. Creti, A. Di Blasi, V. Antonucci, J.Brunea, A. Chapotot, A. Bozzi, and J. Schoemans, *J. Power Sources* **123** (2003) 107.
153. P. Argyropoulos, W. M. Tamma, and K. Scott, *J. of Membrane Science* **171** (2000) 119.
154. T. Hatanaka, N. Hasegawa, A. Kamiya, M. Kawasumi, Y. Morimoto, and K. Kawahara, *Fuel* **81** (2002)2173.
155. B. Bae and D. Kim, *J. Membr. Sci.* 220 (2003) 75.
156. H. Sawa and Y. Shimada, *Electrochemistry* **72** (2004) 111.
157. R. Carter, R. Wycisk, H. Yoo, and P. Pintauro, *Electrochem. Solid State Lett.* **5** (2002) A195.
158. J. Weston, E. Chalkova, M. Hofmann, C. M. Ambler, H. R. Allcock, and S. N. Lvov, *Electrochim. Acta* **48** (2003) 2173.
159. L. Li, L. Xu, and Y. X. Wang, *Chem. J. Chin. Univ.-Chin.* **25** (2004) 388.
160. H. Y. Chang and C. W. Lin, *J. Membr. Sci.* **218** (2003) 295.
161. Th Dippel, K. D. Kreuer, J. C. Lassegeus, and D Rodriuez, *Solid State Ionics* **61** (1993) 41.
162. L J. Hobson,Y Nakano, H. Ozu, and S. Hayase, *J. Power Sources* **104** (2002) 79.
163. X. Z. Fu, J. Li, C. H. Li, and D. W. Liao, Prog. *Chem.*16 (2004) 77.
164. E Yu, Ph.D. Thesis, University of Newcastle upon Tyne, UK, 2003.
165. T. N. Danks, R. C. T. Slade, and J. R. Varcoe, *J. Mater. Chem.* **12** (2002) 3371.
166. H. Herman, R. C. T. Slade, and J. R. Varcoe, *J. Mater. Chem* **218** (2003).
167. S. Surampudi, S. R. Narayanan, E. Vamos, H. Frank, G. Halpert A., LaConti, J. Kosek G. K., SuryaPrakash, and G A. Olah, *J. Power Sources* **47** (1994) 377.
168. H. Dohle, T. Brewer, J. Mergel, J. Neitzel, and D. Stolten D. *Fuel Cell Seminar*, Portland, Oregon, 2000.
169. H. Dohle, J. Divisek, and R. Jung, *J. Power Sources.* **86** (2000) 469.
170. J. Cruickshank and K. Scott, *J. Power Sources* **70** (1998) 40.
171. H. Grune, G. Kruftand, M. Waidaus, in *Fuel Cell Seminar*, San Diego, CA, Nov. 28-Dec 1 (1994), Abstracts pp. 474-478.
172. M. Baldauf and W. Preidel, *J. Power Sources* **84** (1999) 161.
173. S. Hikita, K. Yamane, and Y. Nakajima, *Japan Society of Automotive Engineers* review **22** (2001) 151.
174. X. Ren, M. S. Wilson, and S. Gottesfeld, *J. Electrochem. Soc.* **143** (1996) L12.
175. M. Hogarth, P. Christensen, A. Hamnett, and A. Shukla, *J. Power Sources* **9** (1997) 125.

[176] A. K. Shukla, M. K. Ravikumar, A. S. Arico, G. Candiano, V. Antonicci, N. Giargano, and A. Hamnett *J. Appl. Electrochem.* **25** (1995) 528.
[177] A. S. Arico, P. Creti, P. L. Antonucci, and V. Antonucci, *Electrochem. and Solid State letters* **1** (1998) 66.
[178] S. R. Narayanan A. Kindler, W. Chun, B. Jeffries-Nakamura, H. Frank, G Halpert, and S Surampudi , in *Proton Conducting Membrane Fuel Cells I*, Ed. by S. Gottesfeld, G. Halpert, and A. Langrebe, The Electrocehm. Soc. Proceedings Series, Pennington, NJ, 1999, p261.
[179] X. Ren, P. Zelenay, S. Thomas, J. Davey, and S. Gottesfeld, *J. Power Sources* **86** (2000) 111.
[180] X. Ren, S. C. Thomas P., Zelenay, and S. Gottesfeld,. *Fuel Cell Seminar*, Portland, Oregon 2000.
[181] H. Dohle J. Mergel H. Scharmann, and H. Schmitz, in *Direct Methanol Fuel Cells*, Ed. by S. R. Narayanan, S. Gottesfeld, and T. Zawodzinski, Electrochem Soc., NJ, 2001.
[182] M. P. Hogarth and G. A. Hards, *Platinum Metals Rev.* **40** (1996) 150.
[183] A. S. Arico, V. Baglio, P. Creti and V. Antonucci. *Fuel Cell Seminar*, Orlando 2000.
[184] D. P. Wilkinson and A. E. Steck, in *Proceedings of 2nd Int. Symp. New Materials for Fuel Cell and Modern Battery Systems,* Montreal, July 6-10, 1997, p. 27.
[185] S. R. Narayanan, T. I. Valdez and F. Clara, *Fuel Cell Seminar*, Portland, Oregon, 2000, p. 795.
[186] T. I. Valdez, S. R. Narayanan, H. Frank and W. Chun, in *Proc. 12th Annual battery Conference on Applications and Advances*, Long Beach CA 14-17, Jan1997, p. 239.
[187] S. Gottesfeld, X. Ren, P. Zelenay, H. Dinh, F. Guyon, and J. Davey, *Fuel Cell Seminar* Portland, Oregon, 2000, p. 799.
[188] R. G. Hockaday, M. DeJohn, C. Nava, P. S. Turner, H. L. Vaz, and L. Luke Vazul, *Fuel Cell Seminar*, Portland, Oregon, 2000,p. 791.
[189] H. Y. Ha, S. Y. Cha, I. H. Oh and S. A. Hong, *Fuel Cell Seminar*, Portland, Oregon, 2000, p. 175.
[190] L. Liu, C. Pu, R. Viswanathan, Q. Fan, R. Liu, and E. S. Smotkin. *Electrochimica Acta*, **43** (1998) 3657.
[191] A. K. Shukla, M.K. Ravikumar, M. Neergat ,and K. S. Gandhi, *J. Appl. Electrochem.* **29** (1999) 129.
[192] D. H. Jung, C. H. Lee, C. S. Kim, and D. R. Shin, *J. Power Sources* **71** (1998) 169.
[193] D. Buttin, M. Dupont, M. Straumann, R. Gille, J-C.Dubois, R. Ornelas, G. P. Fleba, E. Ramuni, V. Antonucci, A. S. Arico, P. Creti, E. Modica, M. Pham-Thi, and J-P. Ganne, *J Appl. Electrochem.* **31** (2001) 275.
[194] K. Scott, P. Argyropoulos, and W. M. Taama, *Trans. I. Chem. E.* **78** (2000) 881.
[195] M. Baldauf and W. Preidel, *J. Power Sources* **84** (1999) 161.
[196] K. G. Stanley, Q. M. J. Wu, T. Vanderhoek, and I. Nikumb, in *IEEE CCEC 2002: Canadian Conference on Electrical and Computer Engineering*, Vols. 1-3, Ed. by W. Kinsner, A. Sebak, and K. Ferens, May 12-15, 2002.
[197] D. J. Kim, E. A. Cho, S. A. Hong, I. H. Oh, and H. Y. Ha, *J. Power Sources* **130** (2004) 172.
[198] J. Mex, L. Sussiek, and M. Muller, *J Chem. Eng. Commun.* **190** (2003) 1085.
[199] E. Reddington, A. Sapienza, B. Gurau, R. Viswanathan, S. Sarangapani, E. S. Smotkin, and T. E. Mallouk, *Science* **280** (1998) 1735.
[200] A. S. Arico, P. Creti, P. L. Antonucci, J. Cho, H. Kim, and V. Antonucci, *Electrochimica Acta*. **43** (1998) 3719.

[201] P. L. Antonucci, A. S. Arico, P. Creti, E. Ramunni, and V. Antonucci, *Solid State Ionics* **125** (1999) 431.
[202] C. Yang, S. Srinivasan, A. S. Arico, P. Creti, and V. Baglio,. *Electrochem. And Solid State Letters* **4** (2001) A31.
[203] E. Peled, T. Duvdevani, A. Aharon and A. Melman, *Electrochem. and Solid State Letters* **1** (1998) 210.
[204] J. T. Wang, S. Wasmus, and R. F. Savinell, *J. Electrochem. Soc.* **142** (1996) 1233.
[205] J. Kerres, A. Ullrich, Th. Haring, M. Baldauf, U. Gebhardt, and W. Preidel, *J. of New Materials for Electrochemical Systems* **3** (2000) 129.
[206] M. Shen, K. Scott, J. W. Kuhlmann, S. Roy, J. Horsfall, N. M.Walsby, K. Lovell, *J. Membrane Science.* (2004) in press.
[207] G. Q. Sun, *J. Appl. Electrochem.* **28** (1998) 1087.
[208] R. W. Reeve, P. A. Christensen, A. J. Dickinson, A. Hamnett, and K. Scott, *Electrochimica Acta* **45** (2000) 4237.
[209] E. Peled, T. Duvdevani, A. Aharon, and A. Melman, *Electrochem. and Solid State Letters* **3** (2000) 525.
[210] R. R. Adzic, M. C. Aramor and A. V. Tripkovic, *Electrochemica Acta*, **29**(1984) 907.
[211] J. Taraszewska and G. Roslonek, *J. Electroanal. Chme.* **364** (1994) 209.
[212] J. D. Genders, E. L. George and D. Pletcher, *J. Electrochem Soc.* **143** (1996) 177.
[213] E. Yu and K. Scott, *J Power Sources* (2004) in press.
[214] D. N. Prater and J. Rusek, *J Applied Energy.* **74** (2002) 135.
[215] P. G. Grimes, B. Fielder, and J. Adam, *Proc. Annual Power Sources Conf.* **15** (1961) 29.
[216] G. Grunenberg, W. Wicke, and E. Justi, British Patent 994448 (1961).
[217] G. Goebel, B. D. Struck, and W. Vielstich, in *Fuel cells- Modern processes for the electrochemical production of energy,* W. Vielstich (translated by D. J. G. Ives), Wiley, New York, 1965.
[218] W. van Gool, *Philps Res. Reports* **20** (1965) 81-93.
[219] G. A. Louis, J. M. Lee, D. L. Maricle, and J. C. Tocciola, US Patent 4248941 (1981).
[220] C. K. Dyer, *Nature* **343** (1990) 547.
[221] T. Hibino and H. Awahara, *Chem. Letts.* (1993) 1131.
[222] I. Riess, P. J. van der Put, and J. Schoonman, *Solid State Ionics* **82** (1995) 1.
[223] T. Hibino, K. Asano, and H. Iwahara, *Nippon Kagaku Kaishi* **7** (1994) 600.
[224] T. Hibino, A. Hashimoto, T. Inoue, J. I. Tokuno, S. I. Toshida, and M. Sano, *Science* **288** (2000) 2031.
[225] T. Hibino, K. Ushiki, and Y. Kuwahara Japanese Patent 02910977 B2 (1995/98).
[226] M. Joerger, *Joint meeting of the 192nd Electrochem Soc. Meeting and 48th Annual Meeting of the Int. Soc. of Electrochem.*, Paris 31/8-5/9.
[227] K. Asano, T. Hibino, and H. Iwahara. *J. Electrochem. Soc.***142** (1995) 3241.
[228] T. Hibino, A. Hashimoto, T. Inoue, J. I. Tokuno, S. I. Toshida and, M. Sano, *J. Electrochem. Soc.* **147** (2000) 2888.
[229] M. A. Priestnall, V. P. Kotzeva, D. J. Fish, and E. M. Nilsson, *J. Power Sources* **106** (2002) 21.
[230] S. Popovici, W. Leyffer, and R. Holze, *J. Porphyrins Phthalocyanines* **2** (1998) 249.
[231] T. Okada, M. Yoshida, T. Hirose, K. Kasuga, and T. Yu, *Electrochim Acta* **45** (2000) 4419.

[232] G. Q. Sun, J-T. Wang, S. Gupta, and R. F. Savinell, *J. Appl. Electrochem.* **31** (2001) 1025.

[233] P. Convert, C. Coutanceau, P. Crouïgneau, F. Gloaguen, and C. Lamy, *J. Appl. Electrochem.* **31** (2001) 945.

[234] D. Chu and R. Jiang, *Solid State Ionics* **148** (2002) 591.

[235] P. S. Kauranen and E. Skou. *J. Electroanalytical Chemistry* **408** (1996) 189.

[236] K. Sundmacher, T. Schultz, S. Zhou, K. Scott, M. Ginkel, and E. D. Gilles,. *Chem. Eng. Sci.* **56**, (2001) 333.

[237] B. D. McNichol *J. Electroanal. Chem.* **118** (1981) 71.

[238] T. Freelink, W. Visschev, and J. A. R. van Veen, *Langmuir* **12** (1996) 3702.

[239] J Nordlund and G Lindbergh, in *Direct Methanol Fuel Cells,* Vol. 2001-4, Ed. by S. R. Narayanan, S.Gottesfeld, and T. Zawodzinski, Proceeding of The Electrochemical Soc. Inc. (2001) 331.

[240] J. P. Meyers and J. Newmann, *J. Electrochem. Soc* **149** (2002) A710.

[241] J. P. Meyers and J. Newmann, *J. Electrochem. Soc* **149** (2002) A 729.

[242] H. A. Gasteiger, N. Markovic, P. N. Ross Jr., and E. J. Cairns, *J. Electrochem Soc.* **141** (1995) 1795.

[243] K. Scott and P. Argyropoulos, *J. Elctroanalytical Chem.* **567** (2004)103.

[244] J. Kim, S. Lee, S. Srinivansan, and C. E. Chamberlin, *Journal of Electrochemical Soc.* **142** (1995) 2670.

[245] P. Argyropoulos, K. Scott, A. K. Shukla, and C. Jackson, *J. Power Sources* **123** (2003) 190.

[246] J. Nordlund, The anode of the direct methanol fuel cell. *Doctoral Thesis*, KTH, Stockholm, Sweden (2003).

[247] A. Simoglou, P. Argyropoulos, A. J. Morris, K. Scott, E. B. Martin, and W. M. Taama, *Chem. Eng. Sci.* **56** (2001) 6761.

[248] A. Simoglou, P. Argyropoulos, A. J. Morris, K. Scott, E. B. Martin, and W. M. Taama, *Chem. Eng. Sci.* **56** (2001) 6773.

[249] C. H. Wirguin. *J. Membrane Sci.* **120** (1996) 1.

[250] J. A. Kerres, *J. Membrane Sci.* **185** (2001) 3.

[251] T. E. Springer, T. A. Zawodzinski, and S. Gottesfeld, *J Electrochem Soc.* **138** (1991) 2334.

[252] D. M. Bernardi and M. W. Verbrugge, *J. Electrochem. Soc* **139** (1992) 2477.

[253] S. Um, C. Y. Wang, and K. S. Chen, *J. Electropchem. Soc.* **147** (2000) 4485.

[254] V. Gurau, H. Liu, and S. Kakac, *A. I. Ch. E. J.* **44** (1998) 2410.

[255] M W Verbrugge. J. Electrochem. Soc **136** (1989) 417.

[256] K. Scott, W. M. Taama, and J. Cruickshank, *J. Power Sources*, **65** (1998) 40.

[257] P. S. Kauranen and K. Skou, *J. App. Electrochem.* **26** (1996) 909-917.

[258] X Ren, T. E. Springer, T. A. Zawodzinski and S. Gottesfeld, *J. Electrochem. Soc.* **147** (2000) 466.

[259] S. R. Narayana, A. Kindler, B. Jeffries-Nakamura, W. Chun, H. Frank, M. Smart, T. I. Valdez. S. Surampudi, and G. Halpert, *11th-Annual Battery Conference on Applications and Advances*, IEEE Industry Applications Society, (ISBN 0780329945), 1996, p. 113.

[260] H. Dohle, J. Divisek, and R. Jung, *J Power Sources.* **86** (2000) 469.

[261] A. A. Kulikovsky, *J. Appl. Electrochem.* **30** (2000) 1005.

[262] K. Sundmacher and K. Scott, *Chem. Eng. Sci.* **54** (1999) 2927.

[263] M. K. Ravikumar and A. K. Shukla, *J. Electrochem. Soc.* **143** (1996) 2601.

[264] K. Scott, W. M. Taama, and J. Cruickshank, *J. Power Sources*, **65** (1997) 159.

[265] S. R. Narayana, A. Kindler, B. Jeffries-Nakamura, W. Chun, H, Frank, M. Smart, T. I. Valdez, S. Surampudi and G. Halpert. *11th-Annual Battery*

[266] X. Ren, W. Henderson, and S. Gottesfeld, *J. Electrochem. Soc* 144 (1997) L267.
[267] K. Scott, W. M. Taama, S. Kramer, P. Argyropoulos, and K. Sundmacher. *Electrochimica Acta* **45** (1999) 945.
[268] K. Scott, P. Argyropoulos, and K. Sundmacher *J. Electroanal. Chem.* **477** (1999) 97.
[269] K. Scott, P. Argyropoulos, P. Yiannopoulos, and W. M. Taama, *J. Appl. Electrochem.* **31** (2001) 823.
[270] P. S. Kauranen K. Skou, and J. Munk *J. Electroanal. Chem.* **404** (1996) 1.
[271] K. Scott, Mass transfer in flow fields, in *Handbook of Fuel Cell: Fundamentals Technology and Applications*, Vol. 4, Ed. by W. Vielstich, A Lamm, and H. A. Gasteiger, Wiley and Sons, England, 2003, Ch. 64.
[272] K. Scott, S. Kraemer, and K. Sundmacher, *5th-European Symposium on Electrochemical Engineering*, I. Chem. E. Symp., Ser. no 145, Exeter. 24-26, March 1999, p. 11.
[273] Z. H. Wang and C. Y. Wang, in *Direct Methanol Fuel Cells*, Proc. Vol. 2001-4, Ed. by S. R. Narayanan, S. Gottesfeld, and T. Zawodzinski, The Electrochemical Soc. Inc., Pennington NJ, USA, 2001, p. 286.
[274] J. Divisek, J. Fuhrmann, K. Gartner, and R. Jung. *J. Electrochem. Soc.* **150** (2003) A811.
[275] G. Murgia, L. Pisani, A. K. Shukla, and K. Scott, *J. Electrochem. Soc.* **150** (2003) A1213
[276] S. F. Baxter, V. S. Battaglia and R. E. White, *J. Electrochem. Soc.* **146** (2000) 437.
[277] A. A. Kulikovsky, J. Divisek, and A. A. Kornyshev, *J. Electrochem. Soc.* **147** (2000) 953.
[278] A. A. Kulivosky, *J. Appl. Electrochem.* **30** (2000) 1005.
[279] T. I. Valdez, S. R. Narayanan, H. Frank, and W. Chun, *Annual Battery Conference on Application and Advances*, Long Beach, USA. 1997.
[280] P. Argyropoulos, K. Scott, and W. M. Taama. *J. Power Sources.* **87** (2000) 153.
[281] P. Argyropoulos, K. Scott, and W. M. Taama. *J. Appl. Electrochem.* **31** (2001) 13.
[282] T. Schultz, S. Zhou, and K Sundmacher, *Chem. Eng. Technol.* **24** (2001) 12.
[283] P. Argyropoulos, K. Scott, and W. M. Taama *Chem. Eng. Technol.* **23** (2000) 985.
[284] P. Argyropoulos, K. Scott, and W. M. Taama, *The Chemical Engineering Journal* **78** (2000) 29.
[285] T. Schultz, Experimental and model-based analysis of the steady stae and dynamic operating behaviour of the direct methanol fuel cell, Ph.D. Thesis, Otto-von-Guericke University, Magdeburg, Germany, 2004.

5

Review of Direct Methanol Fuel Cells

Brenda L. García and John W. Weidner

University of South Carolina, College of Engineering and Information Technology, Swearingen Engineering Center, Department of Chemical Engineering,, Columbia, South Carolina

I. INTRODUCTION

Direct Methanol Fuel Cells (DMFCs) are galvanic electrochemical flow systems that convert chemical energy from methanol and oxygen directly to electrical energy. This direct conversion avoids the limitations of the Carnot cycle that constrains the performance of engines and gives DMFCs a higher theoretical efficiency. DMFCs were first investigated in the 1950's and research on these systems had a renaissance in the 1990's with the advent of Nafion® use for fuel cell applications.[1-3] DMFCs are akin to Proton Exchange Membrane Fuel Cells (PEMFCs) with the main difference being that DMFCs oxidize methanol (1) on the anode to produce protons while PEMFCs oxidize hydrogen (2) to produce protons.

$CH_3OH + H_2O \rightarrow CO_2 + 6\ H^+ + 6\ e^-$ ($E° = 0.02$V vs. SHE)   (1)

$3\ H_2 \rightarrow 6\ H^+ + 6\ e^-$ ($E° = 0.00$V vs. SHE)   (2)

Both DMFCs and PEMFCs reduce oxygen on the cathode according to (3):

$$\frac{3}{2} O_2 + 6 H^+ + 6 e^- \rightarrow 3 H_2O \quad (E°=1.23V \text{ vs. SHE}) \quad (3)$$

This makes the over all,

$$CH_3OH + \frac{3}{2} O_2 \rightarrow CO_2 + 3 H_2O \quad (E°_{cell}=1.21V \text{ vs. SHE}) \quad (4)$$

$$2 H_2 + O_2 \rightarrow 2 H_2O \quad (E°_{cell}=1.23V \text{ vs. SHE}) \quad (5)$$

The theoretical potentials of both reactions are nearly equal and looking at logistical issues appears to give DMFCs many advantages over PEMFCs and lithium ion batteries.

The theoretical specific energy density of methanol (~ 20 MJ/kg) is an order of magnitude higher than lithium ion battery systems (~ 2 MJ/kg), and DMFCs are in early stages of commercialization for applications that compete with lithium ion batteries.[4,5] Methanol has a significantly lower energy density than hydrogen (~ 120 MJ/kg)[4,5] based on mass, but the volumetric energy density of methanol is higher than that of gaseous hydrogen. And, because methanol is a liquid it can be stored in lightweight plastic tanks, while hydrogen usually is gaseous and has to be kept in high-pressure cylinders or hydride beds. This gives DMFC systems an even larger advantage when the storage efficiency of the fuel (kg of fuel / total kg of storage system + fuel) is also considered in the calculation.[4] Methanol is also fairly abundant in supply. The American Methanol Institute (AMI) estimates the annual global methanol production capacity to be around 11 billion gallons with a utilization of only 80 percent. The AMI also contends that trillions of cubic feet of natural gas are flared annually and that conversion of a fraction of this to methanol could power millions of fuel cells.[2] If DMFCs could provide performance high enough to be considered for vehicular transportation, the cost of converting existing infrastructure to distribute methanol would be significantly lower than trying to build a system to store and distribute hydrogen gas. Thus, on paper it would seem like DMFCs are the natural choice over PEMFCs. However, DMFCs suffer from two 2 major drawbacks: slow anode kinetics and methanol crossover to the cathode.

The oxidation rate of methanol is much lower than the rate for hydrogen at cell temperatures between 50 °C to 150 °C and operating pressures between 1 atm and 3 atm on the anode. It is estimated that even with the best DMFC anode catalysts around

0.2 V or 16% of the theoretical cell potential are lost to anode activation.[6] Methods to improve methanol oxidation in DMFCs have generally focused on improving catalyst design or decreasing activation losses by operating at higher temperatures using novel membrane materials. Also, platinum loadings in DMFCs are usually around 2 mg/cm^2 on the anode and 2 mg/cm^2 and the platinum loadings in PEMFCs are around 0.4 mg/cm^2 on the anode and cathode. This is a significant difference since the price of platinum has increased about 50% since 2002 to ~ \$27 per gram.[7]

Methanol crossover has a three-fold effect on DMFC systems: it decreases the cell potential due to a mixed potential reaction on the cathode, lowers the overall fuel utilization, and contributes to long-term membrane degradation. The performance loss due to methanol oxidation on the cathode is estimated at 0.1V.[6] For 1 M methanol most researchers measure crossover of methanol of 100 mA/cm^2 of "leakage current" with traditional Nafion® 117 membranes at open circuit conditions. Efforts to reduce the impact of methanol crossover involve reducing the amount of crossover by functionalizing acid sites in the Nafion® membrane, developing novel membrane materials, or using methanol tolerant catalysts on the cathode and recovering cathode methanol.

One method of understanding DMFC performance is by means of mathematical models. These models are developed in order to capture the physical processes occurring within the cell. When these models are combined and validated with experimental data they provide an important tool to get an insight of the full cell behavior. Processes such as mass transfer, kinetics and the electrochemistry are considered in these models from different perspectives. These models are useful, not only to acquiring knowledge for improving fuel cells, but also to incorporate them into high system levels in order to understand the effects in the whole process.

The focus of this review will be to give an overview of advances in strategies to improve DMFC performance and mathematical models for understanding the impact of these advances. Innovations that have been proven in DMFC testing are going to be reviewed more than developing technologies. The first two Sections of this review focus on efforts to improve anode and cathode kinetics. The third Section details research on achieving higher operating temperatures in DMFCs. Section four examines methods of reducing losses from methanol crossover. The final

Section gives an in depth analysis of advances in modeling DMFC systems.

II. ANODE KINETICS

The electrochemical oxidation of methanol has been studied extensively since the 1960's.[8] Iwasita[9] recently wrote an extensive review of the methanol oxidation literature and a more DMFC specific review of methanol oxidation was published by Wasmus and Kuever.[10] The review of methanol oxidation presented here will focus on outlining the important aspects of DMFC catalysts with emphasis on *in-situ* results, but providing insight into new developmental catalysts that could improve DMFC performance.

1. Reaction Mechanism

Figure 1 shows a proposed reaction network for methanol oxidation.[11] The mechanism for methanol oxidation has been debated significantly over the years,[8,9] with research continuing to define the most significant pathways for different reaction conditions both experimentally[12,13] and numerically.[14] There are multiple paths by which methanol oxidation can proceed. The selectivity of the reaction to formaldehyde, formic acid, and CO_2 depends on the reaction conditions, the catalyst used and the electrolyte. The selectivity of methanol oxidation over Pt to CO_2 is very high and formation of other byproducts will be neglected in this discussion. The electrolyte in a DMFC is generally a Nafion® membrane and can be characterized as a weak acid where the sulfonic acid sites ($R\text{-}SO_3^-$) are fixed in the membrane and do not poison the catalyst. Chlorine and sulfur containing electrolytes

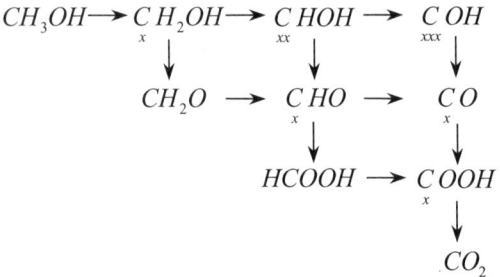

Figure 1. Reaction network for methanol oxidation. Reprinted from Ref. 11, (© 2000), by permission of John Wiley and Sons.

such as H_2SO_4 and $HClO_4$ are known to adversely affect electrode performance. This fact must be considered when reviewing studies of methanol oxidation in different electrolytes.

It should be noted in Figure 1 that all reaction pathways to form CO_2 involve formation of adsorbed CHO or CO. It is believed that these species poison the catalyst surface and that their oxidation is the rate-limiting step for methanol oxidation. The mechanism shows that oxidation of the surface intermediates involves the addition of a hydroxyl group before the final oxidation to CO_2.

The mechanism for the oxidation of CO is very important in the reaction mechanism for methanol oxidation and this step. This means that with a platinum catalyst electro-adsorption of water is fairly slow and therefore addition of an OH^- group to the CO adsorbed on the reaction site must come from an attack of a liquid phase water molecule. However, it is believed that on platinum alloy catalysts the reaction involves the electro-adsorption of water on the alloying metal such as ruthenium and subsequent addition of the adsorbed hydroxyl group to the CO molecule as shown in Eqs. (8) and (9):

$$H_2O + Ru(s) \rightarrow Ru\text{–}OH + H^+ + e^- \quad (8)$$

$$Ru\text{–}OH + Pt\text{–}CO \rightarrow Pt\text{–}COOH + Ru(s) \quad (9)$$

An illustration of the activation of water from Desai and Neurock[14] is shown in Figure 2 and shows the adsorption of water on Ru and the removal of a proton with the help of a neighboring Pt atom. Their calculations show a reduction of 37 kJ/mol in the activation energy of water from a Pt surface to a Pt_2-Ru surface. Figure 3 shows the mechanism for CO oxidation on an active site of a Pt_2-Ru surface. This reaction scheme is often referred to as the bi-functional mechanism. The implication of this mechanism for catalyst design is that the reaction surface must have arrangements of atoms where Pt and Ru sites are adjacent or in clusters. Structure and composition of effective catalysts will be examined and discussed in the following Section.

2. Methanol Oxidation Catalysts

(*i*) *Platinum and Platinum Catalyst Structure*

Platinum alloy electrocatalysts that form surface oxygen species are advantageous for promoting methanol oxidation via the

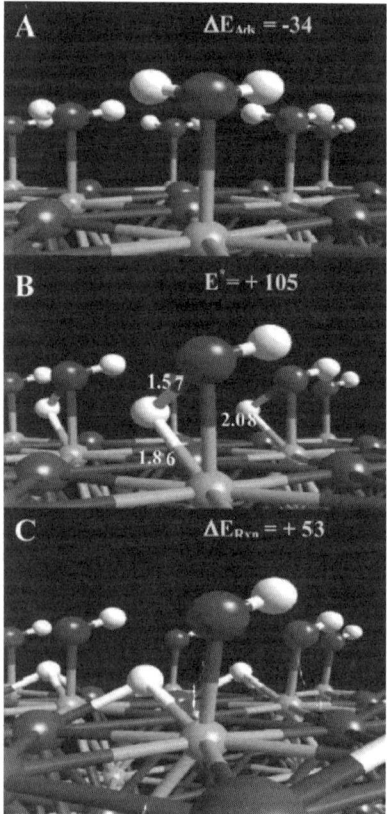

Figure 2. Mechanism for the activation of water predicted from Density Functional Theory (DFT). Reprinted from Ref. 14, (© 2003), with permission from Elsevier.

bifunctional mechanism. For this reason many researchers have synthesized bimetallic catalysts using various alloying metals such as Rh, Sn, Ru, W, Mo.[8-10,16-39] These metals are known to form surface oxide species that could promote the oxidation of methanol. However, only Pt-Ru alloy catalysts are known to be more active than Pt over a wide range of conditions.[8-10]

Gastieger et al.[40] provided the first analysis of Pt-Ru catalysts over well-characterized surfaces and used a statistical model to provide insight in the morphology of active sites. They used Low Energy Ion Scattering (LEIS) to assess the surface composition of

Direct Methanol Fuel Cells

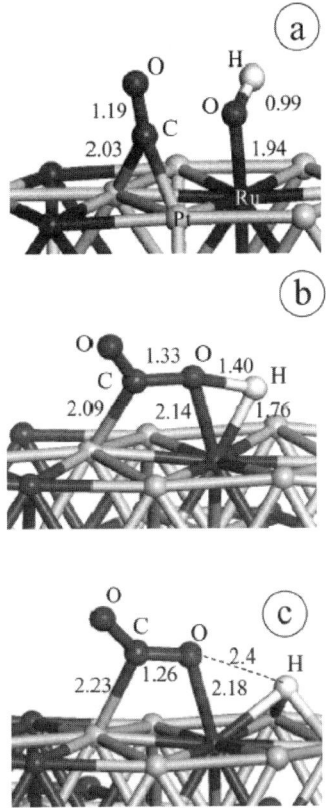

Figure 3. Mechanism for the oxidation of CO on a Pt2-Ru surface. Reprinted from Ref. 14, (© 2003), with permission from Elsevier.

alloys with a specific bulk composition. Their analysis showed that a surface concentration of ≈ 10 atomic % Ru was optimum for methanol oxidation. This analysis countered popular theory at the time because it had been believed that 50–50 Pt-Ru alloys gave the highest activity. However, the effects of surface preparation had not been considered in previous analysis. Gasteiger et al.[40] argued that the cleaning process of cycling the voltage to 1.3 or 1.5 V that was commonly used caused dissolution of Ru and thus lowered the surface concentration from the bulk composition. This work also hypothesized that the active sites consisted of a cluster of three Pt atoms adjacent to a Ru atom. Their illustration of the active sites in catalysts with different surface concentrations is shown in Figure 4. They next calculated the probability of finding the active sites on surfaces with different crystal structure and surface composition. Their probability distributions show that the most active sites will be found when the surface concentration of ruthenium is around 10 atomic %. Their work also showed that surfaces with 50–50 Pt-Ru composition had a very low activity for methanol oxidation as predicted by their statistical model.

The fundamental experiments provided by Gasteiger et al.[40] provide a solid foundation for understanding methanol oxidation on smooth Pt-Ru alloy electrodes, but DMFCs generally use gas diffusion electrodes that are a porous mixture of electrolyte, carbon, and catalyst particles. The structure and surface

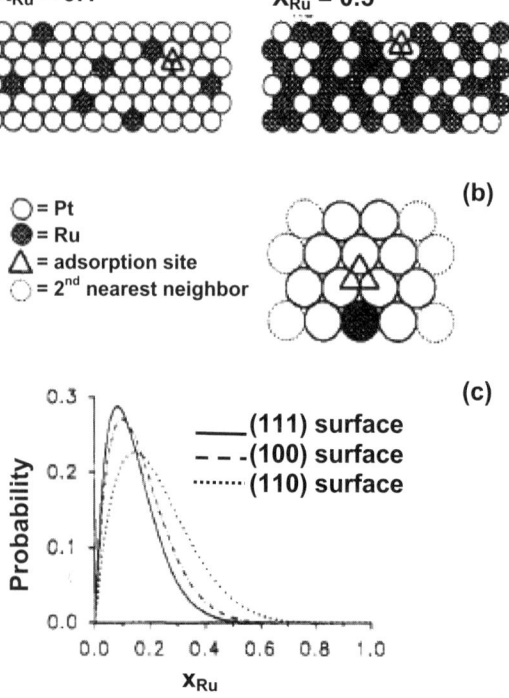

Figure 4. Illustration of Pt_3-Ru methanol oxidation active sites and the probability of finding the sites for various surface compositions. Reprinted with permission from Ref. 40, (© 2003), American Chemical Society.

composition of the catalysts are significantly affected by electrode preparation techniques. A recent survey of catalyst synthesis techniques for low temperature fuel cells by Chan[18] gives an insightful review of catalyst and catalyst support synthesis methods, electrode performance measurement, and structure and stability of nanoparticles. One conclusion of this review was that measuring catalyst activity in gas diffusion electrodes similar to those used in DMFCs is difficult *ex-situ*. For this reason, only *in-situ* performance of catalysts is examined in this review.

Hogarth and Ralph[41] gave insight into their DMFC catalyst development process at Johnson Matthey. This paper is an excellent example of the full catalyst development process for DMFC catalysts and is currently available from the Johnson Matthey website (http://www.platinum.matthey.com/publications). It should also be noted that Johnson Matthey platinum website has a number of interactive features that are very useful for understanding the platinum market. As the DMFC industry is currently very dependent on the price of this commodity (similar to airlines and oil) understanding the platinum market is important.

To investigate the reaction mechanism for the Methanol Oxidation Reaction (MOR) and the promotional effects of Pt-alloys Hogarth and Ralph perform Cyclic Voltammetry (CV) experiments with supported and unsupported Pt and Pt-Ru catalysts. Pt-Ru catalysts showed a peak for methanol oxidation, indicating adsorbed species, at the same potential that oxidation of gas phase CO as shown in Figure 5. This supports the hypothesis that adsorbed CO is the surface poisoning species in the MOR. They also calculated the Electrochemical Platinum Surface Area for oxidizing CO ($EPSA_{CO}$) and $EPSA_{MeOH}$ by integrating the CO adsorption peak measured by CV and compared the value to the total Pt surface area as measured by chemisorption. This comparison gave them the fraction of the total Pt surface area accessible to CO that is available to electrochemically oxidize CO and methanol. Their results showed that around 60% of the Pt surface area accessible to gas phase CO could be used for electrochemical oxidation of adsorbed CO from gas or methanol. The experiments also showed a lowering of the CO oxidation potential with Pt-Ru catalysts. This indicates that the Ru promotes the MOR as shown by previous researchers[8-10] and has been explained by the bifunctional mechanism.[8-10,14] They showed that Pt-Ru had greater activity than pure Pt or other Pt-Alloys for methanol oxidation in rotating disk electrode experiments.

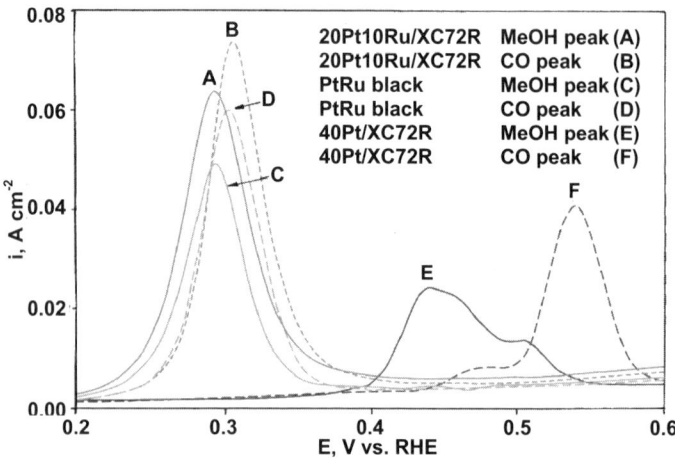

Figure 5. Methanol and CO stripping peaks for Pt and Pt-Ru catalysts as measured by Cyclic Voltammetry (CV). Reprinted from Ref. 41, (© 2002), with permission Johnson Matthey PLC.

Rotating Disk Electrode (RDE) experiments were also performed to screen alloys for the ability to oxidize CO. After normalizing the results for the Electrochemical Metal Area (ECA), measured by CO stripping voltammetry, the Pt-Ru catalyst shows a lower overpotential than other bimetallic and ternary catalysts for the MOR. Hogarth and Ralph note that RDE experiments similar to the ones in this paper using H_2SO_4 or another strong liquid electrolyte access all of the platinum surface area as measured by CO chemisorption or ECA instead of just the $EPSA_{CO}$ or $EPSA_{MeOH}$ accessed in full cell experiments. Ternary metal alloys with a Pt-Ru base showed performances slightly lower than Pt-Ru. This led to the conclusion that only slight improvements in performance could be found by examining ternary catalyst alloys and a decision by the authors that complications involved in ternary metal catalyst production were not worth any modest gains in catalyst performance.

The third step in developing the methanol oxidation catalyst was to determine the optimal Pt to Ru ratio. This was accomplished by synthesizing catalyst with different atomic compositions and examining their performance in RDE experiments. Also, the bulk composition of the alloys was examined using X-Ray Diffraction (XRD). Results showed that Pt-Ru compositions between $Pt_{70}Ru_{30}$

and $Pt_{30}Ru_{70}$ had similar activity in the MOR and that the surface area measured by ECA and XRD was highest for these catalysts. XRD data showed that the bulk alloy became Pt-rich for atomic compositions of Ru above $Pt_{80}Ru_{20}$. This result points to formation of segregated Ru on the surface of the catalyst.

For the final step in anode characterization, Hogarth and Ralph examined the effect of the catalyst support on the activity and tested *in situ* performance. They found that in RDE experiments supported catalysts performed equally to unsupported catalysts and that supported catalysts with different wt% Pt performed equally. However, half-cell tests (MeOH/H_2) of the anode catalysts demonstrated that the composition of the supported catalysts significantly affected polarization. Results showed that 40 wt% Pt and 20 wt% Ru supported on carbon (Vulcan XC-72) showed better performance than unsupported Pt-Ru on carbon (JM HiSPEC™ 6000). This effect was attributed to thickness and porosity effects in the catalyst layer because if the wt% of Pt in the electrode is small then the electrode generally has to be thicker to have equivalent Pt loadings. There are also effects due to transport of protons to the membrane, but these were not discussed in this paper.

Using the information gathered from the anode characterization, they constructed MEAs to see how these catalysts performed in full DMFC experiments. An MEA with a Nafion® 117 membrane, 1:1 atomic ration Pt-Ru anode (1 mg/cm² Pt loading), and a 4-mg/cm² Pt cathode produced a 100-mW/cm² power density at 90 °C with 0.75-M methanol. The cell was operated at 4.5/5 (A/C) stoich with 4-bar air. These operating conditions give anode and total specific powers of 100 W/g and 20 W/g. They showed that using a Nafion® 112 membrane under these conditions give almost equal performance to a Nafion® 117 membrane at voltages above 0.5 V. They were also able to reduce the cathode flowrate to 2 stoich with no performance loss. Increasing the cell temperature to 130 °C improved activation and resulted in a 60% increase in power density.

Hogarth and Ralph reduced the Pt loading to 1 mg/cm² on the cathode to see the effect on performance. They found that performance could be maintained at the level of a 4-mg/cm² Pt cathode by increasing the cathode flowrate to 10 stoich and decreasing the methanol concentration to 0.5 M. These modifications increase their total specific power to 100-W/g Pt.

(ii) Platinum and Platinum Alloy Catalyst Performance

In order for DMFCs to compete with reformate/air PEMFCs then DMFCs need to be capable of producing 400–500 mA/cm^2 at 500 mV (200–250 mW/cm^2) with only 50 mA/cm^2 of leakage current at this condition (~ 90% methanol utilization).[42,43] These conditions are strenuous to meet with current technologies and will be used as a benchmark throughout this review. The reason to look at the power density at 500 mV is because power-conditioning hardware generally favors high-voltage, low-current applications to low-current, high-voltage applications. In addition to these standards, the platinum loading will also be considered in examining the platinum efficiency (watts per gram of platinum). No penalty shall be assessed for operation of the cell at high pressure, but it should be noted that compressors generally subtract around 20% from the total system efficiency. The amount of additional metals used shall also be neglected because most alloying metals are significantly less expensive than platinum (i.e., Ru costs $2.6 per gram).

Surampudi et al.[44] published one of the first studies of a high-performance DMFCs. Their cell produced 140 mW/cm^2 at 95 °C with 3-bar oxygen on the cathode and 2-M methanol. They used a Nafion® 117 membrane and 5-mg/cm^2 Pt-Ru anode catalyst loading. This gives an anode catalyst activity of about 55-W/g Pt. Increasing methanol concentration improved performance up to 2 M, but above this decreasing performance was seen due to methanol crossover.

Ren et al.[45] used commercially available materials to construct a DMFC that produced 340 mW/cm^2 at 130 °C with 5-bar oxygen and 1-M methanol. This performance was obtained using a Nafion® 112 membrane and 2.2 mg/cm^2 anode and cathode loadings. The anode catalyst was unsupported Pt-RuO$_x$ from E-TEK mixed with Pt-black from Johnson-Matthey with a specific area of 20 m^2/g. The anode and total specific powers were 300-W/g Pt and 100-W/g Pt. The authors saw a performance increase using thinner membranes such as Nafion® 112 even though they are known to have higher rates of methanol crossover. They also noted that at lower temperatures their catalysts performed similar to those of Surampudi et al.[44]

Scott et al.[46] demonstrated a cell giving a power density of 200 mW/cm^2 at 80 °C, 1-bar oxygen and 1-M methanol. The membrane was Nafion® 117 and the catalyst loadings were 2 and

1 mg/cm^2 on the anode and cathode respectively. The catalysts were prepared using a colloidal method and were mixed with ketjenblack 600 carbon. Under these conditions the authors see anode and total specific powers of 200-W/g Pt and 100-W/g Pt. The authors contend that with 5-bar pressure on the cathode they can achieve similar performance to than Ren et al.[45] If this were true the activity of the colloidal catalyst would probably exceed that of the "off-the-shelf" catalyst from Ren et al. Scott et al. showed that performance of a DMFC can be improved more than a 30% by pressurizing the oxygen cathode. This increase in performance seems to justify pressurizing the DMFC cathode.

Witham et al.[47] sputter depositied anode catalysts for DMFCs. The performances of their catalysts were below 100 mW/cm^2, but they demonstrated an anode with a loading of 0.03 mg/cm^2 of catalyst with an anode activity of 4000 W/g Pt. The catalyst activity from this method dropped significantly with increasing the anode catalyst loading. The authors thought that the decrease in activity when the loading was increased may have been caused by spraying Nafion® solution onto the membrane over the sputter deposited catalyst before attaching the electrode. A subsequent study of sputter deposited Pt on DMFC electrodes[48] found that sputtering catalyst onto the membrane did not enhance performance, but sputtering Pt on the catalyst layer appeared to help.

Baldauf and Preidel[49] used Johnson Matthey catalysts with 1.3- and 4-mg/cm^2 (A/C) catalyst loadings showed a power density of 250 mW/cm^2 for a single cell. Operating conditions were 110 °C cell temperature, 3-bar oxygen, and 1-M methanol. The anode and total specific activities for this catalyst were 360 and 60-W/g Pt. Similar MEAs used by the authors in a -cm^2 stack showed a power density of only 180 mW/cm^2 at 110 °C and 1.5 bar of air at 30 L/min. They concluded that the high cathode flowrate needed to give the peak performance was not justifiable because of compressor cost and inefficiency and that much of the need for high flowrate on the cathode could be solved by reducing methanol crossover.

Thomas et al.[50] showed improvements on the earlier kinetic data of Ren et al.[45] as well as investigating DMFC performance stability and fuel utilization. Using Johnson Matthey Pt-Ru catalysts and a total Pt loading of 2.6 mg/cm^2 they reached 200 mW/cm^2 at 100 °C with an air cathode operating a 3-bar and 0.5-M methanol on the anode. The membrane used in this study was Nafion® 117. The calculated total specific power was 77-W/g Pt at 0.5V. They showed a 2000-hour life test with a power density over 150 mW/cm^2 and fuel utilization better than 90% for this MEA. To improve fuel utilization they suggested operating the anode with lower concentrations of methanol or operating at higher current densities so that more of the methanol is consumed in the reaction. In tests at different temperatures it was shown that the flux of 1-M and 0.5-M methanol through a Nafion® 117 membrane nearly doubled (from 90 to 175 mA/cm^2 and from 50 to 100 mA/cm^2) when the cell temperature was increased from 60 °C to 120 °C. In this study the error in crossover due to CO_2 diffusion through the membrane was not considered. For temporary improvement in performance opening the cell circuit or stopping the methanol flow for ~ 30 minutes was shown to give a bump in performance. The overpotential of a cathode with high catalyst loading due to methanol crossover was shown to be less than 20 mV at 100 mA/cm^2.

Shukla et al.[51] extended previous work[52] on the performance of a DMFC at atmospheric pressure. They showed that a cell with a 1 and 4.6-mg/cm^2 (A/C) Pt loadings on a Nafion® 117 MEA could produce 160 mW/cm^2 with an oxygen cathode and 90 mW/cm^2 with an air cathode at 90 °C. They also performed a durability test at 100 mA/cm^2 for oxygen and air cathodes at cell temperatures of 60 °C and 90 °C. The results of these durability tests, shown in Figure 6 and Figure 7, illustrate the significant performance decay experience by DMFCs operating under atmospheric conditions. For

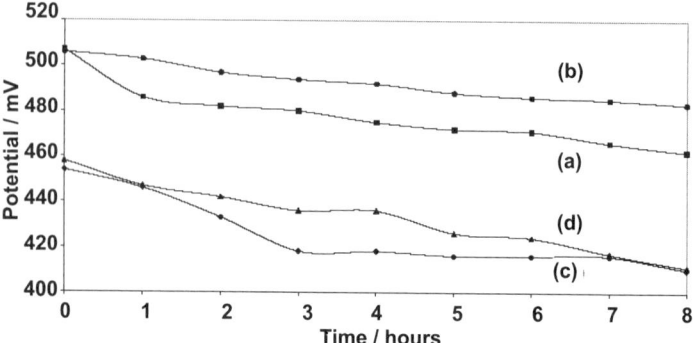

Figure 6. DMFC durability at 90°C with 1-M methanol on the anode, an atmospheric pressure on the cathode and (a) 1-L/min oxygen; (b) 2-L/min oxygen; (c) 1-L/min air; (d) 2-L/min air. Reprinted from Ref. 51, (© 2002), with permission from Elsevier.

Direct Methanol Fuel Cells

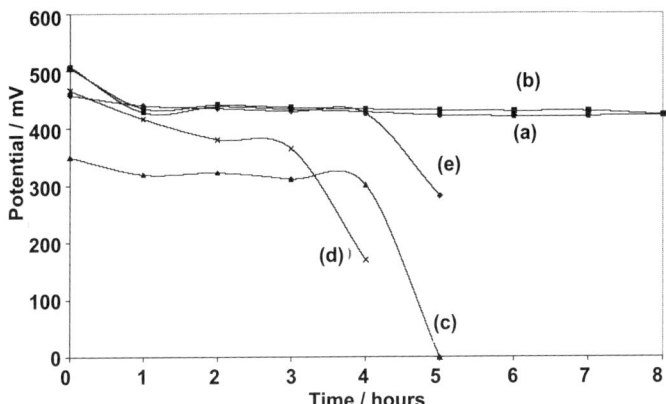

Figure 7. DMFC durability at 60 °C with 1-M methanol on the anode, an atmospheric pressure on the cathode and (a) 1-L/min oxygen; (b) 2-L/min oxygen; (c) 1-L/min air; (d) 2-L/min air. Reprinted from Ref. 51, (© 2002), with permission from Elsevier.

the oxygen cathode at 2 L/min the power density decreased by 4% and at 1 L/min the decrease was about 8% in only 8 hours. An 8% loss in power density over 8 hours was also seen using an air cathode at 90 °C. At 60 °C the oxygen cathode performed steadily for 1- and 2-L/min flowrates, while the air cathode lost all performance after 4–6 hours with flow rates of 1, 2 and 5 L/min. The decay in performance shown by Shukla et al. was more than 50 times as fast as that shown by Thomas et al.[50] when viewing the decay in terms of W/cm^2-hr. These results justify pressurizing the DMFC cathode in order to prolong durability.

Arico et al.[53] synthesized anodes with different atomic Pt-Ru compositions and evaluated the activity of these catalysts at 130 °C. The cell used a Nafion® 112 membrane and electrodes with 2-mg/cm^2 Pt loading prepared by a colloidal method. Running with 2 M methanol, 2.5 atm on the anode and 3 atm on the cathode they achieved 275 mW/cm^2 with a 1:1 atomic ratio Pt-Ru catalyst. This gives total and anode specific power densities of 138 and 69 W/g Pt. Table 1 shows the DMFC power density at 0.5 V for anodes with different Pt-Ru compositions. The most active catalyst had a 1:1 atomic ratio of Pt and Ru.

Arico et al.[54] published results showing the performance of a DMFC with a 0.1 mg/cm^2 anode Pt loading. The method of reducing the loading is to form a submonolayer of Pt on top of Ru

Table 1
Power Densities for Pt-Ru Alloy Electrodes Tested at 130 °C Cell Temperature, 2/2-mg/cm^2 (A/C) Pt Loading, 2.5 atm Methanol, and, 3-atm Oxygen. Reprinted from Ref. 53, (© 2002), with permission from Elsevier.

Sample	XRD-derived composition at.% (± 2%)	XPS-derived composition at.% (± 5%)	Power density at 0.5 V (mW cm^{-2})	Maximum power density (mW cm^{-2})
DMFC1	Pt	---	20	75
DMFC2	Pt$_{72}$Ru$_{28}$	Pt$_{90}$Ru$_{10}$	85	190
DMFC3	Pt$_{64}$Ru$_{36}$	---	155	250
DMFC4	Pt$_{52}$Ru$_{48}$	Pt$_{60}$Ru$_{40}$	275	390
DMFC5	Pt$_{38}$Ru$_{62}$	Pt$_{35}$Ru$_{65}$	175	260

particles. The catalyst particles were formed by impregnation of amorphous Ru-oxide with H_2PtCl_6 followed by reduction in hydrogen. The membrane used in this experiment was Nafion® 117 and the cathode Pt loading was 2 mg/cm^2. The test was conducted with 1-M methanol on the anode at 3 bar and air on the cathode at 3.5 bar in the cathode. The anode and total specific power densities were 1000 and 50 W/g Pt.

Several authors have attempted novel synthesis methods that show improved electrode performance over commercial catalysts, but lower performance than is found by many researchers in literature. Choi and Woo[55] synthesized a Pt-Ru nanowire electrode. The nanowire electrode showed lower activation energy than an electrode with a Johnson Matthey catalyst prepared by a similar method. However, the maximum power density was only around 50 mW/cm^2 at 80°C and atmospheric pressure. Coutanceau et al.[56] demonstrated improved performance of Pt-Ru electrodes synthesized by a galvanostatic pulse method. But only a 100-mW/cm^2 power density.

Table 2 summarizes the MEA preparation, performance, and activity for test of fuel cells with Pt cathode catalysts and Nafion® membranes. State-of-the-art DMFC catalysts are currently able to deliver 250 mW/cm^2 at 0.5 V at high temperature, high oxygen pressure, and with large Pt loadings.[45,49,53] This performance goal was also attained under similar conditions with high pressure air.[41] Most research groups after attaining this performance level have looked for ways to maintain power density while decreasing the Pt loading of the electrodes, lowering the anode and cathode back pressure and lowering temperature. The specific activity of catalysts that achieve these goals is around 90–50-W/g Pt for the

Direct Methanol Fuel Cells

Table 2
Summary of Operating Conditions and MEA Performance for DMFC Tests Using Nafion® Membranes

Paper	Pt loading (mg/cm²) A	Pt loading (mg/cm²) C	T_{cell} (°C)	Pressure (bar) A	Pressure (bar) C	MeOH conc. (M)	Cathode gas	Anode stoich. @ 0.5 V	Membrane type	Power density (mW/cm²) @ 0.5 V	Specific activity (W/g Pt) Anode	Specific activity (W/g Pt) Total	Kilowatt Pt cost ($/kW) Anode	Kilowatt Pt cost ($/kW) Total
Arico et al. (2002)	2.0	2.0	130	2.5	3.0	2.0	O_2	14	Nafion® 112	275	138	69	199	398
Arico et al. (2002)	2.0	2.0	130	2.5	3.0	2.0	O_2	22	Nafion® 112	175	88	44	312	625
Arico et al. (2002)	2.0	2.0	130	2.5	3.0	2.0	O_2	25	Nafion® 112	155	78	39	353	705
Arico et al. (2002)	2.0	2.0	130	2.5	3.0	2.0	O_2	45	Nafion® 112	85	43	21	643	1286
Arico et al. (2002)	2.0	2.0	130	2.5	3.0	2.0	O_2	193	Nafion® 112	20	10	5	2733	5466
Baldauf and Preidel (2001)	0.7	4.0	130	1.0	3.0	1.0	O_2		Nafion® 117	250	357	53	77	514
Ren et al. (1996)	1.4	2.2	130	3.0	5.0	1.0	O_2	5.8	Nafion® 112	330	236	92	116	298
Ren et al. (1996)	1.4	2.2	130	3.0	5.0	1.0	O_2	6.4	Nafion® 115	300	214	83	128	328
Ren et al. (1996)	1.4	2.2	130	3.0	5.0	1.0	O_2	8.4	Nafion® 117	230	164	64	166	428
Scott et al. (1998)	1.0	1.0	80	1.0	3.0	1.0	O_2		Nafion® 117	175	175	88	156	312
Shukla et al. (2002)	1.0	4.6	90	1.0	1.0	1.0	O_2		Nafion® 117	50	50	9	547	3061
Surampudi et al. (1994)	2.5		95		1.4	2.0	O_2		Nafion® 117	135		54		506
Witham et al. (2000)	0.91	12.0	90	1.0	1.4	1.0	O_2		Nafion® 117	101	111	8	247	3501
Witham et al. (2000)	0.02	12.0	90	1.0	1.4	1.0	O_2		Nafion® 117	74	3714	6	7	4423
Witham et al. (2000)	0.14	12.0	90	1.0	1.4	1.0	O_2		Nafion® 117	74	528	6	52	4486
Witham et al. (2000)	0.05	12.0	90	1.0	1.4	1.0	O_2		Nafion® 117	57	1140	5	24	5777
Arico et al. (2002)	0.1	2.0	130	3.0	3.5	1.0	air		Nafion® 117	100	1000	48	27	574
Hogarth and Ralph (2002)	1.0	4.0	130	1.0	3.0	0.75	air	4.5	Nafion® 117	265	265	53	103	516

Table 2. Continuation

Paper	Pt loading (mg/cm^2) A	Pt loading (mg/cm^2) C	T_{cell} (°C)	Pressure (bar) A	Pressure (bar) C	MeOH conc. (M)	Cathode gas	Anode stoich. @ 0.5 V	Membrane type	Power density (mW/cm^2) @ 0.5 V	Specific activity (W/g Pt) Anode	Specific activity (W/g Pt) Total	Kilowatt Pt cost ($/kW) Anode	Kilowatt Pt cost ($/kW) Total
Hogarth and Ralph (2002)	1.0	4.0	90	1.0	3.0	0.75	air	4.5	Nafion® 117	164	165	33	166	828
Hogarth and Ralph (2002)	1.0	4.0	90	1.0	3.0	0.75	air	4.5	Nafion® 112	143	143	29	191	956
Ren et al. (1996)	1.4	2.2	130	1.8	5.0	1.0	air	15	Nafion® 115	125	89	35	306	787
Ren et al. (1996)	1.4	2.2	130	1.8	5.0	1.0	air	15	Nafion® 112	125	89	35	306	787
Ren et al. (1996)	1.4	2.2	130	1.8	5.0	1.0	air	21	Nafion® 117	90	64	25	425	1093
Shukla et al. (2002)	1.0	4.6	90	1.0	1.0	1.0	air		Nafion® 117	25	25	4	1093	6122
Thomas et al. (2002)		2.6	120	1.0	3.0	0.5	air		Nafion® 117	125		48		568
Thomas et al. (2002)		2.6	110	1.0	3.0	0.5	air		Nafion® 117	115		44		618
Thomas et al. (2002)		2.6	100	1.0	3.0	0.5	air		Nafion® 117	100		38		711
Thomas et al. (2002)		2.6	80	1.0	3.0	0.5	air		Nafion® 117	60		23		1184

[1] A = Anode; C = Cathode.

entire cell. At the average 2004 market price of $27/g Pt this translates to a platinum cost of 300–500 $_{Pt}$/kW. To be competitive in the transportation market (the most demanding sector) total system DMFC costs per kilowatt would have to be around 300–500 $/kW.

Although many different catalysts have been tested for their characteristics in the MOR, the only Pt-Ru catalysts have been extensively tested in DMFCs. Most of the variations in catalyst layer construction has been in the Pt:Ru ratio,[33,53,54] Nafion® loading in the electrode,[57] Pt on carbon loading[41], and catalyst synthesis method.[55,56,58] Current electrode synthesis standards to support commericially available or colloidally synthesized 1:1 atomic ratio Pt-Ru on a high surface are carbon black such as acetylene black, Vulcan XC-72 or Ketjen Black at 40–60 wt%. Nafion® is usually included into the electrodes at 5–15 wt% to improve three-phase contact. These electrodes are then hot pressed onto Nafion® membranes with 1100 equivalent weight and 50, 90, 130, or 180-μm thickness.

The most effective efforts to increase Pt activity have been the sputter deposition of the Pt-Ru catalyst onto the membrane and electrodes[47] and the use of a Ru catalyst with small amounts of surface Pt.[54] However, both of these methods are not yet able to yield catalyst layers that have a high power density. There have been several studies to optimize electrodes by trying different Nafion® loadings[57,59,60] that recommend anode Nafion® loadings between 0–30 wt%, but it appears that the Nafion® content must be optimized for each electrode synthesis technique individually.

III. OXYGEN REDUCTION REACTION CATALYSTS

Researchers who have obtained power densities of 250 mW/cm^2 have had a minimum Pt loading on the cathode of 2 mg/cm^2. This Pt loading is around 5 times as high as the cathode loadings used in PEMFCs. The high Pt loading in DMFC cathodes is necessary because of methanol crossover.

The voltammagrams for Pt catalysts in Figure 5 show that they are poisoned by methanol oxidation and that the adsorbed CO is only oxidized at high potentials. If methanol permeates a DMFC membrane, then a Pt catalyst used on the cathode will be poisoned and the result would be a loss of active area for oxygen reduction. The combination of Pt active site loss with the parasitic consumption of oxygen during methanol oxidation on the cathode results in a "mixed potential" effect on the cathode.[61-63] One method for improving cathode performance is the use of catalysts that are more selective toward the Oxygen Reduction Reaction (ORR) and are thus not poisoned by CO.

Catalysts that are known to be selective towards oxygen reduction are macrocycles (i.e., porphyrins) with complexed iron or cobalt,[64] platinum alloys such as Pt-Cr, Pt-Co, and Pt-Fe,[6,65-68] small-size Pt nanoparticles,[69] and bimetallic RuSe catalysts.[6,70] These catalysts generally show lower activity toward oxygen reduction than Pt/C in systems that do not contain methanol. However, when methanol is present the performance decrease for methanol-tolerant catalysts is much lower than Pt electrodes and the methanol-tolerant catalysts show higher polarization performance.

Shukla and co-workers[66] demonstrated a Pt-Fe/C cathode together with a Pt-Ru/C anode that produced 100 mW/cm^2 at 0.5 V and only 85 °C. Novel prussian blue analog macrocycle catalysts have been shown to be very promising in the ORR during kinetic studies and are tolerant to methanol.[71,72] These catalysts have been shown to be more active than Pt catalysts in PEMFC electrodes at low overpotentials.[71] However, these catalysts have not been tested in DMFCs. RuSe and RuMoSeO$_4$ catalysts showed very low performance in oxygen reduction tests, but this class of catalysts is interesting because they have no Pt and no activity toward the MOR.

If catalysts can be developed that are active only for the MOR or the ORR then a mixed-reactants DMFC could be designed where methanol and oxygen flow together through stacked cells and the anodic and cathodic reactions take place on the respective electrodes due to catalyst selectivity.[6] Figure 8 shows a schematic of a mixed-reactants DMFC. The mixed reactant DMFC design is advantageous over conventional designs because it does not require Nafion® membranes and would simplify gasketing inside the cell. A mixed-reactants DMFC was shown by Shukla et al.[6] to produce a power density of 45 mW/cm^2 at 0.2 V with a 2.5-mg/cm^2 RuSe/C cathode and a Pt-Ru anode.

IV. HIGH TEMPERATURE MEMBRANES

Finding better catalysts is one answer to improving catalyst activation in DMFCs. The second approach is to raise the operating

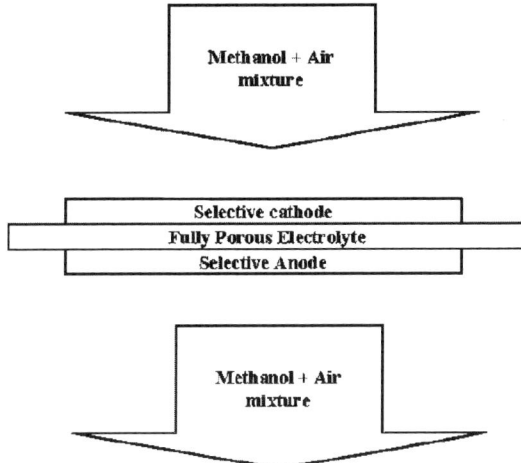

Figure 8. Schematic of a mixed –reactants DMFC. Reprinted, with permission, from the Annual Review of Materials Research Vol. **33**, © 2003 by Annual Reviews, www.annualreviews.org.[6]

temperature. Raising the operating temperature activates both the anode and cathode reactions and decreases the effect of adsorbed CO from methanol oxidation. In addition to improving reaction kinetics, operating at elevated temperature helps with thermal management. At higher temperatures less cooling of the stack is needed and cell temperatures above 100 °C would mean that waste heat from the stack could be used to generate steam that could be used elsewhere in the system. The problem with operating at temperatures above the boiling point of water is due to membrane hydration issues.

Figure 9 shows the maximum activity of water and the maximum membrane conductivity for a Nafion® membrane as a function of cell temperature for different operating pressures. Membrane conductivity increases with temperature up to the boiling point of water at the given operating pressure. Above the boiling point of water there is a significant drop in membrane conductivity because of decreasing water activity. The membrane conductivity at 1 bar decreases around 66% in the first 15 °C above the boiling point. This drop in conductivity is the reason that high-temperature DMFCs operating at ambient pressures and high temperatures will require membranes that can operate under dry conditions. Figure 10 shows the membrane conductivity of a

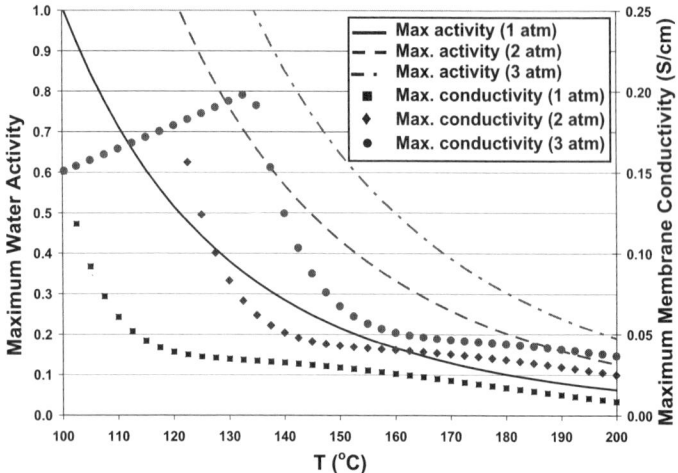

Figure 9. Maximum water activity and membrane conductivity for different cell temperature.

saturated Nafion® membrane for a range of temperatures from 60 °C to 200 °C and the operating pressures that would be needed to maintain saturation. As considered by Kallo,[73] the dilute two-phase methanol-water system will have a slightly different boiling point than water.

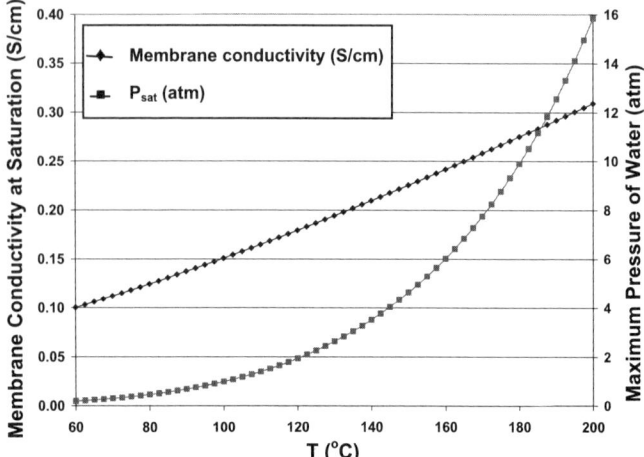

Figure 10. Theoretical conductivity of a saturated Nafion® membrane with temperature and the saturation pressure of water needed at each temperature.

Li et al.[43] recently reviewed developments in membranes for high-temperature fuel cells. They compare the different materials and techniques used to construct novel membranes and characterize them according to the conductivity. In this Section, how the different membranes improve *in situ* DMFC performance will be discussed.

One common method of constructing high temperature membranes is to incorporate a hygroscopic material into the membrane structure that will retain water at high temperatures. The two most common classes of materials that have been used are Zr compounds (i.e., ZrO and $ZrPO_4$) and silicon containing compounds (i.e., SiO_2, tetraethoxysilane (TEOS), or other organic silicates).

Nafion®/$ZrPO_4$ membranes for DMFC uses have been investigated by Yang et al.[74,75] To incorporate $ZrPO_4$ into the membranes they ion-exchanged acid sites in the membrane with Zr^{4+} and then quenched the membrane in H_3PO_4. These modifications increased the weight and thickness of the membrane around 25%.[74] They used electrodes with 2.3 and 2.5-mg/cm^2 (A/C) Pt loadings. The 5-cm^2 cell was operated with 2-M methanol at 4 bar and 2.8 mL/min on the anode and humidified oxygen and air on the cathode at 4 bar and 500 mL/min. The authors show that the Nafion® 115/$ZrPO_4$ membrane produces 300 and 210 mW/cm^2 in oxygen and air respectively at 145 °C. The calculated total specific activities were 63-W/g Pt with oxygen and 41-W/g Pt with air. They also show that the Nafion® 115/$ZrPO_4$ performs better than a humidified Nafion® 115 membrane at 90 °C and high temperatures. However, no results are shown for the membranes operating under the same conditions. Arico, one of the authors, publishes a paper[53] about 1 after the work of Yang et al. where he achieves nearly equivalent results using a lower Pt loading at only 3-atm back pressure on the cathode and no humidification. The Nafion®/$ZrPO_4$ membrane, thus, does not appear to deliver the decrease in operating pressure needed for high-power DMFCs or a significant catalyst activation increase that would allow for lower catalyst layer loading.

Arico et al.[76] investigated ZrO_2 modified membranes and found that these membrane produced a power density of 250 mW/cm^2. They made the membranes by recasting Nafion® where some acid sites had been exchanged with ZrO_2. The membranes had a thickness of 100μm with a 3-wt% ZrO_2 content. For DMFC tests the cell was operated with 2M methanol at 145 °C. With 3.5-bar oxygen the Pt specific activity was 63-W/g Pt, the same as found for $ZrPO_4$ in earlier studies.[74] At 1.5 bar the power density decreased to around 120 mW/cm^2 at 0.5 V, thus decresing the specific power to 30-W/g Pt. The methanol crossover current was found to be 40 mA/cm2. This value is less than half the typical value for a Nafion® 117 membrane that is twice as thick under the same conditions. The authors reported a doubling of the membrane resistance over the 2-bar drop in pressure. Using analysis similar to that in Figure 9, the Springer et al. model of Nafion® conductivity shows that a pure Nafion® membrane should have shown a 4 times higher resistance for the same change in pressure at 145 °C, so some improvement on the ability of the membrane to self-hydrate appears to be made using ZrO_2. However, the overall decrease in performance during the pressure decrease was more than 5 times the decrease predicted by the change in membrane conductivity. This indicates that the pressure has a significant influence on the kinetics of the cathode and/or the hydraulic permeation of methanol through the membrane. The promotional effect from ZrO_2 was slightly less that other additives tried by Arico et al.,[76] but the performance at 0.5 V was similar.

In a manner similar to ZrO_2, the group of the CNR-TAE incorporated SiO_2 and Si with phophotungstic acid (Si-PWA) in DMFC membranes to act as hygroscopic agents.[77,78] The amount of inorganic material added to the membrane was always 3 wt%. The membranes modified with SiO_2 showed peak performances of 150 and 75 mW/cm^2 in 5.5-atm oxygen and air respectively. The specific activity for these catalysts was fairly low at 38- and 19-W/g Pt in oxygen and air respectively.[77] Crossover current densities of 40 mW/cm^2 similar to ZrO_2 modified membranes were also measured. The Si-PWA modified Nafion® membranes in[78] had power densities of 240 and 125 mW/cm^2 with oxygen and air cathodes for 2.2/2- mg/cm^2 (A/C) Pt loadings. The anode feed was 2-M methanol and cathode pressure in the cells was 4 atm. The total specific activities of 55- and 28-W/g Pt were slightly increased from the work done on SiO_2 membranes in,[77] but were nearly equal to SiO_2 modified membranes investigated in this work. The difference between the two papers is probably due to the small differences in electrode structure or operating conditions. The incorporation of heteropolyacids into the membrane appeared to give more improvement in performance in the mass transfer limited region of the polarization curves instead of improving activation of the catalyst.

Another method to make a high temperature membrane for DMFCs is to graft acid groups onto active sites using radiation. Florjanczyk[79] attached vinylphosphonic acid (VPA) and 2-acrylamido-2methyl-1-propanesulfonic acid (AMPSA) into the membrane structure. Using typical electrodes with a 2.3 mg/cm^2 of Pt they were able to achieve 200 mW/cm^2 at 130 °C, while a Nafion® 115 membrane produced only 150 mW/cm^2. The cell was operated with 2-M methanol and 3-bar oxygen. This gives a Pt efficiency of 43-W/g Pt for the VPA modified membrane. At 0.5 V the AMPSA modified membrane produced 100 mW/cm^2 under the same test conditions.

Membranes designed specifically to enhance DMFC performance by allowing operation at higher operating temperatures have only shown moderate increases in the range of operating temperatures to around 145 °C. However, this increase in operating temperature was almost always accompanied by an increase in the operating pressure and high performing cells were never significantly above the boiling point of water. Table 3 summarizes the MEA performance for DMFC tests using modified Nafion® membranes. In the only test where the temperature was significantly above the boiling point of water the cell performance was very low.[76]

V. METHANOL CROSSOVER

The mixed potential effect on the cathode and the decrease in methanol efficiency caused by methanol crossover are serious problems for DMFC systems. This Section will examine the magnitude, causes, and effects of methanol crossover in DMFCs. Then studies that examine how membranes can be modified to decrease methanol crossover will be reviewed.

1. Magnitude of Crossover

Heinzel and Barragan[80] reviewed early work done on understanding methanol crossover in DMFCs. Table 4 shows data collected from early methanol crossover research on the diffusion coefficient of methanol through a Nafion® 117 membrane.[81-84] General trends that were observed in early research is that increasing cell temperature and anode operating pressure increased

Table 3
Summary of Operating Conditions and MEA Performance for DMFC Tests Using Modified Nafion® Membranes for High Temperature Operation.

Paper	Pt loading (mg/cm²) A¹	Pt loading (mg/cm²) C¹	T_{cell} (°C)	Pressure (bar) A¹	Pressure (bar) C¹	MeOH conc. (M)	Cathode gas	Anode stoich. @ 0.5 V	Membrane type	Power density (mW/cm²) @ 0.5 V	Specific activity (W/g Pt) Anode	Specific activity (W/g Pt) Total	Kilowatt Pt cost ($/kW) Anode	Kilowatt Pt cost ($/kW) Total
Antonucci et al. (1999)	2.0	2.0	145	3.0	5.5	2.0	O_2		Nafion® SiO_2 (80μm)	150	75	38	$364	$ 729
Arico et al. (2004b)	2.0	2.0	145	1.0	1.5	2.0	O_2		Nafion® ZrO_2 (100μm)	120	60	30	$456	$ 911
Arico et al. (2004b)	2.0	2.0	145	1.0	3.5	2.0	O_2		Nafion® ZrO_2 (100μm)	250	125	63	$219	$ 437
Florjanczyk (2001)	2.3	2.3	130	2.0	3.0	2.0	O_2	23.2	Nafion® 117 / VPA	200	87	43	$314	$ 629
Florjanczyk (2001)	2.3	2.3	130	2.0	3.0	2.0	O_2	46.3	Nafion® 117 / AMPSA	100	43	22	$629	$1,257
Staiti et al. (2001)	2.2	2.2	145	3.0	4.0	2.0	O_2		Nafion® Si-PWA (80μm)	240	109	55	$251	$ 501
Yang et al. (2001)	2.3	2.5	145	4.0	4.0	2.0	O_2	18.0	Nafion® 115 / $ZrPO_4$	300	130	63	$210	$ 437
Antonucci et al. (1999)	2.0	2.0	145	3.0	5.5	2.0	air		Nafion® SiO_2 (80μm)	75	38	19	$729	$1,458
Staiti et al. (2001)	2.2	2.2	145	3.0	4.0	2.0	air		Nafion® Si-PWA (80μm)	125	57	28	$481	$ 962
Yang et al. (2001)	2.3	2.5	145	4.0	4.0	2.0	air	25.7	Nafion® 115 / $ZrPO_4$	210	91	44	$299	$ 625

Table 4
Estimates of Methanol Transport Properties Through Nafion® 117 Membranes.

	Diffusion Coefficient $\times 10^{-5}$ (cm^2/s)	MeOH Permeability $\times 10^{-5}$ (cm^2/s)	Measurement Temperature (°C)	Activation energy (kJ/mol)
Verbrugge (1989)	1.2		25	
Kauranen and Skou (1996)		0.5	60	12
Cruickshank and Scott (1998)	1.0		70 – 96	
Tricoli et al. (2000)		0.34	60	18

methanol crossover, while increasing membrane thickness and equivalent weight decreased methanol crossover. The catalyst loading had a neutral effect on methanol crossover.

Hikita et al.[85,86] investigated the effect of current density, membrane thickness, operating pressure, and catalyst loading on the crossover of methanol. They showed that methanol efficiency near 80% could be achieved independent of membrane thickness by operating at concentrations of methanol below 1.5 M and at 0.5 A/cm^2. At lower current densities methanol consumption in the anode decreases and methanol crossover is increased due to the higher methanol concentration. To reduce the effect from methanol crossover at low current densities a thick membrane (i.e., Nafion® 117) could be used or the concentration of methanol can be reduced. Ren et al.[87] noted that for making these adjustments at low performance the methanol concentration could be varied as low as 0.1 M without a severe impact on performance. Hitika et al.[85] also found no effect from catalyst loading on methanol crossover.

Ren et al.[88] measured the water uptake and methanol crossover in Nafion® 117 and Nafion® 1210 membranes. They showed that the water uptake (λ) by Nafion® 1210 is 15.3 (H$_2$0/S0$_3^-$) and the value for Nafion® 117 is 20.8. Figure 11 shows the methanol crossover current for the membranes with different concentrations of methanol at open circuit and limiting current. It can be seen that the crossover rate is lower for the Nafion® 1210 membrane. To understand the impact this difference in methanol crossover has on DMFC performance the researchers investigated the polarization, membrane resistance, anode overpotential, and cathode overpotential. The electrodes used had 1.3 and 2.0-mg/cm^2 (A/C)

Pt loadings and the cell was operated at 80 °C with a 2.1-bar humidified air cathode. The power density for the Nafion® 117 membrane was 25 mW/cm^2 at 0.5 V and was obtained using 0.75 or 1-M methanol. For the Nafion® 1210 membrane the power density at 0.5 V was 50 mW/cm^2. The membrane resistance for the 1210 membrane was 0.36 Ω cm^2 and for the Nafion® 117 membrane was 0.16 Ω cm^2. The trend in resistance was supported by Kallo[73] who found that 1000 equivalent weight Nafion® membrane had a higher conductivity than a 1100 equivalent weight membrane. The anode over potentials were nearly equal for both membranes against a dynamic hydrogen electrode while the Nafion® 117 membrane produced 350 mA/cm^2 at 0.8 V while the Nafion® 1210 produced 200 mA/cm^2 at the same condition. This means that the enhanced crossover from the anode to the cathode was caused by the increased water uptake of Nafion® 117 membranes. The relationship between water transport and methanol transport elucidated in this work is helpful for analyzing crossover results.

Figures 12 and 13 show the effect of back pressure and oxygen flow rate on the flux of water across a Nafion® 117 membrane at

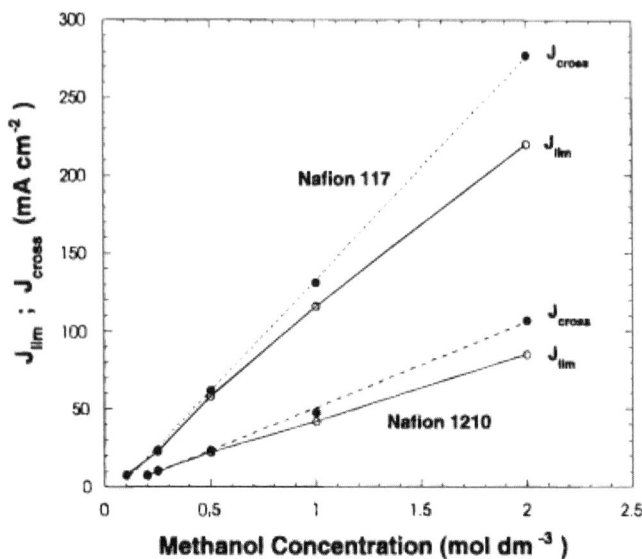

Figure 11. Methanol crossover at open circuit and at the limiting current for a Nafion® 117 and a Nafion® 1210 membrane. Reproduced from Ref. 88, (© 2000), by permission of The Electrochemical Society, Inc.

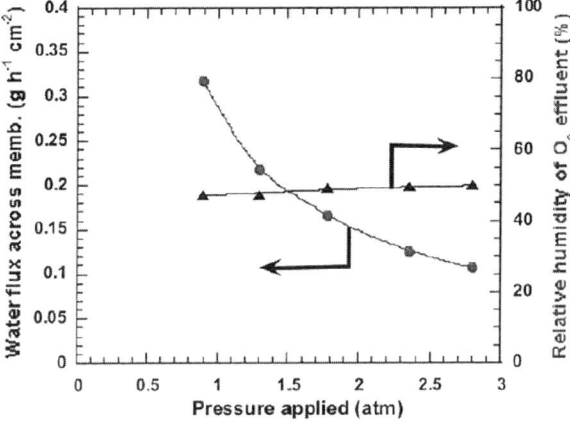

Figure 12. Effect of back pressure on the water flux across a Nafion® 117 membrane at open circuit conditions. Reproduced from Ref. 91, (© 2001), by permission of The Electrochemical Society, Inc.

60 °C and open circuit conditions. The sharp decay in water permeation with increasing back pressure is most likely due to the transport driven by a hydraulic pressure difference anode and cathode. The hypothesis that hydraulic pressure is responsible for

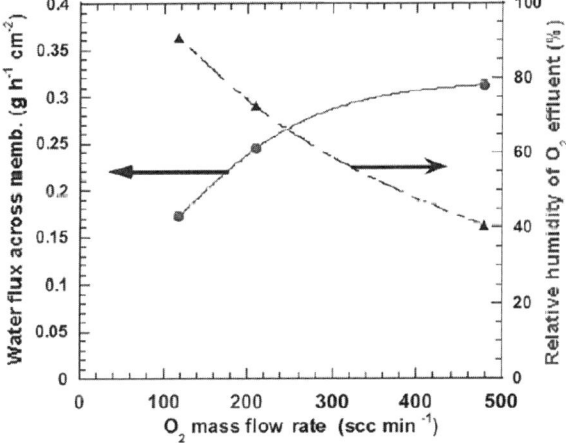

Figure 13. Effect of back pressure on the water flux across a Nafion® 117 membrane at open circuit conditions. Reproduced from Ref. 91, (© 2001), by permission of The Electrochemical Society, Inc.

part of water and methanol permeation seems to explain the results of Hikita[85] at different back pressure on the anode and cathode. The effect of oxygen flow rate in the cathode on water flux and relative humidity also supports experimental results that show increasing performance with high cathode flow rates. The high cathode flow rates would increase the methanol crossover, but would reduce the partial pressure of methanol. Fundamental and empirical insight into water and protonic transport processes in Nafion® in the literature is extensive[88-93] and should be consulted for more insight into these phenomena.

2. Effect of CO_2 Crossover

Most early methods of measuring methanol crossover used an IR-sensor or a GC on the cathode to measure the CO_2 content of the cathode outlet gas. The main assumption of this method for measuring crossover is that the only source of CO_2 in the cathode exhaust is methanol oxidation on the cathode. Recent work has been done to quantify the amount of CO_2 that permeates the membrane due to the concentration gradient between the anode and cathode.[94-97]

Dohle et al.[94] measured the effect of CO_2 crossover at a variety of DMFC operating conditions. Their measurements examined the difference between the CO_2 amounts in the cathode stream for a MeOH/H_2 half-cell experiment and a normal DMFC experiment. In the half-cell experiment the methanol that crosses the membrane is not oxidixed and thus is not detected by an IR-CO_2 sensor, thus only the CO_2 crossover flux is measured. In the typical DMFC experiment both fluxes are measured and a catalytic burner was used on the cathode to fully oxidize MOR intermediates. Figure 14 shows a typical MeOH crossover profile without considering CO_2 crossover and after correcting for CO_2 crossover. At 200 mA/cm^2 the authors show that the flux of CO_2 crossing the membrane is equivalent to the CO_2 that would be produced from a crossover current of 40 mA/cm^2 of methanol. At this condition the flux of CO_2 crossing the membrane is equal to the flux of CO_2 from methanol crossover. With thinner Nafion® 112 membranes this effect is more pronounced and the fluxes from CO_2 and methanol crossover are shown to be equal at 125 mA/cm^2. It was shown that CO_2 permeation increased with increasing methanol concentration, increasing anode pressure, and decreasing temperature. Using a

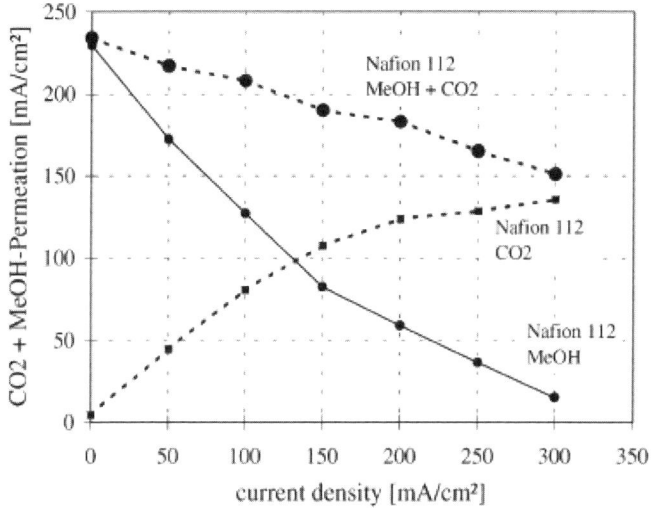

Figure 14. The effect of CO_2 crossover on methanol crossover experiments for a Nafion® 117 membrane at 85°C, 1M methanol, and 3 bar O_2 pressure. Reprinted from Ref. 94, (© 2002), with permission from Elsevier.

simple model for analysis of the contributions from CO_2 crossover due to electroosmotic drag and diffusion, the authors predicted that transport of CO_2 dissolved in water and dragged by electroosmosis was only 5–10% of the total amount measured.

Drake et al.[95] used a GC to measure CO_2 in a DMFC cathode when a mixture of H_2 and CO_2 or H_2O and CO_2 was used in the anode at ambient conditions. For currents up to 100 mA/cm², they found that crossover of methanol could cause up to a 20% error in the estimated methanol crossover. They showed an increase in methanol flux with current and calculated that if 1 molecule of CO_2 per 580 protons was dragged electroosmotically through the membrane then this mechanism could explain the increase in CO_2 flux with DMFC current density.

Jiang and Chu[96] analyzed CO_2 crossover by doing a complete carbon balance on an full DMFC. To measure the amount of CO_2 on both sides of the cell they reacted the exit gases with $Ba(OH)_2$ to precipate $BaCO_3$ and then the amount of CO_2 leaving the cell on each side could be analyzed gravimetrically. This method did not include a catalytic burner to ensure complete oxidation on the cathode, so some partial oxidation products of methanol may be overlooked by this type of analysis. The results shown by Jiang and

Chu agree well with the conclusions of Dohle et al.[94] The results showed that above 200 mA/cm^2 the flux of CO_2 from the anode could exceed the amount produced by the MOR on the cathode.

Gogel et al.[97] attempted to reproduce the experiments of Dohle et al.,[94] but used a Nafion® 105 membrane in place of the Nafion® 117 membrane by Dohle et al. The contradiction of their results is thus very interesting. The 105 membrane shows high rates of methanol crossover as would be expected[88] from a membrane with lower equivalent weight. However, the rate of CO_2 crossover measured by the authors is only around 4 mA/cm^2 at 250 mA/cm^2 but does not increase significantly with a doubling of the cell current density. This result appears to indicate that the crossover of CO_2 occurs mostly in the hydrophobic regions of the membrane while methanol crossover occurs in hydrophilic regions. Since the hydrophobic regions are reduced by the higher water uptake by lower equivalent weight membranes the CO_2 crossover is reduced. When the contention of Dohle et al. that crossover of dissolved CO_2 by electroosmosis contributes to only 5–10% of the total, then the results of Gogel et al. agree with the amount Dohle et al. predict is transported with the aqueous phase.

3. Mixed-Potential Effects

The crossover of methanol to the cathode results in a "mixed-potential" on the cathode due to the parasitic consumption of oxygen on the cathode. For a classical "mixed-potential" to occur, the reactions should take place independently on the electrode surface. Under these conditions the total current of the mixed reaction system should be the sum of the partial currents from the individual reactions.

Chu and Gilman[61] investigated the MOR and ORR on a rotating disk electrode with an H_2SO_4 electrolyte. By comparing polarization results from the ORR in O_2 saturated H_2SO_4, the MOR in H_2SO_4 with 1M methanol, and the MOR and ORR in O_2 saturated H_2SO_4 with 1-M methanol it was observed that the reactions did not conform to the classical "mixed-potential" definition. At potentials > 0.8 V, where CO is oxidized quickly, the MOR and ORR currents summed to the total current. At potentials < 0.4 V the mass transfer of oxygen is limiting and the surface is mostly covered by adsorbed CO. In the range 0.4V < V_{cell} < 0.8 V, the authors conclude that the presence of adsorbed CO from methanol oxidation significantly affects the ORR partial current, but the ORR intermediates do not appear to affect the adsorbed CO. This result follows the conclusion of Chu and Gilman.

Kauranen and Skou[62] carried out similar experiments to[61] on a 40-wt% Pt/C electrode. They also developed a model of the electrode kinetics to try and explain the mix-potential effects. They assumed that the CO adsorbtion step was the limiting anode process and that the kinetics followed a Langmuir type mechanism. On the cathode, Tafel kinetics was used to describe the kinetics. The Tafel kinetic expression for the ORR kinetics gave predictions of the cathodic current that was too large. The authors found that they needed to include a term in the cathode kinetics to describe the poisoning of the surface CO. From a fit of their data they found that the correction term needed was constant with voltage and equivalent to the coverage of CO at high potential. The authors speculated that the potential independent CO poisoning effect in the ORR is due to surface defects in the PT particles. They contend that the MOR generally takes place on the edges of Pt particles where Ru atoms are present and the ORR takes place on the flat surfaces of particles. At defects near the edges they believe the conditions exist where the intermediates could interact.

Jusys and Behm[63] used Differential Electrochemical Mass Spectrometry (DEMS) to study reaction products from methanol oxidation and oxygen reduction on a 20-wt% Pt/C electrode in a thin-layer flow cell. They showed there was evidence of increased methanol partial oxidation products in the presence of the ORR. This result is contrary to the hypothesis of other researchers[61,62] who believed that the ORR did not have an effect on the MOR. The presence of O_2 in solution caused a -50 mV shift in the methanol oxidation peak.

4. Novel Membranes to Reduce Methanol Crossover

Researchers are currently investigating methods of reducing methanol crossover through DMFC membranes by modifying existing membranes or developing novel membrane materials. The main methods to modify existing Nafion® 117 membranes are incorporating inorganic molecules into the structure and extending the polymer structure. While most modifications are effective at reducing the methanol crossover through the cell, there are very few techniques that also maintain good membrane conductivity. To compound this problem, techniques for synthesizing MEAs with Nafion® membranes do not necessarily work well with modified membranes. Poor lamination of the electrode to the membrane can cause a significant contact resistance and lead to early degradation caused by delamination.[98] Developing novel membrane materials has proven more promising than membrane modification. This Section will give an overview of both approaches to reducing crossover with an emphasis on techniques that show high performance.

Modification of Nafion® 117 membranes with SiO_2 is a technique used by researchers to improve the methanol crossover performance in addition to the high temperature characteristics mentioned in Section IV. Miyake et al.[99,100] investigated Nafion®/Silica membranes and found that to decrease methanol crossover by 10 mA/cm^2 over 10 wt% silica was needed in the membrane. At this loading of silica the membrane conductivity was lowered by around 25% from the pure Nafion® value. The membrane conductivity continued to decrease with increased silica loading. Nunes et al.[101] found that a $ZrO_2/ZrPO_4$/Nafion® membrane had a 30-times lower crossover current than Nafion®, with only a 10% decrease in conductivity, but in-situ performance was not investigated. Investigations using organic silica and organic silica mixed with heteropoly acids[102,103] show that the open circuit voltage is slightly higher than pure Nafion® for the membranes, but then performance drops rapidly. This is indicative of a slight inhibition of the methanol crossover and lower membrane conductivity, similar to the predictions of Miyake et al.

Impregnating or sputtering Pd metal into the membrane has been investigated by some researchers as a method to reduce crossover.[104-106] Since Pd is hydrogen permeable, it should impede the transport of methanol molecules without significantly reducing the proton conductivity of the membrane. Given that the 2004 average price of Pd is $7.50 per gram and that Pd metal is electronically conductive it seems expensive and risky to use this method. The equation to calculate the price per kilowatt given the metal loadings in the electrodes and the market prices of the metals is shown below in Eq. (10),

$$\frac{\$}{\text{kW}} = \frac{1000}{31.1p} \sum_i \sum_j Z_i \rho_{ij} \quad (10)$$

where p is the power density in mW/cm^2, Z_i is the market price for metal i in \$/oz$_{\text{troy}}$, and ρ_{ij} is the loading of metal i in DMFC region j. Using this equation, if the Pd loading of 0.5 mg/cm^2 and the Pt loading of 2 mg/cm^2 used in each electrode of[106] this only comes to a 3% increase in the price of precious metal per kW. At 0.5 V in cell operating at 40 °C and with oxygen at atmospheric pressure the Pd sputtered membrane showed a doubling of performance at 0.5 V from 10 mW/cm^2 to 20 mW/cm^2. The membrane that produced this performance was prepared with a water solvent using tetraamminepalladium(II) chloride hydrate and was the only sample that had a conductivity only 5% below that of Nafion®. The open circuit voltage for this sample was 70 mV higher than the value for pure Nafion®. Similar results were seen in other studies.[104,105] Although results have shown that Pd sputtered membranes can reach up to 100 mW/cm^2, experiments have not been conducted that show the durability of Pd altered membranes. If particles agglomerate with time and short the cell this method may of reducing crossover may not be worth the risk.

A radiation grafting technique similar to that presented by Florjanczyk et al.[79] was used by Hatanaka et al.[107] In their experiments, they grafted styrene into poly(ethylene-tetrafluroethylene) (ETFE) membranes. They found that under normal operating conditions the methanol permeability of the grafted membrane was about half of the Nafion® permeability. Polarization results showed that the grafted membrane performance was below that of Nafion® 115 or Nafion® 117 membranes. The authors attributed the lower performance to delamination of the membranes, but did not investigate the membrane conductivity. The work of Hatanka et al.[107] is fairly unique in the field of membrane development because their Nafion® membranes perform at a power density of 100 mW/cm^2 under normal fuel cell conditions. Many other researchers in this field do not fabricate MEAs operating near main stream literature values for comparing novel membrane materials.

Composite polymer materials such as polypyrrole/Nafion®,[98] sulfonated polyimide blend membranes,[108] and sulfonated polyetheretherketone (sPEEK) membranes[42,109,110] show promise in reducing the crossover of methanol, but have shown decreased proton conductivity that appears to cancel any benefit from the crossover reduction. This decrease in proton conductivity measured in in-situ experiments could possibly be accentuated by lamination problems.

One of the most promising membrane materials currently being invested are composites of poly(vinylidene fluoride) (PVdF) with Nafion®[111,112] and its composites with poly(styrenesulfonic) acid (PVdF-PSSA).[113] PVdF is not intrinsically conductive to protons and has a generally hydrophobic structure,[114] but when used as a copolymer gives membranes a more hydrophobic nature and thus reduces methanol crossover. Song et al.[111] showed that chemically modified blends of PVdF and Nafion® give equal performance to Nafion® membranes. Si et al. extended this work to show that a PVdF layer could be sandwiched between two Nafion® coatings to make a trilayer membrane with advantageous methanol crossover properties and good conductivity. The layers of pure Nafion® on top of the PVdF layers would also improve the lamination of the electrodes to the membranes. The trilayer membrane was shown to have a power density of 25 mW/cm^2 with 1-M methanol, atmospheric air and 4/4-mg/cm^2 Pt (A/C) loadings at 60 °C. Prakash et al.[113] demonstrated a DMFC with a PVdF-PSSA membrane at 55 °C with 8/8-mg/cm^2 Pt (A/C) loadings operating with 0.5-M methanol and atmospheric air. This DMFC had a power density of 25 mW/cm^2 under these conditions. The specific power of these cells is not high, but the fact that they are able to produce a reasonable amount of power at low temperatures and pressures is promising. The next step for these materials should be to investigate operation at high temperatures to see if the catalysts activate better and if the methanol crossover is still low under more demanding high temperature conditions.

VI. DMFC MODELING REVIEW

Many direct methanol fuel cell (DMFC) models have been developed since the 90's in order to capture and explain the transport and kinetic phenomena within the cell. DMFCs can be operated with vaporized or liquid fuel but the vast majority of the models gives attention to liquid-feed DMFCs since they are operated at temperatures below 100 °C and makes easier the experimental validation. In liquid-feed DMFCs the anode backing layer consists of two-phase flow due to the low solubility of CO_2 which is produced during the methanol oxidation and evolves as bubbles. Although many models assume that CO_2 remains dissolved in solution this effect gains more importance at high currents. Two major problems with DMFCs are the poor kinetics of methanol oxidation and methanol crossover from the anode to cathode. The former is best captured when models employ reaction mechanisms while the later is taken into account in the membrane transport equations and as mixed potential effect by including terms in the cathode kinetic expression. Although methanol crossover is of great importance some vapor-feed DMFC models neglect this effect.

A review of the more important DMFC models found in the literature follows. The models are presented for vapor-feed as well as liquid-feed DMFCs. These models are developed for isothermal and steady state conditions unless otherwise specified. One, two, and three-dimensional models are presented. The models range from simple Tafel kinetics to more complex kinetic expressions based on reaction mechanism and from easy analytical solutions to very difficult numerical solutions.

1. One-Dimensional Models

One of the firsts published models in DMFCs was presented by Verbrugge.[81] He developed a very simple model to study methanol diffusion through Nafion®-117 membranes. However, the cell used in this study does not correspond to an actual fuel-cell geometry since the electrodes are placed in the center of a well-stirred reservoir instead of being pressed against the membrane. The transport equations for the membrane were developed based on the dilute solution theory. The resulting set of dimensionless and linear equations with constant coefficients was solved by applying the Laplace-transform technique. This model when compared with experimental data showed that methanol diffuses through the membrane easily.

Baxter et al.[115] proposed a model for a DMFC anode. Their model treats the anode as a porous structure coated with a polymer electrolyte where the void spaces are flooded with aqueous methanol solution. Also, the model assumes that CO_2 remains dissolved in solution thereby neglecting the gas phase. The transport equations are based on dilute solution theory and the kinetics of methanol oxidation is described by the Butler-Volmer expression. A sensitivity analysis was used to determine the model parameters from experimental data obtained for a methanol-feed, polymer-electrolyte fuel cell operated with a hydrogen-producing electrode where the measured voltages were corrected to reflect the anodic overpotential.

The experimental data from Baxter et al.[115] was also used to validate the DMFC anode model developed by Jeng and Cheng.[116] This model incorporates diffusion and convection in the backing layer. Although carbon dioxide evolution is not considered in the backing layer, the gas bubble release into the flow channel is considered to hinder the methanol transport to the backing layer. Thus, the methanol transport to the backing layer in their model is represented by a mass transfer coefficient. The methanol oxidation rate is described by Butler-Volmer which is then simplified to a Tafel type equation. The set of nonlinear first-order equations were solved by the shoot-and-correct technique in terms of two initial values.

Models for the DMFC anode have been proposed based on the methanol oxidation reaction mechanism. Nordlund and Lindbergh[117] describe the structure of the electrode by a homogeneous agglomerate model where each agglomerate is assumed spherical. They used a simplified reaction mechanism from the mechanism presented by Hamnett.[15] The elementary steps for methanol oxidation are described as

$$CH_3OH + Site_1 \leftrightarrow CH_3OH_{ad,1} \tag{11}$$

$$CH_3OH_{ad,1} \rightarrow CO_{ad,1} + 4H^+ + 4\,e^- \tag{12}$$

$$H_2O + Site_2 \leftrightarrow OH_{ad,2} + H^+ + e^- \tag{13}$$

$$CO_{ad,1} + OH_{ad,2} \rightarrow CO_2 + H^+ + e^- + Site_1 + Site_2 \tag{14}$$

The kinetics is then described based on this mechanism coupled with the Butler-Volmer kinetics. The transport properties in the pore structure and the agglomerates are described by effective parameters. The carbon dioxide is assumed to be dissolved in solution and the Stefan-Maxwell equations are used to describe the mass transfer in the backing layer and the porous anode. The set of nonlinear equations is solved using a shooting method. Also, Scott and Argyropoulos[118] developed an analytical DMFC anode model based on a simplification to the methanol oxidation reaction mechanism from Frelink.[119] The steps follow Eqs. (11) to (14), where step (12) is considered in equilibrium similar to steps (11) and (14) leading to the following kinetic expression:

$$r = \frac{k' c_{MeOH} e^{\beta \eta}}{1 + k'' c_{MeOH} e^{\beta \eta}} \tag{15}$$

where c_{MeOH} is the methanol concentration, η is the anode overpotential, and k', k'' and β are constants. The expression for the voltage is then developed by applying Ohm's law through the membrane and combining it with the kinetic overpotential. A similar approach was used by Kulikovsky[120] who presented an analytical model for the DMFC anode. This model uses the kinetic rate expression for methanol oxidation developed by Meyers and Newman[121] which was derived from a mechanism proposed by Gasteiger.[40] The reaction at the anode is

$$CH_3OH + H_2O \leftrightarrow 6\ H^+ + 6\ e^- + CO_2 \qquad (16)$$

which consist in the following elementary steps

$$CH_3OH_{ad,1} + Site_1 \rightarrow CH_3OH_{ad,1} \qquad (17)$$

$$CH_3OH_{ad,1} \leftrightarrow CO_{ad,1} + 4\ H^+ + 4\ e^- \qquad (18)$$

$$H_2O + Site_2 \leftrightarrow OH_{ad,2} + H^+ + e^- \qquad (19)$$

$$CO_{ad,1} + OH_{ad,2} \rightarrow CO_2 + H^+ + e^- + Site_1 + Site_2 \qquad (20)$$

This mechanism led to a kinetic expression in the following form:

$$r = \frac{k' c_{MeOH} e^{\beta\eta}}{c_{MeOH} + k'' e^{\beta\eta}} \qquad (21)$$

The basic difference between Eq. (15) and Eq. (21) is in the adsorption of methanol. In the mechanism of Frenlink, the adsorption of methanol is taken as an equilibrium process while in the mechanism of Gasteiger methanol adsorption is not considered to be in equilibrium. However, this model can be solved only in the range of low or high currents.

The mathematical models discussed above have focused on the DMFC anode. However, this comprises just a portion of the models available in the DMFC literature. Several models to predict full DMFC performance have been attempted. Argyropoulos et al.[122] developed a semi-empirical model for a DMFC. The model describes the mass transport empirically by use of mass transfer coefficients and Tafel kinetics is used for methanol oxidation and oxygen reduction. These equations are then combined to obtain an expression for the cell voltage that is validated with experimental data. Kulikovsky[123,124] was one of the first researchers to introduce an analytical model for a DMFC. The model neglects the formation of CO_2 in the anode and uses simple Tafel kinetic expressions for the anode and the cathode. A simple reaction term is included in the cathode expression to represent the oxygen consumption in reaction with the methanol that crosses the membrane.

Dohle et al.[125] developed a mathematical model for a DMFC to understand the methanol permeation in the cell. The molar flux through the backing layers is described by a superposition of Stefan-Maxwell diffusion, Knudsen diffusion and Darcy permeation. Mass transport in the membrane is considered diffusive and by electroosmosis. The catalyst layers are described by Tafel type expressions where the mixed potential at the cathode due to methanol crossover is taken into account. The numerical solution of this model is based on the finite integration technique

A DMFC two-phase model was developed by Murgia et al.[126, 127] In the liquid phase, solutes such as methanol are described by the Nernst-Planck relationship and water transport is described by Darcy law. The gas mixture is considered ideal and is described by the Stefan-Maxwell equations. Butler-Volmer kinetics is used to express the rate of methanol oxidation and oxygen reduction. The condensation or evaporation of water at the cathode is considered via empirical relations and the effect of capillary pressure is represented by a Gaussian function. The model is solved numerically, but uses an analytical solution for the anode kinetic overpotential to decrease the solution time.

A mathematical model for a DMFC was presented by Scott et al.[128] The model takes into accounts the transport of water, methanol, and carbon dioxide gas. The effect of dissolved carbon dioxide and bubble evolution are lumped into an effective mass transfer coefficient. Tafel kinetics is used to describe the methanol oxidation. Finally, the semi-empirical model for the open circuit potential[129,130] is incorporated together with the Butler-Volmer cathode polarization model to predict overall cell potential.

A DMFC two-phase model that describes many simultaneous phenomena was developed by Sundmacher and Scott.[129] In the anode, the model includes mass transfer from the flow channel to the catalyst layer and describes the mass transport in the backing layer by diffusion and convection. The mass transport in the membrane compartment is described by diffusion and convective flow due to the effects of migration and pressure-driven transport. These equations were developed through mass and charges balances. The gas phase formation due to carbon dioxide evolution is taken into account. The component distributions are described by

K-values. While the gas phase is treated as an ideal gas mixture, the nonideality of the liquid phase is accounted for by using activity coefficients estimated from the UNIQUAC-method. The kinetics of methanol oxidation and oxygen reduction are described by the Butler-Volmer rate expression where the overpotential due to methanol oxidation in the cathode side is included. The resulting nonlinear set of model equations were analyzed in terms of dimensionless parameter groups in which the least-squares method of Marquardt-Levenberg was applied.

The first semi-analytical DMFC model in one dimension to include a kinetic expression derived from a mechanism was developed by García et al.[131] Their model only considers the liquid phase by assuming that CO_2 remains dissolved in solution. The transport equations were developed in terms of mass balances and Fick's law. In the membrane the mass transport of methanol is considered by diffusion and electroosmosis. A macro-homogeneous approach is assumed in the catalyst layers where the expression for methanol oxidation is taken from Meyers and Newman[121] and Tafel kinetics with the mixed potential effect describe the oxygen reduction. Analytical solutions were obtained for the different layers in the cell and combined with the kinetics expression to predict cell performance. The model was validated with experimental data as shown in Figure 15. Parameters were adjusted in order to fit the model with the experimental results for different methanol concentrations. Methanol concentration profiles from model simulations are shown in Figure 16. This model incorporates a complex anode kinetic expression with a very simple development that leads to a very short solution time.

Meyers and Newman[121] presented a mathematical model to describe multicomponent transport in a DMFC. The model includes a rigorous development of the thermodynamic equilibrium equations by considering first-order nonidealities between pairs of species in a charged ion-exchange membrane.[132] The transport equations were formulated on the basis of concentrated solution theory. The reaction kinetics for the methanol oxidation is based on a mechanism described by Gasteiger et al.[40]

Constitutive equations were used to describe the fluxes in the liquid, membrane, and gas phases. Transport in the liquid phase through the anode backing layer and the membrane phase are written for a binary solution of methanol and water. The diffusion of CO_2 out to the anode backing layer is neglected. It is considered

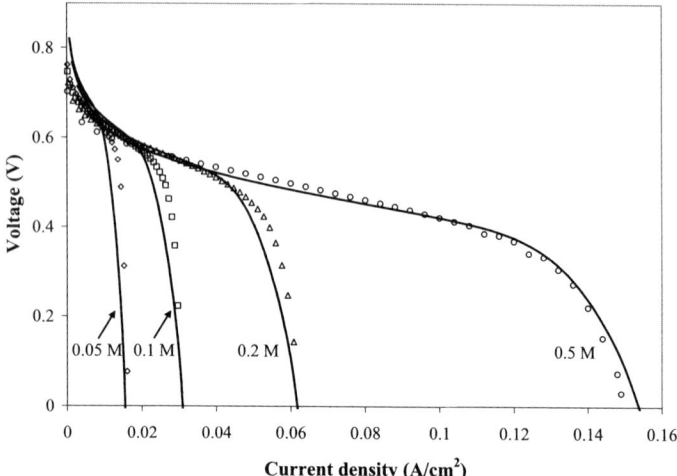

Figure 15. DMFC model against experimental data for different methanol concentrations. Reproduced from Ref. 131, (© 2004), by permission of ASME.

that CO_2 is dilute enough to remain dissolved in the condensed phase derby neglecting its composition in the mixture. The gas phase is expressed in terms of the Stefan-Maxwell equations where the species are treated as ideal gases. Multiple phases are

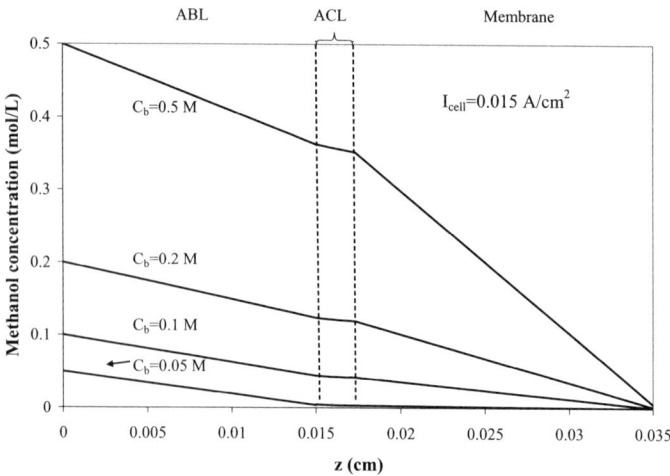

Figure 16. Methanol concentration profiles from the model simulations. Reproduced from Ref. 131, (© 2004), by permission of ASME.

handled by the porous-electrode theory described by Newman and Tiedemann.[133, 134] The rates of reaction for methanol oxidation and oxygen reduction are described by two different kinetic expressions. The rate expression for methanol oxidation is based on a series of elementary steps proposed by Gasteiger et al.[40] as described by Eqs. (17) to (20), which led to the kinetic expression of Eq. (21). The oxygen reduction is described by Tafel kinetics with first-order oxygen dependence.

The model is solved by using a finite difference scheme where the BAND(J) subroutine is implemented. However, due to the amount of details included and the nonlinearities in the equations the model become cumbersome and difficult to run requiring careful initial guesses of the concentration and potential profiles in order to converge. In order to determine transport and kinetic properties the model was adapted to simulate the methanol electrolysis cell described by Ren et al.[135] Unlike the fuel cell model, the electrolysis cell model has a filled-liquid backing instead of a gas backing at the cathode. Figure 17 shows the electrolysis cell simulation with data provided by Los Alamos National Laboratory (LANL) for a variety of methanol

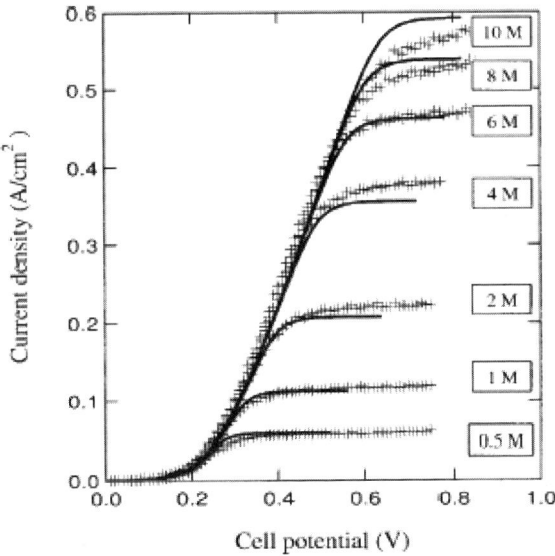

Figure 17. Current density vs. cell potential for methanol electrolysis at 80 °C with data provided by LANL. Reprinted from Ref. 121a, (© 2002), with permission from The Electrochemical Society, Inc.

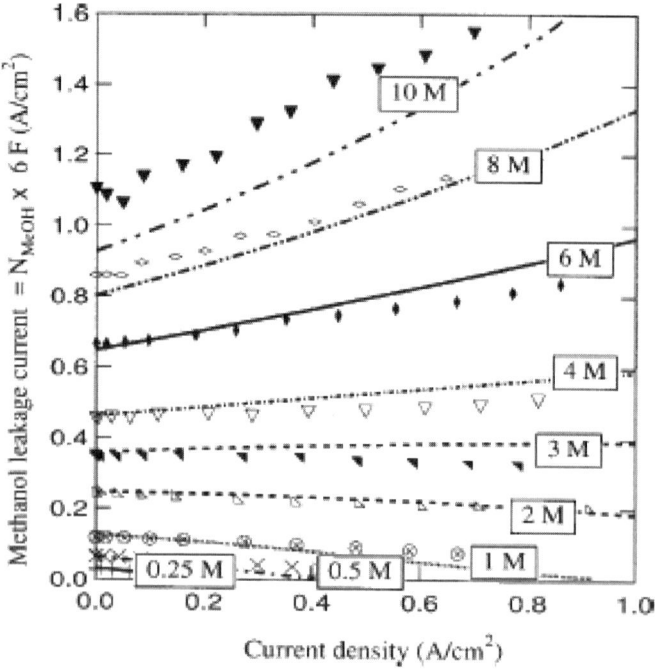

Figure 18. Model simulations for methanol crossover with experimental data from Ren et al. Reproduced from Ref. 121b, (© 2002), by permission of The Electrochemical Society, Inc.

concentrations. Methanol crossover from the fuel cell simulation is shown in Figure 18 also for different methanol concentrations.

Models for vapor-feed DMFCs have been few. Scott et al.[136] were the first to present a semi-empirical model for a vapor-feed DMFC to determine the effect of methanol crossover in the cell performance. The model employs the Stefan-Maxwell equations to describe the diffusion mass transfer in the porous backing layers on the basis that all components are in the gas phase. The electrocatalyst layers are described by Tafel expressions where the cathode overpotential due to methanol crossover is taken into account. The transport in the membrane is described using Fick's law. It is assumed that the membrane is fully hydrated and no pressure drop occurs across the membrane. A similar model was also developed by Scott and Taama,[46] however, the diffusion through the backing layer was neglected by assumed that the effect of diffusion through the highly porous structure is small under realistic operating conditions. The resulting non-linear equations for both models were solved using the finite-difference method and Newman's BAND algorithm. Also, Cruickshank and Scott[83] presented a simple model of methanol diffusion and electroosmosis for a vapor-feed DMFC. The model was used to analyze the effect of methanol crossover on the cell performance.

DMFC models with a greater degree of complexity are available in the literature. So far, most research works have been focused in steady state investigations. However, a few models to analyze the dynamic behavior of a DMFC were published.[137,138] Hydraulic behavior in DMFC stacks has been investigated by Argyropoulos et al.[139] Also, the thermal behavior in DMFC stacks can be found in the literature.[140,141]

2. Two-Dimensional and Three-Dimensional Models

The first attempt to construct a two-dimensional model was presented by Kulikovsky et al.[142] The model describes only the cathode compartment by assuming that the membrane is impermeable to gases. The diffusion flux of gases is defined in terms of the Stefan-Maxwell relation and Knudsen flow. A Tafel type expression is used to describe the oxygen reduction. The model is solved by using finite-difference approximation.

Birgersson et al.[143] developed a two-dimensional model for a DMFC anode. The model considers only liquid phase. The mass and momentum equations are used to describe the flow in the channel and the Darcy's law is used for the porous backing. The kinetic at the anode is described by the following parameter-adapted expression for the local current density:

$$\langle i \rangle = A(\langle c_{MeOH} \rangle)^B \frac{e^{\frac{\alpha_A F}{RT}(E_A - E_o)}}{1 + e^{\frac{\alpha_A F}{RT}(E_A - E_o)}} \quad (22)$$

where $\langle c_{MeOH} \rangle$ is the methanol concentration at the porous backing/active layer interface, E_A is the anode potential at the active layer/membrane interface *vs.* a dynamic hydrogen electrode, α_A is the Tafel slope, and A, B, and E_o are experimental fitted parameters. From these equations a reduced model was derived and the resulting system of nonlinear equations was solved using the Newton-Raphson-based algorithm. The model was not validated against experimental data. Later, this model was extended to considers two phases.[144] A similar development of the equations was presented for the two phases. However, the kinetics at the anode was described by the expression for the local current density given by Nordlund[145] as

$$\langle i \rangle = \frac{e^{\frac{\alpha_A F}{RT}(E_A - E_o)}}{1 + \frac{e^{\frac{\alpha_A F}{RT}(E_A - E_o)}}{i_{\lim}}} \qquad (23)$$

where i_{lim} is the limiting current density and E_o is a linear function of temperature. This model showed that the presence of the gas phase improved the mass transfer of methanol at temperatures above 30 °C.

A pseudo two-dimensional model for the full DMFC was developed by Kulikovsky.[146] The mass transport through the different layers is expressed on the base of the continuity equation. Tafel kinetics is used to describe the methanol oxidation and oxygen reduction. The system of equations was solved by the Runge-Kutta method.

A two-dimensional analytical model for the full DMFC was presented by Guo and Ma.[147] The model assumes that mass transport occurs only across the cell layers except for the flow channel where the transport occurs in the two directions. Also, the velocity in the flow channel is assumed uniform and no change in concentration across the catalyst layer is considered. Mass transport equations were developed based on Fick's law. Kinetics on the catalyst layers was described by Tafel expressions. The analytical solution was obtained by assuming that methanol concentration changes linearly in the PEM and neglecting methanol concentration in the cathode. Validation of the analytical solution is shown in Figure 19 where experimental data for different methanol concentrations are plotted against the model simulation. Two-dimensional distribution of the methanol concentration is shown in Figure 20.

Two-dimensional numerical models with Tafel kinetics expressions for DMFCs have been developed. In the model of Kulikovsky[148] the methanol transport are based on Fick's law and the transport of gases in the cathode follows his cathode compartment model[142] as previously described. Electroosmotic effect and pressure gradient are taken into account in the

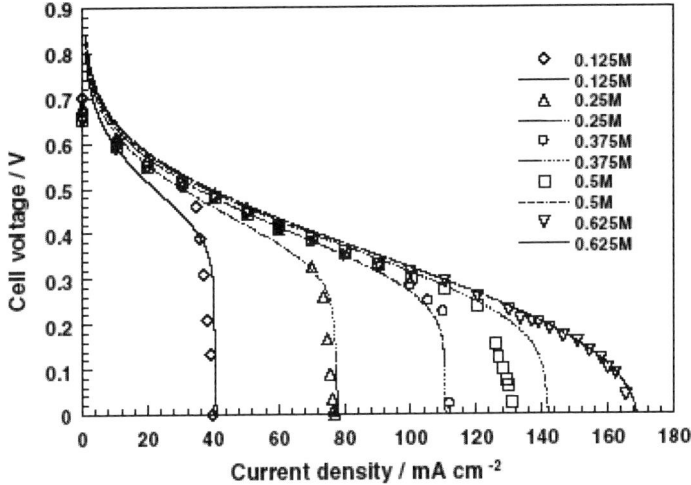

Figure 19. DMFC model against experimental data for different methanol concentrations. Reproduced from Ref. 147, (© 2004), by permission of Elsevier.

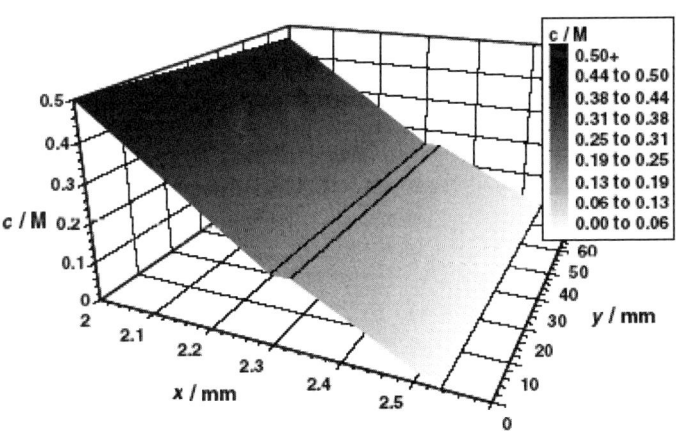

Figure 20. 2D-methanol distribution from the model simulation with 0.5-M methanol bulk concentration. Reproduced from Ref. 147, (© 2004), by permission of Elsevier.

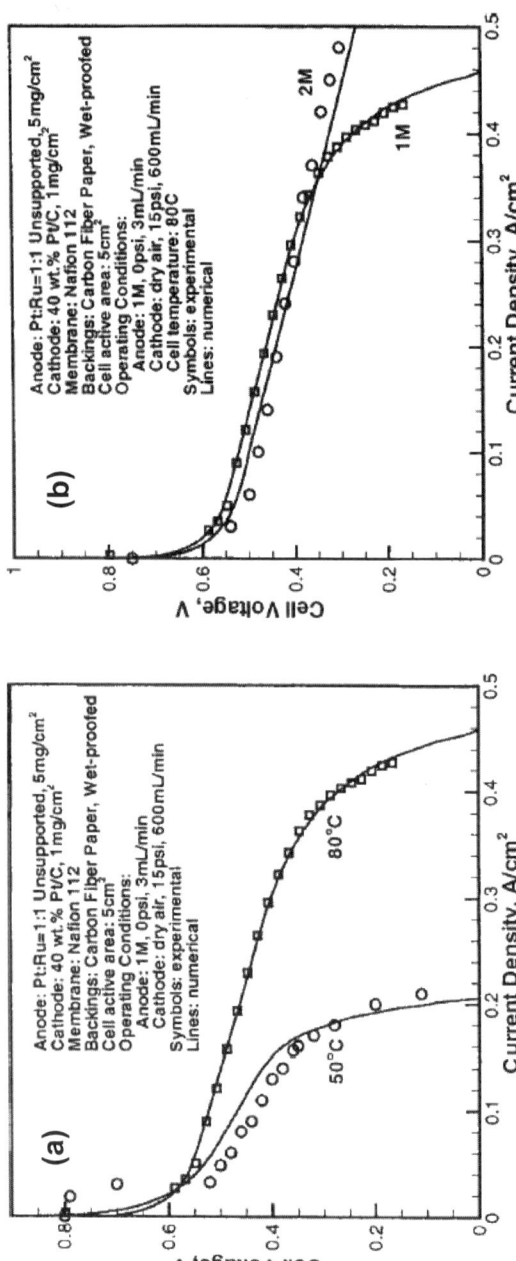

Figure 21. DMFC model against experimental data for (a) two different temperatures and, (b) two different methanol feed concentrations. Reprinted from Ref. 151, (© 2003), by permission from The Electrochemical Society. Inc.

membrane. A model for a vapor-feed DMFC was presented by Kulikovsky et al.[149] Transport of gases is described by Stefan-Maxwell and Knudsen diffusion mechanisms. In this model, however, the membrane is assumed impermeable for gases and fuel penetration thereby neglecting the methanol crossover. A similar model for a liquid-feed DMFC was developed by Fan et al.[150] As in the model of Kulivkosky,[148] this model accounts for the electroosmotic effect and pressure gradient in the membrane. Also, the mixed potential due to methanol crossover is included in the kinetic expression for the cathode. All these models were solved by using finite difference approximation.

Wang and Wang[151] developed a two-phase, 2D along-the-channel multicomponent model for the full DMFC to provide a basic understanding of transport and electrochemical phenomena involved. The catalyst layers are considered as thin interfaces between the backing layer and the membrane. The model is divided into flow channels and porous regions (the remaining fuel cell). For the flow channel a one-dimensional flow pattern is considered and the continuity equations are used to describe the transport. The model for the porous regions is based in species and momentum conservation. Equilibrium conditions are taken into account in the backing layers. Tafel kinetics expressions were used to describe the methanol oxidation and the oxygen reduction. However, a threshold value for the methanol concentration was adapted to split the behavior of the methanol oxidation into zero-order and first-order reaction dependence. First order kinetics was employed to describe the oxygen reduction. Also, a parasitic current term was used in the Tafel expression for the cathode catalyst interface to take into account the mixed potential effect of methanol oxidation due to methanol crossover. The model was solved numerically using computational fluid dynamics (CFD). Figure 21 shows the DMFC simulation for polarization curves compared with experimental data for different temperatures and different methanol feed concentrations.

Another two-dimensional numerical model was developed by Divisek et al.[152] The model describes the mass transport in two phases and describes the exchange between these phases by condensation and evaporation. Water flux includes pressure gradients, gravity, and proton drag effect. The flux of the dissolved species includes dispersion effects and induced flux due to water volumetric flux. The gas transport is governed by the Stefan-

Maxwell equations. Furthermore, the model considers temperature and the catalyst layers kinetics are described by expressions developed from reaction mechanisms. The system of equations was discretized in space and time by the finite volume method and the backward Euler method respectively. However, no transient results were shown.

A quasi-3D numerical model for a vapor-feed DMFC was developed by Kulikovsky.[153] The objective of the model is to demonstrate that a pressure gradient can be employed in a conventional flow field with the flow-through channels. The backing and catalyst layers are described by Stefan-Maxwell and Knudsen diffusion in a 2D scheme while the gas flow in the channels is described in 1D (in perpendicular direction) taking into account mass and momentum flux due to electrochemical reactions. The solution is obtained by assuming uniform distribution of feed gases concentration in the flow channels in order to obtain the solution of the 2D problem which is then used to solve the 1D problem in iterative manner until convergence is reached.[76]

VII. SUMMARY

An approximate target for DMFCs to be competitive with PEMFCs for applications requiring around 50–200 kW of power is to be able to achieve 250 mW/cm^2 above 0.5 V with less than 50 mA/cm^2 of methanol crossover. Current state-of-the-art Pt-Ru catalysts can provide around 300 mW/cm^2 at high temperature with pressurized oxygen on at least the cathode at 0.5 V. Whether the goal of methanol crossover is being met with Nafion® membranes is a current subject of debate in the literature. The method most researchers use to calculate the crossover is by making a carbon balance in the cathode. Several researchers have shown that CO_2 crossover results in overestimation of the amount of methanol crossover in DMFCs. Their results show that at 500 mA/cm^2 the methanol crossover for > 1100 EW Nafion® membranes is near 50 mA/cm^2 at 90 °C. The only dissenting opinion on the magnitude of CO_2 crossover came from researchers who used a 1000-EW membrane.[97] It is possible that lower equivalent weight membranes have a lower proportion of hydrophobic pores where CO_2 could cross thus explaining the difference in results.

The desired performance goals for DMFCs have been obtained for ideal conditions and with large platinum loadings. The 2004 platinum cost for researchers who are achieving the performance goal is around $200-$500 /kW. Goals for total system cost are generally around $300/kW for 5–100 kW stationary power systems, thus making the current platinum cost alone too high to achieve these goals. The need for a pressurized cathode also requires a compressor that both increases system cost and significantly reduces system efficiency. This translates into a need for lower Pt loadings and lower cathode pressure.

The main remedy that has been proposed for most of these problems is operating DMFCs at high temperatures. High temperature operation has thus far been unachievable without pressure on the cathode. Because the methanol diffusion coefficient in the membrane also increases with temperature there are more problems with methanol crossover. These facts have led to research on membrane materials that will be stable and conductive under these conditions. Researchers have tried functionalizing the acid groups in the membrane with ZrO, SiO_2, heteropoly acids, and palladium to reduce methanol crossover and they have tried to develop novel membranes materials that provide superior performance to Nafion®. Functionalizing acid groups led to reduced crossover, but also reduced membrane conductivity. Researchers have not been able to show good performance at temperatures above the boiling point of water at any cathode pressure. This is due to the necessity for membrane hydration using Nafion®. Novel polymers have not yet been developed that perform well in DMFC systems at temperatures above the boiling point of water. Making MEAs from novel membrane materials is also complicated by needing to get a good interface between the ionomer in the membrane and electrode. One of the more interesting solutions to this problem, proposed by Fenton,[112] is to laminate a layer of methanol impermeable polymer between two Nafion® membrane so that the electrodes will make good contact with the membrane. However, this type of solution will still require hydration to perform will at high temperatures.

A variety of one-dimensional and two-dimensional models have been developed to describe DMFC behavior. Due to the complexities of the MOR, most models of the anode kinetics are solved numerically, but can alternatively be simplified and solved analytically for specific cases. Many of the anode kinetics models are developed using multi-step reaction mechanisms where the CO oxidation limits the reaction at high potentials and adsorption of methanol limits the reaction rate at low potential.

Modeling mass transfer in the anode is complex because of the two-phase flow in all regions. Most researchers choose to neglect the CO_2 in the electrode and backing layer due to the complexity of the equations for two-phase flow in porous media. The effect of CO_2 on mass transfer is most often included as a mass transfer coefficient at the interface of the backing layer and flow channel although, Wang[151] has used CFD to include the two phase effects in the backing layer.

The cathode kinetics are usually included using a Tafel expression. Some models solve the concentration profile in the cathode to include the effect of concentration and crossover, but many models consider the oxygen concentration to be a constant. This is probably neglected due to the fact that DMFCs are often operated with a large stoichiometric excess of pure oxygen at high pressures.

Two-dimensional models have been developed to explain the variation in current along the flow path. Transient models have been proposed but not much work has been accomplished in this area.

REFERENCES

[1] B.D. McNicol, , D. A. J. Rand, and K. R. Williams, *Journal of Power Sources* **83** (1999) 15-31.
[2] B.D. McNicol, , D. A. J. Rand, and K. R. Williams, *Journal of Power Sources* **100** (2001) 47-59.
[3] S. Banerjee, and D. E. Curtin, *J Fluorine Chem* **125** (2004) 1211-1216.
[4] J. Larminie and A. Dicks, *Fuel Cell Systems Explained*, 2nd ed. John Wiley & Sons Inc., Hoboken, NJ, 2003.
[5] R. Dillon et al., *J Power Sources* **127** (2004) 112-126.
[6] A. K. Shukla and R.K. Raman, *Annu. Rev. Mater. Res.* **33** (2003) 155-168.
[7] T. Kendall, *Platinum 2004,* Johnson Matthey, 2004.
[8] R. Parsons and T. Vandernoot, *J. Electroanal. Chem.* **257** (1988) 9-45.
[9] T. Iwasita, *Electrochim. Acta* **47** (2002) 3663-3674.
[10] S. Wasmus and A. Kuver, *J. Electroanal. Chem.* **461** (1999) 14-31.
[11] L. Carrette, K. A. Friedrich, and U. Stimming, *Chemphyschem* **1**(4) (2000) 162-193.
[12] E. A. Batista et al., *J. Electroanal. Chem.* **571** (2004) 273-282.
[13] E. A.Batista et al., *Electrochem. Commun.* **5** (2003) 843-846.
[14] S. Desai and M. Neurock, *Electrochimica Acta* **48** (2003) 3759-3773.
[15] A. Hamnett, *Catalysis Today* **38** (1997) 445-457.
[16] S. J. Liao, V. Linkov, and L. Petrik, *Appl. Catal. a-Gen.* **235** (2002) 149-155.
[17] K. W. Park et al., *J. Phys. Chem. B* **106** (2002) 1869-1877.
[18] K. Y. Chan et al., *J. Mater. Chem.* **14** (2004) 505-516.

[19] M. Watanabe, M. Uchida, and S. Motoo, *J. Electroanal. Chem.* **229** (1987) 395-406.
[20] H. N. Dinh et al., *J. Electroanal. Chem.* **491** (2000) 222-233.
[21] H. Hoster et al., *J. Electrochem. Soc.* **148** (2001) A496-A501.
[22] M E. Tess et al., *Inorg. Chem.* **39** (2000) 3942-+.
[23] J. W. Long et al., *J. Phys. Chem. B* **104** (2000) 9772-9776.
[24] S. L. Gojkovic and T.R. Vidakovic, *Electrochim. Acta* **47** (2001) 633-642.
[25] H. Hoster et al., *Phys. Chem. Chem. Phys.* **3** (2001) 337-346.
[26] T. Iwasita et al., *Langmuir* **16** (2000) 522-529.
[27] Z. D. Wei and S.H. Chan, *J. Electroanal. Chem.* **569** (2004) 23-33.
[28] A. J. Dickinson et al., *Electrochim. Acta* **47** (2002) 3733-3739.
[29] A. J. Dickinson et al., *J. Appl. Electrochem.* **34** (2004) 975-980.
[30] Z. J. Jusys, Kaiser, and R.J. Behm, *Langmuir* **19** (2003) 6759-6769.
[31] L. Gao, H.L. Huang, and C. Korzeniewski, *Electrochim. Acta* **49** (2004) 1281-1287.
[32] A. H. C. Sirk et al., *J. Phys. Chem. B* **108** (2004) 689-695.
[33] L. Dubau et al., *J. Appl. Electrochem.* **33** (2003) 419-429.
[34] S. L. Gojkovic, T.R. Vidakovic, and D.R. Durovic, *Electrochim. Acta* **48** (2003) 3607-3614.
[35] A. V. Tripkovic et al., *Electrochim. Acta* **47** (2002) 3707-3714.
[36] M. S. Loffler et al., *Electrochim. Acta* **48** (2003) 3047-3051.
[37] Y. J. Zhang et al., *Catal. Today* **93-95** (2004) 619-626.
[38] B. Yang et al., *Chem. Mater.* **15** (2003) 3552-3557.
[39] L. Dubau et al., *J. Electroanal. Chem.* **554** (2003) 407-415.
[40] H. A. Gasteiger et al., *Journal of Physical Chemistry* **97** (1993) 12020-12029.
[41] M. P. Hogarth and T.R. Ralph, *Platinum Metals Review* **46** (2002) 146-164.
[42] L. Jorissen et al., *J. Power Sources* **105** (2002) 267-273.
[43] Q. F. Li et al., *Chem. Mater.* **15** (2003) 4896-4915.
[44] S. Surampudi et al., *J. Power Sources* **47** (1994) 377-385.
[45] X. M. Ren, M.S. Wilson, and S. Gottesfeld, *J. Electrochem. Soc.* **143** (1996) L12-L15.
[46] K. Scott, W. Taama, and J. Cruickshank, *J. Appl. Electrochem.* **28** (1998) 289-297.
[47] C. K. Witham, et al., *Electrochem. Solid St.* **3** (2000) 497-500.
[48] Y K. Xiu and H. Nakagawa, *J. Electrochem. Soc.* **151** (2004) A1483-A1488.
[49] M. Baldauf and W. Preidel, *J. Appl. Electrochem.* **31** (2001) 781-786.
[50] S. C. Thomas et al., *Electrochim. Acta* **47** (2002) 3741-3748.
[51] A. K. Shukla et al., *Electrochim. Acta* **47** (2002) 3401-3407.
[52] A. K. Shukla et al., *J. Power Sources* **76** (1998) 54-59.
[53] A. S.Arico et al., *Electrochim. Acta* **47** (2002) 3723-3732.
[54] A. S. Arico et al., *Electrochem. Commun.* **6** (2004) 164-169.
[55] W. C. Choi and S. I. Woo, *J. Power Sources* **124** (2003) 420-425.
[56] C. Coutanceau et al., *J. Appl. Electrochem.* **34** (2004) 61-66.
[57] Y. H. Chu et al., *J. Power Sources* **118** (2003) 334-341.
[58] D. X. Cao and S. H. Bergens, *J. Power Sources* **134** (2004) 170-180.
[59] S. C. Thomas, X. Ren, and S. Gottesfeld, *J. Electrochem. Soc.* **146** (1999) 4354-4359.
[60] Z. Wei et al., *J. Power Sources* **106** (2002) 364-369.
[61] D. Chu and S. Gilman, *J. Electrochem. Soc.* **141** (1994) 1770-1773.
[62] P. S. Kauranen and E. Skou, *J. Electroanal. Chem.* **408** (1996) 189-198.
[63] Z. Jusys and R. J. Behm, *Electrochim. Acta* **49** (2004) 3891-3900.

[64]S. Gupta et al., *J. Appl. Electrochem.* **28** (1998) 673-682.
[65]R. K. Raman, G. Murgia, and A. K. Shukla, *J. Appl. Electrochem.* **34** (2004) 1029-1038.
[66]A. K. Shukla et al., *J. Electroanal. Chem.* **563** (2004) 181-190.
[67]H. Yang et al., *J. Phys. Chem. B* **108** (2004) 1938-1947.
[68]H. Yang et al., *J. Phys. Chem. B* **108** (2004) 11024-11034.
[69]F. Maillard et al., *Electrochim. Acta.* **47** (2002) 3431-3440.
[70]T. J. Schmidt et al., *J. Electrochem. Soc.* **147** (2000) 2620-2624.
[71]K. Sawai and N. Suzuki, *J. Electrochem. Soc.* **151** (2004) A2132-A2137.
[72]K. Sawai and N. Suzuki, *J. Electrochem. Soc.* **151** (2004) A682-A688.
[73]J. Kallo, W. Lehnert, and R. von Helmolt, *J. Electrochem. Soc.* **150** (2003) A765-A769.
[74]C. Yang et al., *Electrochem. Solid St.* **4** (2001) A31-A34.
[75]C. Yang et al., *J. Membrane Sci.* **237** (2004) 145-161.
[76]A. S. Arico et al., *J. Power Sources* **128** (2004) 113-118.
[77]P. L. Antonucci et al., *Solid State Ionics* **125** (1999) 431-437.
[78]P. Staiti et al., *Solid State Ionics* **145** (2001) 101-107.
[79]Z. E. Florjanczyk, E. Wielgus-Barry, and Z. Poltarzewski, *Solid State Ionics* **145** (2001) 119-126.
[80]A. Heinzel and V. M. Barragan, *J. Power Sources* **84** (1999) 70-74.
[81]M. W. Verbrugge, *J. Electrochem. Soc.* **136** (1989) 417-423.
[82]V. N. Tricoli, Carretta, and M. Bartolozzi, *J. Electrochem. Soc.* **147** (2000) 1286-1290.
[83]J. Cruickshank and K. Scott, *J. Power Sources* **70** (1998) 40-47.
[84]P. S. Kauranen and E. Skou, *J. Appl. Electrochem.* **26** (1996) 909-917.
[85]S. Hikita, K. Yamane, and Y. Nakajima, *Jsae. Rev.* **23** (2002) 133-135.
[86]S. Hikita, K. Yamane, and Y. Nakajima, *Jsae. Rev.* **22** (2001) 151-156.
[87]X. M. Ren et al., *J. Power Sources* **86** (2000) 111-116.
[88]X. Ren, T. E. Springer, and S. Gottesfeld, *J. Electrochem. Soc.* **147** (2000) 92-98.
[89]K. D. Kreuer et al., *Chemical Reviews* **104** (2004) 4637-4678.
[90]X. Ren, T. Zawodzinski, and S. Gottesfeld, *Abstr. Pap. Am. Chem. Soc.* **217** (1999) U490-U490.
[91]X. M. Ren and S. Gottesfeld, *J. Electrochem. Soc.* **148** (2001) A87-A93.
[92]X. M. Ren, W. Henderson, and S. Gottesfeld, *J. Electrochem. Soc.* **144** (1997) L267-L270.
[93]X. M. Ren et al., *J. Electrochem. Soc.* **147** (2000) 466-474.
[94]H. Dohle et al., *Journal of Power Sources* **105** (2002) 274-282.
[95]J. A. Drak, W. Wilson, and K. Killeen, *J. Electrochem. Soc.* **151** (2004) A413-A417.
[96]R. Jiang and D. Chu, *J. Electroch. Soc.* **151** (2004) A69-A76.
[97]V. Gogel et al., *J. Power Sources* **127** (2004) 172-180.
[98]E. B. Easton et al., *J. Electrochem. Soc.* **150** (2003) C735-C739.
[99]N. Miyake, J. Wainright, and R. F. Savinell, *Journal of the Electrochemical Society* **148** (2001) A898-A904.
[100]N. Miyake, J. Wainright, and R. F. Savinell, *Journal of the Electrochemical Society* **148** (2001) A905-A909.
[101]S. P. Nunes et al., *Journal of Membrane Science* **203** (2002) 215-225.
[102]U. Lavrencic Stangar et al., *Solid State Ionics* **145** (2001) 109-118.
[103]Z. Poltarzewski et al., *Solid State Ionics* **119** (1999) 301-304.
[104]W. C. Choi, J. D. Kim, and S. I. Woo, *J. Power Sources* **96** (2001) 411-414.
[105]Z. Q. Ma, P. Cheng, and T. S. Zhao, *J. Membrane Sci.* **215** (2003) 327-336.

[106] Y. J. Kim et al., *Electrochim. Acta* **49** (2004) 3227-3234.
[107] T. Hatanaka et al., *Fuel* **81** (2002) 2173-2176.
[108] K. Miyatake, H. Zhou, and M. Watanabe, *Macromolecules* **37** (2004) 4956-4960.
[109] L. Li, J. Zhang, and Y. X. Wang, *J. Membrane Sci.* **226** (2003) 159-167.
[110] L. Li, J. Zhang, and Y. X. Wang, *J. Mater. Sci. Lett.* **22** (2003) 1595-1597.
[111] M. K. Song et al., *Journal of Power Sources* **117** (2003) 14-21.
[112] Y. Si et al., *Journal of the Electrochemical Society* **151** (2004) A463-A469.
[113] G. K. S. Prakash et al., *J. Fluorine Chem.* **125** (2004) 1217-1230.
[114] M. A. Navarra et al., *Journal of the Electrochemical Society* **150** (2003) A1528-A1532.
[115] S. F. Baxter, V. S. Battaglia, and R. E. White, *Journal of the Electrochemical Society* **146** (1999) 437-447.
[116] K. T. Jeng and C. W. Chen, *Journal of Power Sources* **112** (2002) 367-375.
[117] J. Nordlund and G. Lindbergh, *Journal of the Electrochemical Society* **149** (2002) A1107-A1113.
[118] K. Scott and P. Argyropoulos, *Journal of Power Sources* **137** (2004) 228-238.
[119] T. Frelink, W. Visscher, and J. A. R. van Veen, *Langmuir* **12** (1996) 3702.
[120] A. A. Kulikovsky, *Electrochemistry Communications* **5** (2003) 530-538.
[121] J. P. Meyers and J. Newman, *Journal of the Electrochemical Society* **149** (2002) (a) A718-A728m and (b) A729-A735.
[122] P. Argyropoulos et al., *Journal of Power Sources* **123** (2003) 190-199.
[123] A. A. Kulikovsky, *Electrochemistry Communications* **4** (2002) 318.
[124] A. A. Kulikovsky, *Electrochemistry Communications* **4** (2002) 939-946.
[125] H. Dohle, J. Divisek, and R. Jung, *Journal of Power Sources* **86** (2000) 469-477.
[126] G. Murgia et al., *Journal of the Electrochemical Society* **150** (2003) A1231-A1245.
[127] R. K. Raman, G. Murgia, and A. K. Shukla, *Journal of Applied Electrochemistry* **34** (2004) 1029-1038.
[128] K. Scott, P. Argyropoulos, and K. Sundmacher, *Journal of Electroanalytical Chemistry* **477** (1999) 97-110.
[129] K. Sundmacher and K. Scott, *Chemical Engineering Science* **54** (1999) 2927-2936.
[130] K. Scott, K, *Journal of Power Sources* **83** (1999) 204-216.
[131] B. L. García, V. A. Sethuraman, J. W. Weidner, R E. White, and R. Dougal, *Journal of Fuel Cell Science and Technology* **1**(1) (2004) 43.
[132] J. P. Meyers and J. Newman, *Journal of the Electrochemical Society* **149** (2002) A710-A717.
[133] J. Newman and W. Tiedemann, *AIChE Journal* **21** (1975) 25.
[134] J. S. Newman and W. Tiedemann, *Advances in Electrochemistry and Electrochemical Engineering* **11** (1978) 353.
[135] X. Ren et al., *The Electrochemical Society Proceedings Series*, 1995. Pennington, NJ.
[136] K. Scott, W. Taama, and J. Cruickshank, *Journal of Power Sources* **65** (1997) 159-171.
[137] S. Zhou et al., *Physical Chemistry Chemical Physics* **3** (2001) 347-355.
[138] K. Sundmacher et al., *Chemical Engineering Science* **56** (2001) 333-341.
[139] P. Argyropoulos, K. Scott, and W. M. Taama, *Journal of Applied Electrochemistry* **30** (2000) 899-913.
[140] P. Argyropoulos, K. Scott, and W. M. Taama, *Journal of Power Sources* **79** (1999) 169-183.
[141] P. Argyropoulos, K. Scott, and W. M. Taama, *Journal of Power Sources* **79** (1999) 184-198.

[142] A. A. Kulikovsky, J. Divisek, and A. A. Kornyshev, *Journal of the Electrochemical Society* **146** (1999) 3981-3991.
[143] E. Birgersson et al., *Journal of the Electrochemical Society* **150** (2003) A1368-A1376.
[144] E. Birgersson et al., *Journal of the Electrochemical Society* **151** (2004) A2157-A2172.
[145] J. Nordlund and G. Lindbergh, *Journal of the Electrochemical Society* **151** (2004) A1357-A1362.
[146] A. A. Kulikovsky, *Electrochemistry Communications* **6** (2004) 1259-1265.
[147] H. Guo and C. Ma, *Electrochemistry Communications* **6** (2004) 306-312.
[148] A. A. Kulikovsky, *Journal of Applied Electrochemistry* **30** (2000) 1005-1014.
[149] A. A. Kulikovsky, J. Divisek, and A. A. Kornyshev, *Journal of the Electrochemical Society* **147** (2000) 953-959.
[150] J. R. Fan et al., *Energy & Fuels* **16** (2002) 1591-1598.
[151] Z. H. Wang and C.Y. Wang, *Journal of the Electrochemical Society* **150** (2003) A508-A519.
[152] J. Divisek et al., *Journal of the Electrochemical Society* **150** (2003) A811-A825.
[153] A. A. Kulikovsky, *Electrochemistry Communications* **3** (2001) 460-466.

6

Direct Numerical Simulation of Polymer Electrolyte Fuel Cell Catalyst Layers

Partha P. Mukherjee,[1] Guoqing Wang,[2] and Chao-Yang Wang[1]*

[1]*Electrochemical Engine Center (ECEC), and Department of Mechanical and Nuclear Engineering, The Pennsylvania State University, University Park, PA*
[2]*Plugpower Inc., Latham, New York*
* Corresponding author

I. INTRODUCTION

Fuel cells, due to their high energy efficiency, zero pollution and low noise, are widely considered as the 21st century energy-conversion devices for mobile, stationary and portable power. Among the several types of fuel cells, polymer electrolyte fuel cell (PEFC) has emerged as the most promising power source for a wide range of applications.

A typical PEFC is schematically shown in Figure 1 and divided into seven subregions: the anode gas channel, anode gas diffusion layer (GDL), anode catalyst layer (CL), ionomeric membrane, cathode CL, cathode GDL, and cathode gas channel. The proton-exchange membrane electrolyte is a distinctive feature of PEFC. Usually the two thin catalyst layers are coated on both sides of the membrane, forming a membrane-electrode assembly

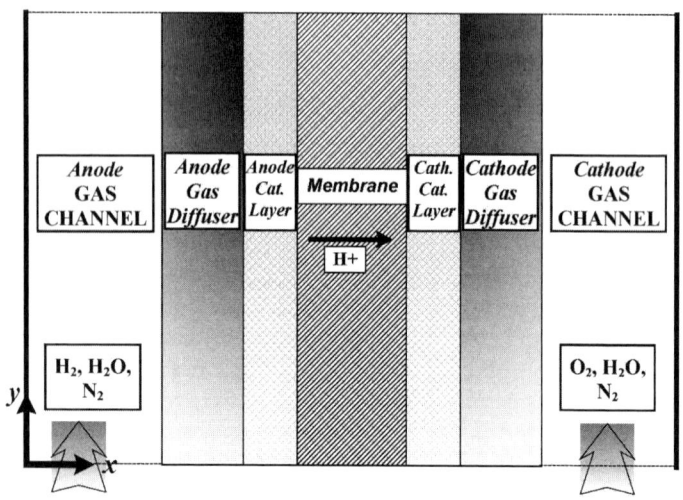

Figure 1. Schematic diagram of a polymer electrolyte fuel cell.

(MEA). The anode feed, generally, consists of hydrogen, water vapor, and nitrogen or hydrogen/water binary gas, whereas humidified air is fed into the cathode. Hydrogen and oxygen combine electrochemically within the active catalyst layers to produce electricity, water and waste heat. The catalyst layer of thickness around 10 μm is, therefore, a critical component of a PEFC and requires extensive treatment. Gottesfeld and Zawodzinski[1] provided a good overview of the catalyst layer structure and functions. The hydrogen oxidation reaction (HOR) occurs at the anode side catalyst layer and protons are generated according to the following reaction:

$$H_2 \rightarrow 2H^+ + 2e^- \qquad (1)$$

The oxygen reduction reaction (ORR) takes place at the cathode catalyst layer and water is produced.

$$O_2 + 4H^+ + 4e^- \rightarrow 2H_2O \qquad (2)$$

Thus, the overall cell reaction is:

$$2H_2 + O_2 \rightarrow 2H_2O \qquad (3)$$

HOR has orders of magnitude higher reaction rate than ORR, which leaves ORR as a potential source of large voltage loss in PEFCs. Due to the acid nature of the polymer membrane and low-temperature operation, Pt or Pt-alloys are the best-known catalysts for PEFCs. For the electrochemical reaction to occur in the cathode catalyst layer, the layer must provide access for oxygen molecules, protons, and electrons. Therefore, the catalyst layer consists of:

1. the ionomer, i.e., the ionic phase which is typically Nafion® to provide a passage for protons to be transported in or out,
2. metal (Pt) catalysts supported on carbon i.e., the electronic phase for electron conduction, and
3. pores for the oxygen gas to be transferred in and product water out. The salient phenomena occurring in the catalyst layer, therefore, include interfacial reaction at the electrochemically active sites, proton transport in the electrolyte, electron conduction in the electronic phase (i.e., Pt/C), and oxygen diffusion through the gas phase, liquid water, and electrolyte phase.

Several modeling approaches have been used for the catalyst layers. In most of the macroscopic models reported in the literature,[2-5] the active catalyst layer was not the main focus but rather treated as a macrohomogeneous porous layer. A few detailed models were specifically developed for PEFC catalyst layers based on the theory of volume averaging and they can be further distinguished as homogeneous model, film model and agglomerate model. Springer and Gottesfeld,[6] Perry et. al.,[7] and Eikerling and Kornyshev[8] presented several analytical and numerical solutions for the cathode catalyst layer under various conditions. Recent reviews by Wang[9] and Weber and Newman[10] provided good overviews of the various catalyst layer models.

However, the above-mentioned macroscopic models do not address localized phenomena at the pore scale. Pisani et al.[11] constructed a pore-level model over an idealized, one-dimensional model geometry of pores and assessed the effects of the catalyst layer pore structure on polarization performance. Recently, a direct numerical simulation (DNS)[12] model has been developed at Penn State Electrochemical Engine Center to describe the oxygen, water and charge transport at the pore level within 2-D and 3-D computer-generated catalyst layer microstructures.

The objective of the current chapter is to present a systematic development of the direct numerical simulation (DNS) model starting with idealized, regular 2-D and 3-D catalyst layer microstructures to a purely random 3-D microstructure and finally to a statistically more rigorous description of a 3-D correlated microstructure. The pore-scale transport of charge and species within a microscopically complex CL microstructure and its importance in the development of high-performance catalyst layers are elucidated.

II. DIRECT NUMERICAL SIMULATION (DNS) APPROACH

As mentioned earlier, traditionally, porous electrodes in fuel cells are modeled using the macrohomogeneous technique, where the properties and variables of each phase are volume-averaged over a representative elementary volume containing a sufficient number of particles.

In such an approach, microscopic details of the pore structure are smeared and the electrode is described using the porosity, interfacial area per unit volume, effective conductivity, diffusivity etc. through a homogenized porous medium. With these volume-averaged variables, macroscopic governing equations are derived from their microscopic counterparts by assuming the uniformity of the microscopic properties within the representative elementary volume, which implies existence of phase equilibrium. In these volume-averaged equations, empirical correlations are used to describe the effective properties as a function of porosity and tortuosity, which are characteristic of the porous structure. Therefore, in the macrohomogeneous model, both the structure and the variables are homogenized microscopically. The effects of the microstructural morphology are ignored and also empirical transport properties need to be introduced. On the microscopic level, however, different modes of oxygen transport are responsible for the reactant supply to the reaction surface in the cathode catalyst layer. The relative importance of each distinct mechanism depends on the specific pore structure and relative volumes of small and large pores. It is also possible that oxygen has to dissolve in and diffuse through the ionomer (i.e., Nafion®) and/or the product liquid water in order to reach the reaction surface. Therefore, using the effective overall diffusion coefficient for oxygen in the porous layer seems to be a gross simplification.

Indeed, a most recent measurement of Stumper et al.[13] indicated that the effective oxygen diffusivity through the composite media of the catalyst layer, microporous layer and gas diffusion layer is only one tenth of its theoretical value in air. A detailed characterization of the pore structure is thus an important prerequisite to describe oxygen transport in a microscopic model. To account for these effects of microphase morphology, a direct numerical simulation (DNS) approach is developed.

The concept of DNS first appeared in the modeling of turbulence in fluid mechanics, where micro-level vortices were traced by solving the Navier-Stokes equations directly, instead of using Reynolds averaging.[14] Another application of DNS is in the study of combustion in porous materials, where the flame thickness is of the same order as the pore size.[15]

The motivation of the DNS modeling is to solve the various point-wise accurate conservation equations on a real microstructure of a porous electrode, instead of the volume-averaged equations based on the homogenized structure. Thus, the geometry of each phase in the porous electrode has to be identified in order to implement the microscopic transport equations for each phase.

1. Advantages and Objectives of the DNS Approach

The advantage of the DNS approach is that the governing equations used are point-wise accurate instead of phase-homogenized equations as in the macrohomogeneous model. Therefore, the DNS model can include the important effects of microstructural morphologies, thus yielding more accurate predictions. In the application to fuel cells, this method can be used to analyze and identify the various losses from the cathode catalyst layer, using computer-generated porous microstructures. Humidity effect of the inlet oxidant can be studied by additionally solving the water transport problem through the catalyst layer. In addition, optimal compositions and structures of various phases in the catalyst layer to achieve minimum voltage loss can be explored using the DNS approach.

Another unique application of the DNS approach is to examine and evaluate the empirical correlations needed in the macrohomogeneous model by building a pore-level database. With the detailed distributions of species concentrations and electrical potentials averaged macroscopically, the effective diffusivity and conductivity can be calculated via Fick's law and Ohm's law, respectively. Furthermore, since the macrohomogeneous model is based on the assumptions of homogeneous structures and local equilibrium, the DNS method is the only approach to simulate the thin and highly heterogeneous electrodes (the catalyst layer of 10 μm in thickness is an example), high-rate or fast-transient operations, where the use of a volume-averaging method in a macrohomogeneous model would lose its physical basis.

In addition, the DNS method could distinguish the performance of electrodes when their microstructures vary from each other, and this eventually will help to engineer various micro-morphologies of porous materials for different applications at the phase interfacial scale. However, the DNS model is computationally much more intensive since it resolves the microstructure of a porous electrode. Especially when the real 3D geometries are to be included, it would cost computational time because extremely fine grids would be required to capture each phase intercrossing in a complex manner. Nonetheless, parallel computation on an inexpensive Linux cluster should significantly increase the computing power for DNS calculations.

2. DNS Model - Idealized 2-D Microstructure

There are essentially two steps in the DNS model. The first step is the catalyst layer microstructure reconstruction. The second step is solving transport equations for protons, electrons and chemical species directly on the reconstructed electrode microstructure. For the illustration of these two steps, a two-dimensional regular microstructure is employed first.

An ideal cathode catalyst layer structure would have the following properties: all Pt catalysts should be dispersed on the surfaces between the electronic phase (i.e., carbon) and the electrolyte, typically Nafion®, forming a catalyzed interface for the ORR. This catalyzed interface must be further accessible by oxygen. According to these basic requirements for the catalyst layer, a 2-D cathode catalyst layer is first generated, as shown in Figure 2. In the schematic diagram, x direction is along the catalyst layer thickness, and y direction represents a periodic repeating unit of a realistic catalyst layer. The physical system includes three phases, an electronic phase, an electrolyte phase and a gas phase. At the interface between the electronic and gas phase, there is a thin electrolyte film, and Pt catalyst is also assumed to exist on the

DNS of PEFC Catalyst Layers

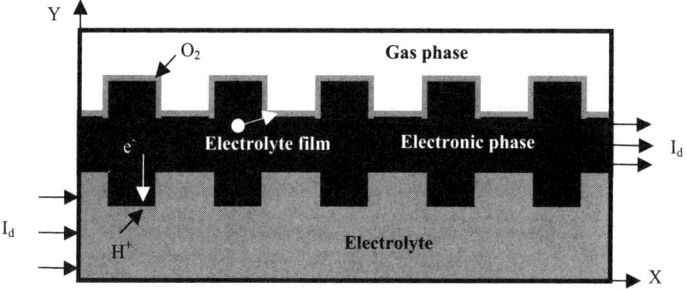

Figure 2. Schematic diagram of the 2-D computational domain.

electronic surface so that ORR takes place on this electrochemically catalyzed interface. In the layer, there are ten particles of 2 μm in size arranged along the x direction with a total thickness of 20 μm. The protons migrate through the electrolyte after crossing the membrane located at the left boundary (i.e., $x = 0$) of the domain as explained in Figure 2 and reach the catalyzed interface. At the same time, O_2 diffuses to the catalyst layer from the cathode backing layer at the right boundary (i.e., $x = x_L$) of the domain, and then dissolves in the electrolyte film, where the ORR occurs.

As a first step toward DNS modeling, the following simplifications and assumptions are made:

1. O_2 diffusion resistance through the polymer electrolyte is ignored due to the small thickness of the film (~5 nm). Thermodynamic equilibrium between the gas phase oxygen concentration and that dissolved in the electrolyte phase is assumed to exist at the reaction interface.
2. The product water mass conservation is not considered, assuming that product water is in the gas phase due to heat generation in the CL and diffuses out of CL sufficiently fast. This assumption may bring considerable error at large current densities, which will be justified later.
3. The proton conductivity in the polymer electrolyte is treated as constant, though it actually depends on the water content in the ionomer. This assumption will be relaxed in Section IV.
4. The cell operation temperature is assumed to be constant and steady state is assumed.

A single set of differential equations valid for all the phases is developed, which obviates the specification of internal boundary conditions at the phase interfaces. The mass balance of O_2, charge conservation and the electrochemical reaction are formulated based on the above assumptions. Due to slow kinetics of the ORR, the electrochemical reaction is assumed to be governed by Tafel kinetics as follows:

$$j = -i_0 \frac{c_{O2}}{c_{O2,ref}} \exp\left(\frac{-\alpha_c F}{RT}\eta\right) \qquad (4)$$

where, i_0 is the exchange current density, c_{O2} and $c_{O2,ref}$ refer to local oxygen concentration and reference oxygen concentration respectively, α_c is the cathode transfer coefficient for ORR, F is the Faraday's constant, R is the universal gas constant, and T is the cell operating temperature. The overpotential, η, is defined as:

$$\eta = \phi_s - \phi_e - U_o \qquad (5)$$

where, ϕ_s and ϕ_e are the electronic and electrolyte phase potentials at the reaction site respectively. U_o is the thermodynamic equilibrium potential of the cathode under the cell operation temperature.

The charge conservation for electron and proton and O_2 conservation can be described by the following equations, respectively:

$$\nabla \cdot (\sigma \nabla \phi_s) + a \int_\Gamma j\delta(x - x_{\text{interface}})ds = 0 \qquad (6)$$

$$\nabla \cdot (\kappa \nabla \phi_e) + a \int_\Gamma j\delta(x - x_{\text{interface}})ds = 0 \qquad (7)$$

$$\nabla \cdot (D\nabla c_{O2}) + a \int_\Gamma \frac{j}{4F}\delta(x - x_{\text{interface}})ds = 0 \qquad (8)$$

where, a represents the specific interfacial area and is defined as the interfacial surface area where the reaction occurs per unit volume of the catalyst layer, s is the interface, Γ represents the interfacial surface over which the surface integral is taken, $\delta(x - x_{\text{interface}})$ is a delta function which is zero everywhere but unity at the interface where the reaction occurs. The transfer current, j, is positive for the electronic phase and negative for the electrolyte since the current is transferred from the electronic phase into the electrolyte. σ and κ represent electronic conductivity and electrolyte conductivity respectively and D refers to the oxygen diffusivity in the gas phase.

At the left boundary, a constant current density, i_d, is applied through the electrolyte phase, while it flows out from the electronic phase on the right boundary. A constant value of oxygen concentration, c_{O_2}, equal to the gas channel inlet value, $c_{O_2,0}$, is assumed at the right boundary. The boundary conditions can then be expressed as:

$$c_{O_2} = c_{O_2,0} \qquad \text{in the gas phase at } x=x_L, \qquad (9)$$

$$-\kappa \frac{\partial \phi_e}{\partial x} = i_d \qquad \text{in the electrolyte phase at } x = 0, \qquad (10)$$

$$-\sigma \frac{\partial \phi_s}{\partial x} = i_d \qquad \text{in the electronic phase at } x = x_L, \qquad (11)$$

and,

$$\frac{\partial c_{O_2}}{\partial n} = 0, \quad \frac{\partial \phi}{\partial n} = 0 \quad \text{everywhere for other boundaries} \qquad (12)$$

The governing equations are discretized by the control-volume-based finite difference method by Patankar,[16] and the resulting sets of algebraic equations are iteratively solved. The rectangular physical domain is divided into uniform grids. The numerical grids used in the following simulations are 160 in the x direction and 24 in the y direction. All the parameters, including the properties of each phase, are given in Table 1. The equations are solved simultaneously, and convergence is considered to be reached when the relative error in each field between two consecutive iterations is less than 10^{-5}.

3. Three-Dimensional Regular Microstructure

The idealized two-dimensional CL microstructure is bound to show some departure from reality. For instance, the reaction area obviously seems to be much less than that in a real 3-D catalyst layer, in which case the reaction area is roughly 100-times larger than the nominal electrode cross-sectional area. However, the main

Table 1
Property Data for the DNS Calculations with the Regular 2-D and 3-D Microstructure

Parameter	Solid phase	Electrolyte	Gas phase
Conductivity, σ (S/cm)	50	0.05	0
O_2 diffusion coefficient, $D^g_{O_2,0}$ (cm^2/s)	0	0	0.01

Pressure at the gas channel inlet, p (kPa)	150
Reference concentration, (mol/cm^3)	51.1×10^{-6}
Temperature, T (°C)	80
Catalyst layer thickness, (μm)	20
Simulation height of layer, (μm)	3
Volume fraction of electronic phase, ε_s	0.5
Volume fraction of electrolyte, ε_e	0.25
Volume fraction of gas phase, ε_g	0.25
Exchange current density, i_0 (A/cm^2)	1.0×10^{-8}
Cathodic transfer coefficient, α_c	1.0
Open-circuit potential, U_0 (V)	1.1

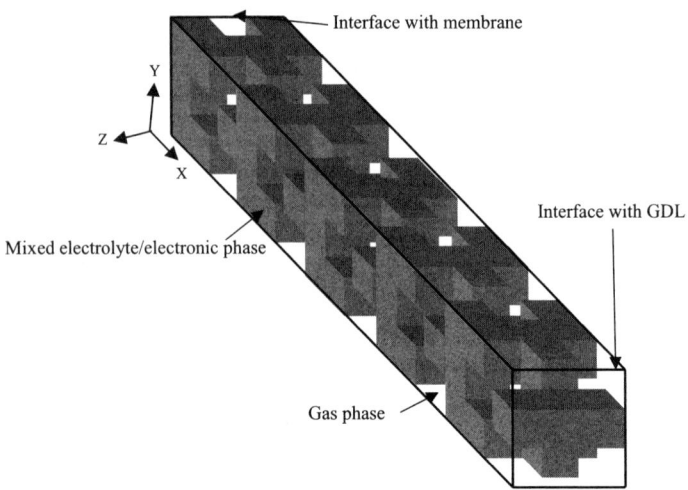

Figure 3. Schematic diagram of the 3-D regular microstructure.

purpose of the 2-D DNS model is to demonstrate the concept and utility of the DNS approach in assessing the effects of the micro-morphology on diffusion and reaction. This model also allows for a fundamental understanding of the physics occurring in the catalyst layer.

As an attempt toward describing a more realistic CL microstructure, the DNS model is now extended to a three dimensional CL microstructure, as shown in Figure 3. Here, the catalyst layer is simplified to contain two phases, the gas phase and a mixed electrolyte/electronic phase. On the left boundary, the protons migrate into the catalyst layer from the membrane. Oxygen and electrons reach the CL through the gas voids and electronic phase, respectively, at the right boundary, attached to the GDL. A typical CL is about 10–20 μm in thickness and the pore size is about one to two order of magnitude smaller. In the y-z directions, the computational domain is assumed to have symmetry boundary conditions such that many repeating units constitute the entire catalyst layer.

A few assumptions are additionally made to the regular three-dimensional catalyst layer structure for simplicity.

- The electronic phase potential is assumed to be uniform because the electrode is very thin and its electronic conductivity is sufficiently high. Under this assumption, the mixed phase is treated as an electrolyte phase by applying an effective ionic conductivity. This correction accounts for the volume fraction of the electrolyte phase with respect to the mixed phase volume fraction and thus assumes a Bruggeman type correlation as:

$$\kappa = \kappa_0 \cdot (\frac{\varepsilon_e}{\varepsilon_e + \varepsilon_s})^{1.5} = \kappa_0 \cdot (\frac{\varepsilon_e}{1-\varepsilon_g})^{1.5} \qquad (13)$$

where, κ_0 is the intrinsic conductivity of the electrolyte, ε_e, ε_s and ε_g are the electrolyte, electronic and gas pore volume fractions, respectively. For simplicity, the mixed electrolyte/electronic phase will be referred to simply as the electrolyte phase in the rest of the chapter.

- The interface between the gas phase and the mixed phase is assumed to be completely catalyzed and activated by platinum nanoparticles. Therefore, the entire interface is active for the ORR.

- The system is also assumed to be isothermal and steady state is considered.

Under the aforementioned assumptions, the governing equations for charge and oxygen conservation can be written, respectively, as:

$$\nabla \cdot (\kappa \nabla \phi_e) + a \int j \delta(x - x_{\text{interface}}) ds = 0 \quad (14)$$

$$\nabla \cdot (D_{O_2}^g \nabla c_{O_2}) + a \int \frac{j}{4F} \delta(x - x_{\text{interface}}) ds = 0 \quad (15)$$

The second term in both the equations represents a source/sink term only at the catalyzed phase interface where the electrochemical reaction takes place.

In order to facilitate numerical solution of Eqs. (14) and (15) without having to resolve the microscopically complex phase interface, the governing equations are extended to the entire computational domain by incorporating a phase function f. The phase function is defined as unity in the electrolyte phase and zero in the gas phase, respectively. The proton conductivity and oxygen diffusivity can be generally expressed, at each cell center, in terms of the discrete phase function as:

$$K(i, j, k) = \kappa \cdot f(i, j, k) \quad (16)$$

$$D(i, j, k) = D_{O_2}^g [1 - f(i, j, k)] \quad (17)$$

The transfer current between the two neighboring cells at the phase interface, shown in Figure 4, is given by the Tafel equation as follows:

$$j = i_0 \frac{c_{O_2}(i+1, j, k)}{c_{g,ref}^{O_2}} \exp[\frac{\alpha_c F}{RT} \phi_e(i, j, k)] \text{ (A/cm}^2) \quad (18)$$

$\phi_e(i,j,k)$ is used to represent the overpotential in the kinetic expression since both the open-circuit potential and the electronic phase potential are constant. It is worth noting that the prefactor, i_0, is the modified exchange current density after replacing overpotential, η, in Eq. (4) with the expression given by Eq. (5). The control volume, with cell center (i,j,k), forms six interfaces with the neighbors where the electrochemical reaction might occur.

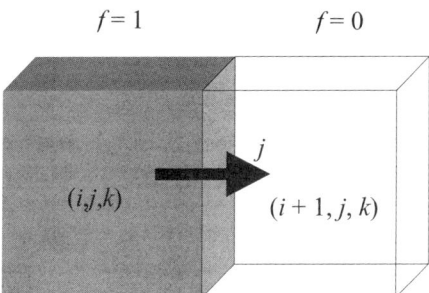

Figure 4. Transfer current between the two adjacent cells.

The sum of the flux from all the reactions can be expressed as the volumetric source term. The corresponding source terms, S_ϕ and S_{O_2}, in the governing equations, Eqs. (14) and (15), respectively, can be expressed in a discretized fashion at the cell center (i, j, k) as:

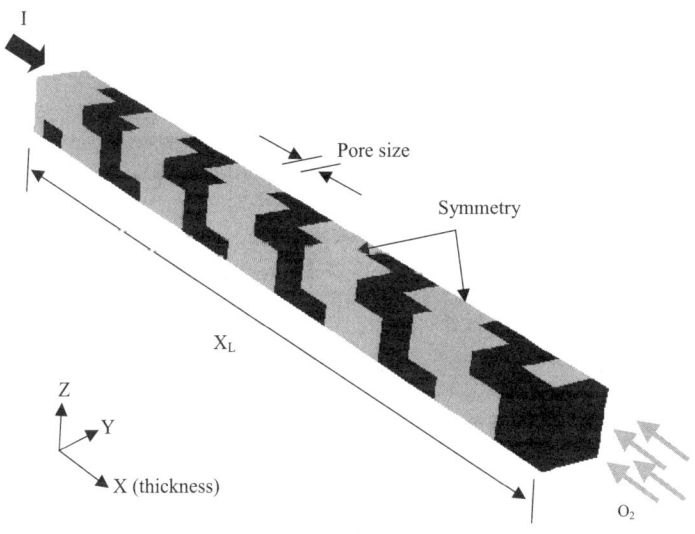

Figure 5. Computational domain for the 3-D DNS model.

$$S_\phi(i,j,k) = -\frac{i_0}{c_{O_2,ref}^g} f(i,j,k) \exp[\frac{\alpha_c F}{RT}\phi_e(i,j,k)] \cdot$$
$$\{[1-f(i-1,j,k)]c_{O_2}(i-1,j,k)/\Delta x$$
$$+[1-f(i+1,j,k)]c_{O_2}(i+1,j,k)/\Delta x$$
$$+[1-f(i,j-1,k)]c_{O_2}(i,j-1,k)/\Delta y \qquad (19)$$
$$+[1-f(i,j+1,k)]c_{O_2}(i,j+1,k)/\Delta y$$
$$+[1-f(i,j,k-1)]c_{O_2}(i,j,k-1)/\Delta z$$
$$+[1-f(i,j,k+1)]c_{O_2}(i,j,k+1)/\Delta z\}$$

$$S_{O_2}(i,j,k) = -\frac{i_0}{4Fc_{O_2,ref}^g}[1-f(i,j,k)]c_{O_2}(i,j,k) \cdot$$
$$\{f(i-1,j,k)\exp[\frac{\alpha_c F}{RT}\phi_e(i-1,j,k)]/\Delta x$$
$$+ f(i+1,j,k)\exp[\frac{\alpha_c F}{RT}\phi_e(i+1,j,k)]/\Delta x$$
$$+ f(i,j-1,k)\exp[\frac{\alpha_c F}{RT}\phi_e(i,j-1,k)]/\Delta y \qquad (20)$$
$$+ f(i,j+1,k)\exp[\frac{\alpha_c F}{RT}\phi_e(i,j+1,k)]/\Delta y$$
$$+ f(i,j,k-1)\exp[\frac{\alpha_c F}{RT}\phi_e(i,j,k-1)]/\Delta z$$
$$+ f(i,j,k+1)\exp[\frac{\alpha_c F}{RT}\phi_e(i,j,k+1)]/\Delta z\}$$

From Eq. (19), it is evident that only the electrolyte phase, with neighboring gas phase cells, has a non-zero source term for charge transport. Likewise, Eq. (19) defines non-zero source term due to oxygen consumption only for that gas phase cell having neighboring electrolyte cells.

The computational domain, as schematically shown in Figure 5, is taken as one quarter of the whole domain, shown in Figure 3, due to symmetry considerations in the y and z directions. The domain size is 20 μm × 3 μm × 3 μm. At the left boundary, through which the protons migrate from the membrane, one layer of electrolyte-only cells is added to the computational domain. The operating current is uniformly applied on this additional layer, making the boundary condition straightforward to be implemented.

For the same purpose, one layer of pore-only cells is applied at the right boundary, which supplies oxygen at a constant concentration. The boundary conditions can be summarized as:

- on the symmetry planes:

$$y = 0, y = y_L, z = 0, z = z_L, \quad \frac{\partial c_{O_2}}{\partial n} = 0, \quad \frac{\partial \phi_e}{\partial n} = 0 \quad (21)$$

- at the left boundary (i.e., CL-membrane interface):

$$x = 0, \quad \frac{\partial c_{O_2}}{\partial n} = 0, \quad -\kappa \frac{\partial \phi_e}{\partial n} = i_d \quad (22)$$

- at the right boundary (i.e., CL-GDL interface):

$$x = x_L, \quad c_{O_2} = c_{O_2,0}, \quad \frac{\partial \phi_e}{\partial n} = 0 \quad (23)$$

The transport properties and the electrochemical kinetic parameters used in this three-dimensional study are mainly taken from Table 1 in order to compare the results with those from the two-dimensional predictions.

A baseline simulation, with nominal porosity of 0.25 and with uniform mesh size of 42x12x12 in the x, y and z directions, respectively, was performed.

4. Results and Discussion

(i) 2-D Model: Kinetics- vs. Transport-Limited Regimes

In this section, the capabilities of the present model are illustrated by comparing the simulation results with some experimental data. Further, the simulation results are analyzed to understand the various voltage losses from the cathode catalyst layer during its operation. Two sets of simulations are carried out using pure oxygen and air as oxidants at various current densities. Oxygen and air are both fed at a pressure of 150 kPa when the cell is operated at 80 °C. In each set of simulations, a special case, in which the diffusion coefficient of O_2 in the gas phase is set to be infinitely large, is simulated to mimic the limiting case without O_2 depletion effect. Then the two groups of results are compared with corresponding experimental measurements from the literature, under similar conditions.

Polarization curves for various simulations are summarized in Figure 6, including an analytical solution for the limiting case with both infinitely large proton conductivity in the electrolyte and O_2 diffusivity in the gas phase. In general, when the conductivities of both electronic phase and electrolyte phase become infinitely large, the overpotential will be uniform across the electrode with a constant open-circuit potential. Then if the mass diffusivity of the reactant is set to infinitely large to have a uniform concentration distribution, the electrochemical reaction rate will be uniform throughout the electrode. In this case with Tafel kinetics, the current balance for the catalyst layer can be written as follows:

$$i_0 \frac{c_{O_2}}{c_{O_2,ref}} \exp(-\frac{\alpha_c F}{RT}\eta).A_{reaction} = i_d.A_{cross} \qquad (24)$$

where $A_{reaction}$ stands for the total reaction area, and A_{cross} represents the cross-sectional area on which the discharge current density i_d is

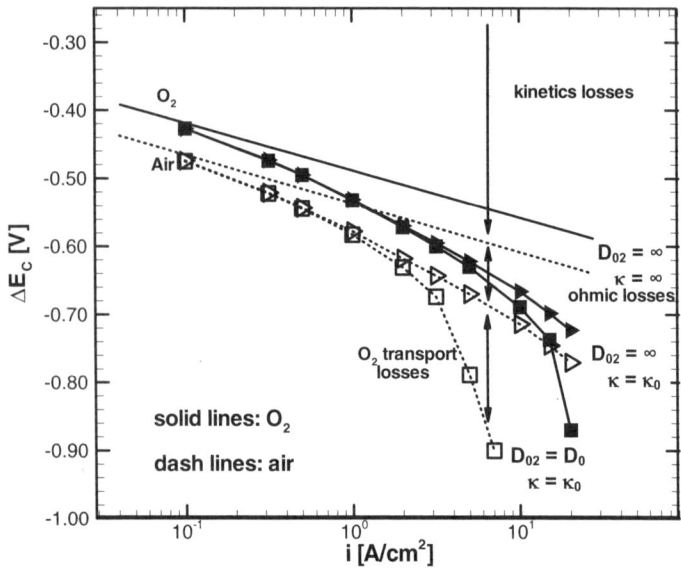

Figure 6. Numerical predictions of the voltage losses in the cathode catalyst layer.

applied. It should be noted that the concentration, c_{O_2} and overpotential, η, are constant in Eq. (24). A new parameter is defined to denote the area ratio, which is given by:

$$A_0 = \frac{A_{reaction}}{A_{cross}} \tag{25}$$

The only voltage loss in this case is the kinetics loss, which can be derived as:

$$|\eta| = 2.303 \frac{RT}{\alpha_c F}[\log i_d - \log(A_0 i_0 \frac{c_{O_2}}{c_{O_2,ref}})] \tag{26}$$

In Eq. (26), the coefficient at the right hand side, $2.303 RT/\alpha_c F$, is the Tafel slope, denoted as b and having the unit of mV/dec. In our simulation, b has the value of 70 mV/dec with α_c of 1.0 and an operating temperature of 80 °C. Eq. (26) states that in the absence of ohmic and transport losses, the cathode voltage drop will increase by 70 mV once the current density i_d increases by a factor of 10 or the concentration, c_{O_2}, the area ratio A_0 decrease by a factor of 10. In Figure 6, the two straight lines show the pure kinetics losses for O_2 and air as oxidant, respectively. Since the mole fraction of O_2 in air is 0.21, there will be about 47 mV more losses when air is used instead of pure O_2 from the following calculation:

$$\Delta F_{O_2 \to air} = 70 \cdot \log\left(\frac{100\%}{21\%}\right) = 47 mV \tag{27}$$

Figure 6 also indicates that there are additional voltage losses when real electrolyte conductivity and O_2 diffusivity are employed. These additional losses have been identified as ohmic losses and O_2 transport losses in the plot. To reveal these losses, the distributions of O_2 concentration and overpotential are plotted for three different cases in Figure 7. The corresponding operating current density is 3.16 A/cm^2 with air as the oxidant. When realistic electrolyte conductivity is used, the overpotential becomes non-uniform as shown by the dash line. The overpotential at the interface with the membrane, η_0, which represents the total voltage loss at the catalyst layer, increases significantly from the dash dot

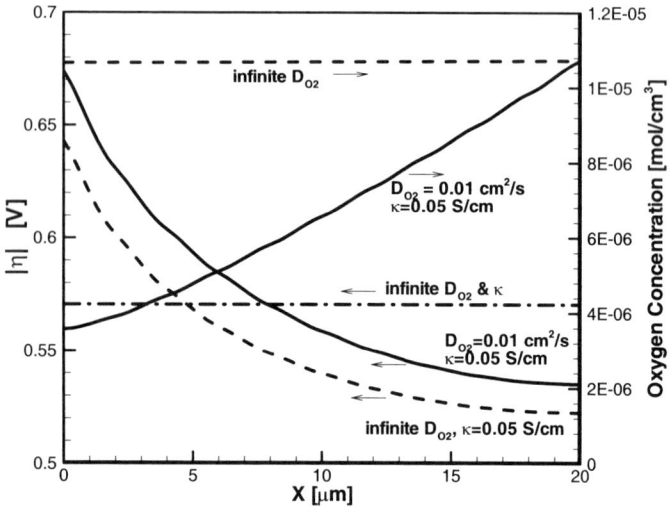

Figure 7. Distributions of the overpotential and oxygen concentration across the thickness of the catalyst layer at current density of 3.16 A/cm^2.

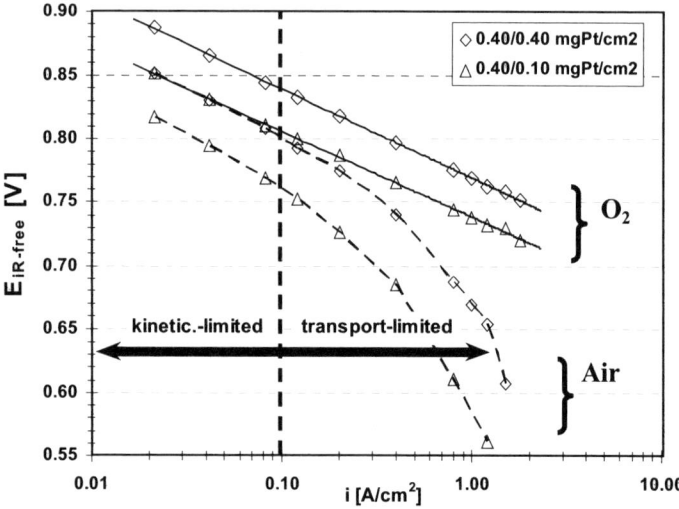

Figure 8. Experimental observations of the voltage losses (Gasteiger et al.[17]).

line to the dash line, when κ changes from infinity to 0.05 S/cm. When realistic diffusivity is incorporated, as shown by the solid line, η_0 becomes even larger due to the O_2 diffusion resistance. The value of η_0 could increase further if O_2 is depleted at the CL-GDL interface, which would occur at a higher current density. Correspondingly, these two additional voltage losses are marked as ohmic losses and O_2 transport losses in Figure 6.

Experimental data by Gasteiger et al.[17] is plotted in Figure 8 in a similar fashion as in Figure 6. The y-axis denotes the cathode IR-free potential by eliminating all ohmic resistances. Different Pt loadings are applied for both O_2 and air as oxidants. It is observed that the curves simply shift down when the lower Pt-loading of 0.10 mg Pt/cm^2 is applied at the cathode. This is because the total reaction area has decreased, subsequently increasing the kinetics losses, as explained by Eq. (26).

Now, comparing the numerical results in Figure 6 with the experimental observations, in Figure 8, reveal that the kinetic losses look different quantitatively. At the current density of 0.1 A/cm^2, the ORR kinetics losses in the experiments are about 370 mV with an equilibrium potential of 1.18V corresponding to the air oxidant and 0.40 mg Pt/cm^2 catalyst loading in cathode. However, the same losses in the numerical simulation are as large as 470mV. This difference is mainly due to the small reaction surface area in the idealized structure used in the simulations. It can be estimated that the total surface area ratio of Pt catalyst at the cathode is about 140 cm^2 Pt/cm^2 (electrode cross-sectional area) in experiments when 0.40/0.40 mg Pt/cm^2 Pt loadings are used with typical dispersion surface area of Pt particles at 35 m^2/g Pt. On the other hand, this value in the model is only 10. Consequently, this difference by an order of magnitude in the surface area results in more voltage losses by as much as 80mV in simulations than that in experiments. Furthermore, the Tafel slope obtained from the experiments is measured to be 66mV/dec, while being 70 mV/dec in simulations. This leads to another 30 mV more kinetics losses in the simulation results than the experimental data.

Secondly, the O_2 transport characteristics appear different between experiments and simulations, though they are in qualitative agreement. In Figure 8, the transport-limited regime is identified when the current density is larger than 0.1 A/cm^2, which means the transport losses begin to appear in that region for air oxidant. In simulations, when air is used as oxidant, the transport of O_2 does not result in additional voltage drop until the current density is increased to 1 A/cm^2, where the two dash lines begin to deviate from each other. For pure oxygen as the oxidant, the diffusion becomes limiting at 5 A/cm^2, although the experiments have not been carried out at such a high rate. There are three factors that could explain why the O_2 transport losses appear later in the simulations than in the experiments. The most important reason is perhaps due to very high diffusion through the idealized geometry of pore spaces, as mentioned earlier. Furthermore, in the present model, the blocking effect of product water has not been considered, which could also help O_2 diffusion to some extent. Another possible explanation is that at the interface between the simulated catalyst layer and the gas diffusion layer (GDL), the O_2 concentration value is assumed to be the same as that at the inlet of the flow channel. Neglecting the diffusion resistance through the GDL could make the concentration, c_{O_2}, on the right boundary significantly larger than the realistic value.

(ii) *Comparison of the Polarization Curves between 2-D and 3-D Simulations*

The polarization curve from the three-dimensional DNS calculation is compared in Figure 9 together with that from the two-dimensional results. These two simulations are carried out under exactly the same conditions, except for the different geometries. It should be noted that the term "polarization curve" refers to the voltage loss vs. current density curve throughout this chapter instead of the standard I-V curve, otherwise used widely in the fuel cell literature. As expected, the effects brought by the three-dimensional geometry can be identified in two different regimes. In the kinetic control regime, voltage loss from the three-dimensional model is about 20 mV less than that for the two dimensional model. Apparently this is due to the increased phase interfacial area, which is one of the purposes to introduce the three-dimensional model. From the calculation, the total interfacial area for the 3-D geometry is around 20 times the electrode cross-sectional area, doubling the total area from the 2-D geometry. Therefore, it results in 20 mV less kinetics losses based on the 70 mV/dec Tafel slope as calculated in Eq. (26). Another influence of the 3-D geometry is reflected on the transport of oxygen. Due to more tortuous path of the 3-D structure, oxygen transport through

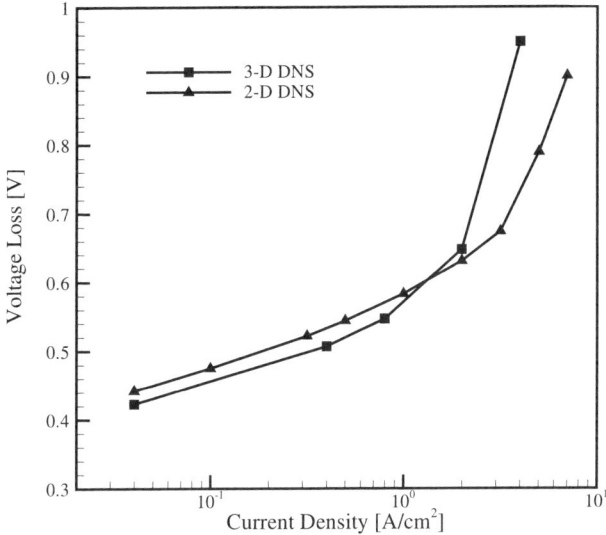

Figure 9. Comparison of the polarization curves from the 2-D and 3-D DNS calculations.

the gas phase is restricted in the 3-D geometry. As a result, oxygen depletion occurs earlier and the mass transport limiting current density decreases.

From the above discussions, it is evident that the DNS approach has been successfully deployed to delineate the various physical processes accounting for the different voltage losses in the cathode catalyst layer. The comparison with experimental results further points to several areas of improvements, which include a better representation of the morphology of the porous microstructure and the consideration of water distribution. These issues are addressed in the subsequent sections.

III. THREE-DIMENSIONAL RANDOM MICROSTRUCTURE

In Section II, a regular microstructure was constructed to represent the simplified three-dimensional cathode catalyst layer for application of the DNS model. The regular 3-D structure, while offering an improvement over the two-dimensional geometrical simplicity, is still plagued with morphological and associated physical limitations. Besides the simplicity of the structure, statistical information still deviates from a realistic porous catalyst layer. For example, the phase interfacial area in the simplified structure needs further enhancement and the constituent phases are not as tortuous as that in a practical porous medium, as well. In this section, a purely random porous microstructure is constructed and the 3-D DNS model is extended accordingly.

The general objective of constructing a random microstructure is to mimic more closely the geometry of a real porous medium so as to preserve its statistical feature. This method is able to generate "digital" microstructures with desired properties. As a general approach, certain low-order statistical properties (e.g., porosity and two-point correlation function) of the real porous medium are measured experimentally first and an artificial medium is reconstructed with the same average parameters.

1. Random Structure

Except for a few man-made microstructures, most of the real porous media are random. However, the word 'random' is quite vague and it can be used to qualify very different situations, such as pure disorder and correlated disorder in a porous medium and thus requires more precise definition.

For an arbitrary piece of porous medium, the pore structure can be completely characterized by a binary phase function, $Z(\vec{r})$, which assumes discrete values in the 3-D space as:[18]

$$Z(\vec{r}) = \begin{cases} 1 & \text{if } \vec{r} \text{ is in the pore space} \\ 0 & \text{otherwise} \end{cases} \quad (28)$$

where, $\vec{r}$ denotes the position with respect to an arbitrary origin. The first two moments of the phase function, the porosity, ε, and the autocorrelation function, $R_Z(\vec{u})$, are defined respectively as:[18]

$$\varepsilon = \overline{Z(\vec{r})} \quad (29)$$

$$R_Z(\vec{u}) = \overline{[Z(\vec{r}) - \varepsilon][Z(\vec{r} + \vec{u}) - \varepsilon]}/(\varepsilon - \varepsilon^2) \quad (30)$$

where, overbar denotes statistical averages. Porosity, ε, is a positive quantity limited to [0, 1] interval and is the probability that a point is in the pore space. $R_Z(\vec{u})$ verifies the general properties of a correlation function and is the probability that two points in the porous medium at a distance, $\vec{r}$, are both in the pore space. For a statistically homogeneous porous medium, ε is a constant and $R_Z(\vec{u})$ is only a function of the lag vector, $\vec{u}$, and does not depend on the spatial coordinates (i.e., independent of $\vec{r}$). Additionally, if the medium is isotropic, then the autocorrelation function does not depend on the direction but only on the norm, u, of the vector, $\vec{u}$. Furthermore, for a purely disordered porous medium, the autocorrelation function is independent of u and identically goes to zero. In such a porous structure, each elementary space, resulting from the discretization of the 3-D continuum space, is occupied at random either with solid or void with a given probability, ε and can be realized, in principle, by throwing a dice. This simplest construction rule is used here to generate a purely random 3-D catalyst layer microstructure for the DNS model.

In the present work, a purely random porous medium is computer-generated by employing a random number generator with porosity, ε and pore size, d as the chosen target geometry features for the reconstruction to match. Specifically, the porous catalyst layer is constructed in a discrete manner. It is considered to be composed of $N_x \times N_y \times N_z$ elementary cubes, each of the same size d, which represents the chosen pore scale. These elementary cubes are filled with either the electrolyte phase or pore phase. During construction, the computer generates a random number uniformly distributed within the interval [0, 1] for each cube. When the random number is lower than the given porosity, ε, the corresponding cube is set to be occupied by the pore space. Otherwise, it is occupied by the electrolyte phase.

2. Structural Analysis and Identification

Once the microstructure is constructed, structural connectivity needs to be imposed by forming pore clusters consisting of a group of connected pores. From structural viewpoint, a group of pores, which are connected with each other, is called a pore cluster. When ε is small, all the pores form small and isolated clusters. When ε is large enough, among all the pore clusters, there would be one that penetrates the entire medium from one end to the other. This kind of pore cluster can be termed as "transport" pore cluster, because it forms a continuous network allowing the fluid to transport across the entire medium. In other words, the porous medium is permeable only if such a "transport" pore cluster exists. A pore belonging to the "transport" pore cluster is called a "transport" pore, otherwise it is called a "dead" pore. Apparently, when ε increases, there would be fewer and fewer "dead" pores. When ε is close to one, all the pores would be "transport" pores.

In the current work, the constructed random cathode catalyst layer structure is depicted in Figure 10. The elementary cube size is 0.25 μm, representing the chosen pore size. Therefore, to simulate the 10 μm-thick catalyst layer, 40 cubes are applied along the thickness. Similar to the approach adopted in the 3-D regular structure, as described in the earlier section, one layer of electrolyte-only and pore-only cells are added to the left and right boundaries of the structure, respectively, for ease of implementation of the boundary conditions.

The numerical approach of identifying the "transport" and "dead" portions for each phase starts with assigning an initial value of a phase function, f, to each elementary control volume in the entire computational domain. On the left boundary of electrolyte-

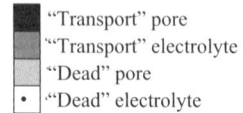

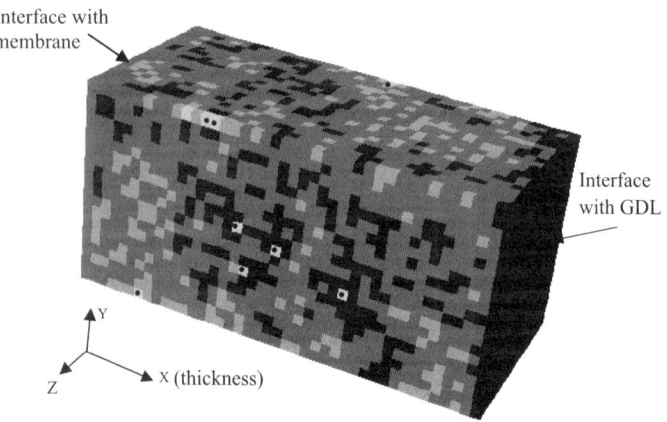

Figure 10. Schematic diagram of the 3-D random catalyst layer microstructure with nominal porosity of 0.36.

only cells, f is set to be one; while f is assigned to be zero within the pore-only cells on the right boundary. Elsewhere within the domain, f is three inside the electrolyte cells and two inside the pore cells. Then beginning from the left boundary, each elementary cell is scanned to identify the "transport" electrolyte. For the cell with f equal to three, if any of its six neighboring cells has f equal to 1, the phase function f of itself is switched to 1. After each scan of the entire domain, the total number of cells with f equal to three is counted. This scan form left to right is repeated until the total number does not vary anymore. Thus cells with f equal to one are identified as "transport" electrolyte, while those with f equal to three represent "dead" electrolyte. Similarly for the pore phase, the scanning process begins from the right boundary to the left and once a cell with f equal to two has any neighboring cell with f equal to zero, the phase function f the cell itself is switched to zero. After sufficient number of scans, when the total number of cells with f equal to two does not change anymore, the cells with f equal to zero and two are identified as "transport" and "dead" pores, respectively.

The "transport" pore and the "transport" electrolyte identified here represent those elementary cells which are accessible for

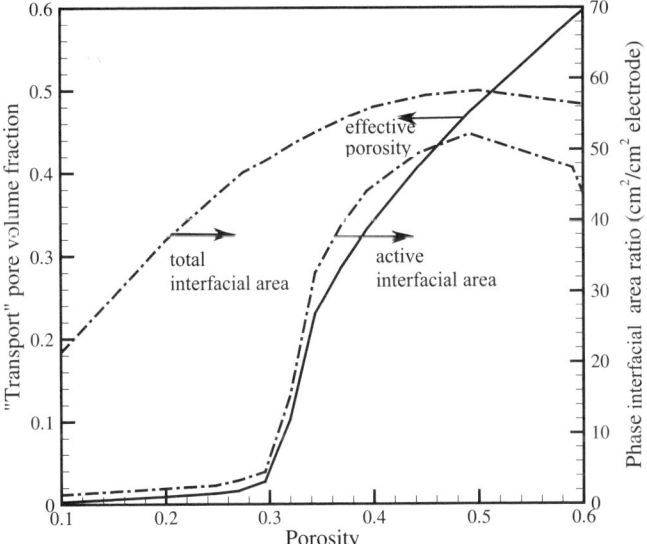

Figure 11. Variation of transport pore volume fraction and phase interfacial area ratio with nominal porosity.

oxygen from the gas diffusion layer (GDL) at the right boundary and the protons from the polymer electrolyte membrane at the left boundary, respectively. In Figure 10, the identified "transport" pore and electrolyte cells, as well as "dead" pore and electrolyte cells are indicated with different gray cubes, corresponding to a natural porosity of 0.36. As expected, there are few "dead" electrolyte cells since the electrolyte volume fraction is relatively large (0.64). On the other hand, only about 70% of total pores are identified as "transport" pores, indicating the effective "transport" porosity is only about 0.26.

Figure 11 shows the effective porosity and phase interfacial area ratio as a function of the natural porosity. The phase interfacial area ratio represents the ratio of the interfacial area in a porous structure to the cross-sectional geometrical area. Here, the total interfacial area ratio represents all interfaces between the electrolyte and pore phases, while the active interfacial area ratio only includes those between "transport" electrolyte and "transport" pores, indicating that the interfacial sites are accessible by protons, electrons and oxygen and hence cause the electrochemical reaction to happen. The figure only shows the natural porosity range up to 0.6 because the effective porosity is nearly identical to the natural porosity when it is larger than 0.6. The interfacial area ratio curve is symmetric around the porosity of 0.5. It is also observed that when the porosity is smaller than 0.35, both effective porosity and active interfacial area ratio are reduced dramatically. This is of significance in practical applications, indicating that 0.35 is the low end of the porosity for the catalyst layer. Furthermore, the random microstructure provides a realistic active interfacial area ratio between 40 and 50. This value of the reaction area ratio corresponds roughly to 0.15 mg Pt/cm^2 catalyst loading with a typical dispersion surface area of 35 m^2/g Pt, which is representative in current applications.

The local profiles of pore and electrolyte volume fractions along the thickness of the catalyst layer are shown in Figure 12. First, the cross-section averaged local natural porosity shows a random fluctuation around the average porosity of the porous structure marked by the horizontal line. Secondly, almost all the electrolyte cells are available for transport, while a considerable portion of the pores are dead pockets. The percentage of "transport" pores, distributes uniformly in most locations except for the front and back end of the structure. At the back end (i.e., near the left

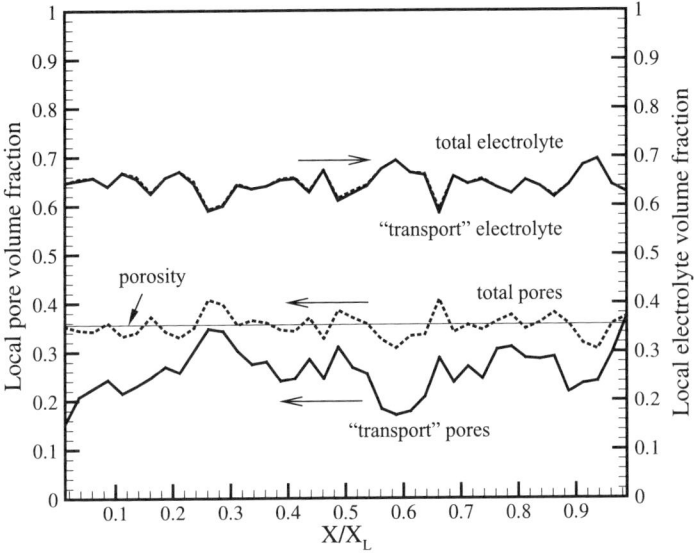

Figure 12. Profiles of pore and electrolyte volume fractions along the thickness of the catalyst layer.

boundary) of the catalyst layer, interfacing with the membrane, more pores are "dead" indicating difficulty in access for the oxygen. On the other hand, near the front end (i.e., adjacent to the right boundary), most of the pores are "transport" pores partly because the structure is open to large pore space in the GDL.

3. Governing Equations

The DNS model developed for the 3-D regular CL microstructure, in the earlier section, is now extended to solve for the conservation equations for charge and oxygen transport on the random 3-D CL structure. The model assumptions remain the same as in the case of the 3-D regular microstructure. The meaning of the symbols can, as well, be traced back correspondingly to the previous sections.

The discrete phase function, f, introduced earlier for the regular microstructure, is redefined for each elementary control volume (i,j,k) within the entire domain as follows:

$$f(i,j,k) = \begin{cases} 0 & \text{"transport" pores} \\ 1 & \text{"transport" electrolytes} \\ 2 & \text{"dead" pores} \\ 3 & \text{"dead" electrolytes} \end{cases} \quad (31)$$

Correspondingly, the proton conductivity and oxygen diffusivity can be expressed as:

$$K(i,j,k) = \kappa \cdot f(i,j,k) \cdot [2 - f(i,j,k)] \cdot [3 - f(i,j,k)]/2 \quad (32)$$

$$D_{O_2}(i,j,k) = D_{O_2}^g \cdot [1 - f(i,j,k)] \cdot [2 - f(i,j,k)] \cdot [3 - f(i,j,k)]/6 \quad (33)$$

For simplicity, we define two new phase functions, f_ϕ and f_{O_2}, as follows:

$$f_\phi(i,j,k) = f(i,j,k) \cdot [2 - f(i,j,k)] \cdot [3 - f(i,j,k)]/2 \quad (34)$$

$$f_{O_2}(i,j,k) = [1 - f(i,j,k)] \cdot [2 - f(i,j,k)] \cdot [3 - f(i,j,k)]/6 \quad (35)$$

Obviously, from the above definitions, it can be seen that f_ϕ is non-zero only within "transport" electrolytes; while f_{O_2} is non-zero only within "transport" pores.

Based on the newly introduced phase functions, Eqs. (32) and (33) can be simplified as:

$$K(i,j,k) = \kappa_e \cdot f_\phi(i,j,k) \quad (36)$$

$$D_{O_2}(i,j,k) = D_{O_2}^g \cdot f_{O_2}(i,j,k) \quad (37)$$

The above expressions indicate that both proton conductivity and oxygen diffusivity are set to be zero in "dead" pores and "dead" electrolytes. Now the governing differential equations for charge and oxygen transport, as detailed in the previous section, can be readily extended to be valid in the entire domain using the discrete phase function, $f(i,j,k)$. Since the electrochemical reaction only occurs at active phase interfaces, only those "transport" pores and "transport" electrolytes next to each other have source terms, S_{O_2} and S_ϕ respectively. These source terms can be expressed in discretized form as:

$$S_\phi(i,j,k) = -\frac{i_0}{c_{O_2,ref}^g} f(i,j,k) \exp[\frac{\alpha_c F}{RT}\phi_e(i,j,k)] \cdot$$
$$\{[1-f(i-1,j,k)]c_{O_2}(i-1,j,k)/\Delta x$$
$$+ [1-f(i+1,j,k)]c_{O_2}(i+1,j,k)/\Delta x$$
$$+ [1-f(i,j-1,k)]c_{O_2}(i,j-1,k)/\Delta y \qquad (38)$$
$$+ [1-f(i,j+1,k)]c_{O_2}(i,j+1,k)/\Delta y$$
$$+ [1-f(i,j,k-1)]c_{O_2}(i,j,k-1)/\Delta z$$
$$+ [1-f(i,j,k+1)]c_{O_2}(i,j,k+1)/\Delta z\}$$

$$S_{O_2}(i,j,k) = -\frac{i_0}{4Fc_{O_2,ref}^g}[1-f(i,j,k)]c_{O_2}(i,j,k) \cdot$$
$$\{f(i-1,j,k)\exp[\frac{\alpha_c F}{RT}\phi_e(i-1,j,k)]/\Delta x$$
$$+ f(i+1,j,k)\exp[\frac{\alpha_c F}{RT}\phi_e(i+1,j,k)]/\Delta x$$
$$+ f(i,j-1,k)\exp[\frac{\alpha_c F}{RT}\phi_e(i,j-1,k)]/\Delta y \qquad (39)$$
$$+ f(i,j+1,k)\exp[\frac{\alpha_c F}{RT}\phi_e(i,j+1,k)]/\Delta y$$
$$+ f(i,j,k-1)\exp[\frac{\alpha_c F}{RT}\phi_e(i,j,k-1)]/\Delta z$$
$$+ f(i,j,k+1)\exp[\frac{\alpha_c F}{RT}\phi_e(i,j,k+1)]/\Delta z\}$$

Effective proton conductivity is employed since the simulated electrolyte phase also includes the electronic phase as assumed. Similar to the treatment in the earlier sections, a Bruggeman correction is applied:

$$\kappa = \kappa_0 \cdot (\frac{\varepsilon_e}{\varepsilon_e + \varepsilon_s})^{1.5} = \kappa_0 \cdot (\frac{\varepsilon_e}{1-\varepsilon_g})^{1.5} \qquad (40)$$

where, ε_e, ε_s and ε_g are the volume fractions of the electrolyte, electronic and gas phases respectively. The intrinsic ionic conductivity, κ_0, is considered to be constant in the present simulations as the membrane is assumed to be fully hydrated.

4. Boundary Conditions

The boundary conditions used for the DNS simulation on random structure remain the same as those described for the regular structure except for the oxygen concentration, $c_{O_2,0}$, at the CL-GDL interface. In the previous simulations, oxygen mass transport resistance through the GDL was ignored and the oxygen concentration at the gas channel inlet was applied directly at the right boundary (CL-GDL interface) of the catalyst layer. In this section, the oxygen concentration drop across the GDL is further included in order to provide a more realistic boundary condition.

As shown schematically in Figure 13, the oxygen concentration in the gas channel is assumed to be uniform, which is physically corresponding to a large stoichiometric flow rate. Through the GDL, an effective diffusion coefficient $D_{O_2,GDL}^{g,eff}$ is applied and the oxygen flux at the CL-GDL interface can be written as:

$$N_{O_2} = D_{O_2,GDL}^{g,eff} \cdot \frac{(c_{O_2,inlet} - c_{O_2,0})}{\Delta X_{GDL}} \tag{41}$$

where, ΔX_{GDL} represents the thickness of the GDL. The porosity ε_{GDL} and tortuosity τ_{GDL} are employed to obtain the effective diffusivity, given as follows:

$$D_{O_2,GDL}^{g,eff} = D_{O_2}^g \cdot \frac{\varepsilon_{GDL}}{\tau_{GDL}} \tag{42}$$

$D_{O_2}^g$ is the oxygen diffusivity which in turn depends on the specified pressure and temperature[19] as:

$$D_{O_2}^g = D_{O_2,0}^g \left(\frac{T}{T_0}\right)^{3/2} \left(\frac{p_0}{p}\right) \tag{43}$$

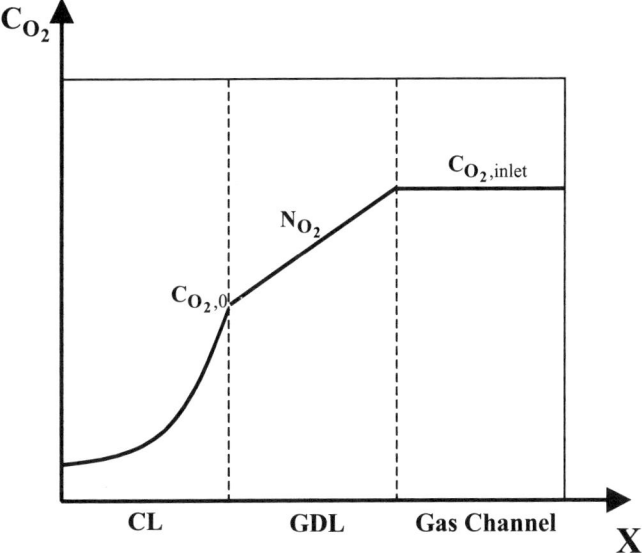

Figure 13. Schematic diagram of the oxygen concentration profile in the cathode.

where, T_0 and p_0 are the reference temperature and pressure respectively; $D_{O_2,0}^g$ is the oxygen diffusivity at the reference condition, T is the fuel cell operating temperature and p is the cathode side inlet gas feed pressure.

At steady state, the total flux through the GDL should be equivalent to the oxygen consumption rate at the catalyst layer, that is:

$$N_{O_2} = \frac{I}{4F} \qquad (44)$$

Thus, the oxygen concentration at the CL-GDL interface can be derived by combining Eqs. (41) and (44) and is given by:

$$c_{O_2,0} = c_{O_2,inlet} - \frac{I \cdot \Delta X_{GDL}}{4F \cdot D_{O_2,GDL}^{g,eff}} \qquad (45)$$

It is evident that the corrected oxygen concentration depends not only on the inlet oxygen concentration but also on the concentration drop through the GDL. At large operating current densities, there could be a considerable drop across the GDL due to the large oxygen flux.

The model input parameters including the properties of the GDL are summarized in Table 2.

5. Results and Discussion

One of the advantages of constructing the random microstructure is that it provides the phase interfacial area and tortuosity that are comparable with the real catalyst layer microstructures. Hence, using this realistic computer-generated random structure, we can evaluate the Bruggeman correlation required for the macro-homogeneous models using the DNS data. Bruggeman exponent factor ξ is commonly applied to determine the effective property as follows:

$$\Gamma_k^{eff} = \Gamma_k \cdot \varepsilon_k^{\xi} \qquad (46)$$

In the 1-D macrohomogeneous model, the same specific surface area a (cm^2/cm^3) as that in the constructed random structure is used in the Butler-Volmer equation to represent the volumetric reaction current, that is

Table 2
Property Data for the DNS Calculations with the 3-D Random Microstructure

Parameter	Value
Oxygen diffusivity in gas phase, $D_{O_2}^g$ (m^2/s)	1.9×10^{-5}
Water vapor diffusivity in gas phase, $D_{H_2O}^g$ (m^2/s)	2.6×10^{-5}
Pressure at the gas channel inlet, p (kPa)	150
Operating temperature, T (°C)	80
GDL thickness, ΔX_{GDL} (μm)	300
GDL porosity, ε_{GDL}	0.4
GDL tortuosity, τ_{GDL}	4
Natural porosity of the catalyst layer, ε_g	0.36
Electrolyte volume fraction in the catalyst layer, ε_e	0.3

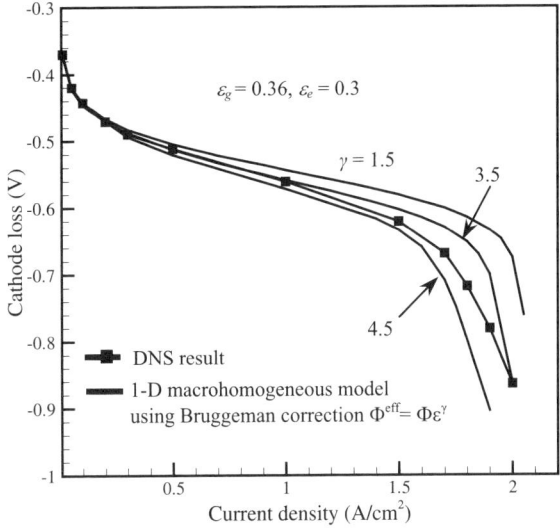

Figure 14. Comparison between the polarization curves from the DNS calculation and the 1-D macrohomogeneous model.

$$j = a \cdot i_0 \left[\exp\left(\frac{\alpha_a F}{RT}\eta\right) - \exp\left(-\frac{\alpha_c F}{RT}\eta\right) \right] \text{ (A/cm}^3\text{)} \quad (47)$$

The comparison of polarization curves predicted by DNS and 1-D macrohomogeneous models is shown in Figure 14. Three different Bruggeman factors, 1.5, 3.5 and 4.5 were attempted. It can be seen that at small current densities (up to 1 A/cm^2), a factor of 3.5 gives a good match to the DNS model; while in the large current density regime, the Bruggeman factor is suggested to be between 3.5 and 4.5. More elaborate comparisons at the current density of 1.5 A/cm^2, such as the oxygen concentration distribution, cathode overpotential and local reaction current distributions are depicted in Figures 15, 16 and 17 respectively.

In the case of the oxygen concentration (Figure 15), the DNS result is in good agreement with the 1-D macrohomogeneous model with the Bruggeman factor of 4.5. However, Figure 16 shows the factor of 3.5 gives a better match for the shape of the overpotential curve except that the DNS result is about 12 mV higher consistently. The higher surface overpotential stems from the lower active interfacial area in the DNS model. It can be seen from Figure 11 that there are only about 65% of the total interfacial area that is active for the electrochemical reaction. Combination of the findings from Figure 15 and Figure 16 shows that the phase with low volume fraction, i.e., the gas phase in the present study, prefers a higher Bruggeman factor not only because of less tortuosity but also because of a lower effective porosity than the natural porosity. If using the effective porosity of 0.26 in the Bruggeman correlation instead of the natural porosity 0.36, the Bruggeman factor would be about 3.4 (i.e., $0.26^{3.4} \approx 0.36^{4.5}$), very close to that (3.5) for the electrolyte phase. Another point worth noting is that the constructed 3-D microstructure stresses the influence of local variation in the effective porosity on the reaction current distribution; while the macrohomogeneous model only uses a constant natural porosity. As shown in Figure 17, the DNS model generates a more uniform reaction distribution than those of macrohomogeneous models using both Bruggeman factors of 3.5 and 4.5. As displayed in Figure 12, although the natural pore volume fraction of this porous medium distributes uniformly around the average porosity, the effective porosity varies across the

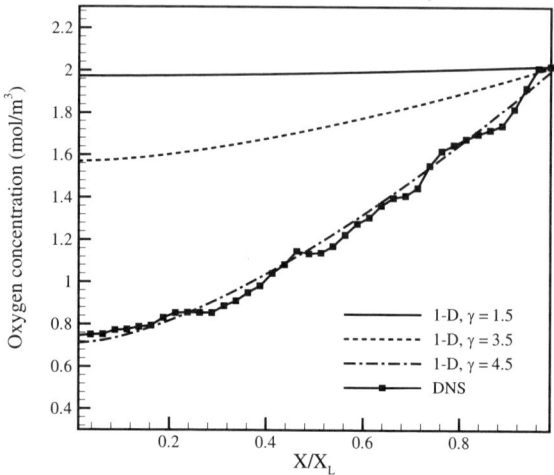

Figure 15. Comparison between the cross-sectional averaged oxygen concentration profiles from the DNS and 1-D macrohomogeneous model predictions.

DNS of PEFC Catalyst Layers

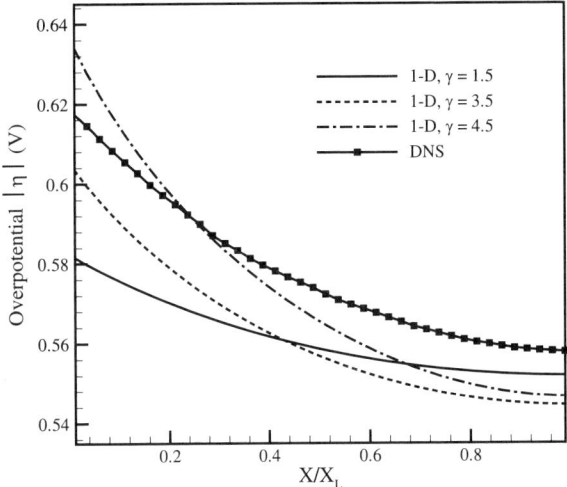

Figure 16. Comparison between the cross-sectional averaged overpotential profiles from the DNS and 1-D macrohomogeneous model predictions.

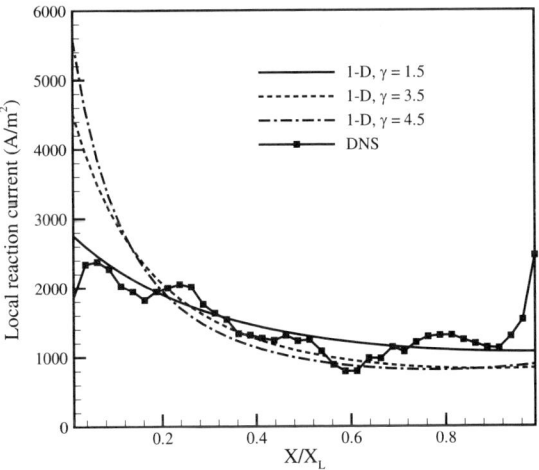

Figure 17. Comparison between the cross-sectional averaged reaction current distributions across the thickness of the catalyst layer from the DNS and 1-D macrohomogeneous model predictions.

thickness of the catalyst layer. More "transport" pores at the front end than at the back end produces a unique reaction current distribution that cannot be captured by the macrohomogeneous model using any Bruggeman correction factor.

IV. DNS MODEL – WATER TRANSPORT

Water management is a central issue in PEFCs. It is referred to as balancing two conflicting requirements: hydration of the polymer electrolyte membrane and avoidance of flooding in porous electrodes and GDL for reactant/product transport. Water management is also a key to high performance and longevity of the polymer membrane. This is because currently available membranes such as Nafion® require water in order to exhibit good proton conductivity. If water is insufficient, the ionomer becomes dehydrated and the proton transport resistance would dramatically increase. On the other hand, liquid water tends to accumulate inside the cathode catalyst layer due to water being produced from the ORR and also water migrating from the anode side via electro-osmotic drag. If water is not removed sufficiently, cathode flooding may occur, resulting in the gas pores being filled with the condensed liquid water, which, in turn, hampers the oxygen diffusion to reaction sites. Generally, fuel and oxidant feed streams are humidified externally to provide adequate water to the polymer membrane. To avoid either membrane dehydration or cathode flooding, it is of great importance to investigate the effect of the inlet humidity on water distribution throughout a cell, particularly inside the cathode catalyst layer, which consists of both the ionomer and gas pores.

Several groups have modeled water transport in PEFCs at various levels of complexity in recent years. Various water transport models for the catalyst layer have been employed in the general framework of computational fuel cell dynamics (CFCD). Notable works include Wang and co-workers,[5,20] Dutta et al.,[21,22] Berning et al.,[23] and Mazumder and Cole.[24] In their work, Dutta et al.[21,22] used an approximate analytical solution for water transport through the membrane. However, they did not consider the MEA in the computational domain. The model of Berning et al.[23] treated the catalyst layer as an interface between the membrane and the GDL. While Mazumder and Cole[24] supposedly ignored any water transport through the membrane, Wang and co-workers,[5,20] in contrast, developed a comprehensive water-transport model applicable throughout a PEFC including the MEA. A recent overview of the various water transport models for a PEFC is provided by Wang.[9] However, all the aforementioned works are based on the macroscopic description. No model has been attempted at the pore-level for water distribution and its intimate interactions with proton transport within the cathode catalyst layer.

In this section, the 3-D DNS model is further extended to include water transport in the cathode catalyst layer. In the previous sections, the proton conductivity of the electrolyte phase was taken to be constant assuming the ionomer to be fully humidified. In the present water transport model, the net water flux through the membrane from the anode side is considered to account for the combined effects of the electro-osmotic drag and back diffusion. Implementation of the various modes of water transport in the DNS model with the 3-D random CL microstructure, generated in the previous section, is described. The effects of humidity and microstructure composition on the cathode performance are investigated in the present section. The importance of the DNS model to optimize the catalyst layer composition is also demonstrated.

1. Water Transport Mechanism

General features of water transport through a PEFC are explained in Figure 18, where a MEA is sandwiched between two gas diffusion layers on the anode and cathode, respectively. To ensure membrane hydration, water is delivered to the fuel cell via humidified fuel and oxidant streams from an external humidifier. Water is transported to the cathode CL from the anode through the membrane by the electro-osmotic drag, expressed by:

$$N_{w,drag} = n_d N_{H^+} = n_d \cdot \frac{I}{F} \tag{48}$$

where, the electro-osmotic drag coefficient, n_d, denotes the number of water molecules carried by each proton across the membrane as current is passed and N_{H^+} is the proton flux. n_d varies in a wide range depending on the degree of membrane hydration according to the experimental measurements by Zawodzinski et al.[25] For a fully hydrated membrane immersed in liquid water, 2.5 water molecules are dragged per H^+ transported, while for a partially

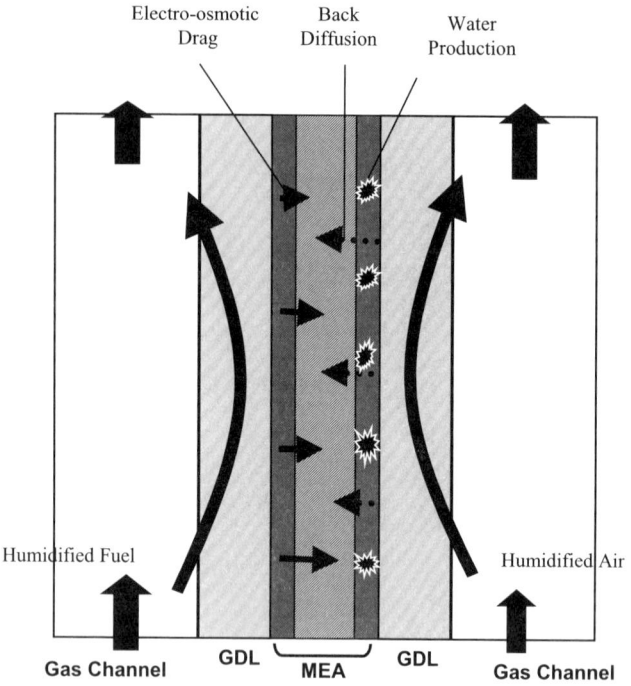

Figure 18. Schematic diagram of the water transport and distribution in a PEFC.

hydrated membrane corresponding to the water content up to 14, the drag coefficient is relatively constant at 1.0. In the present study, a constant drag coefficient of unity is used because the water content of interest ranges from zero to 14.

On the cathode side, water is generated by the ORR. The increase of water concentration from production will result in the back diffusion of water to the anode across the membrane. The back diffusion helps to hydrate the membrane on the anode side and thus partly compensates for the water loss by electro-osmotic drag. At higher current densities, the excessive water produced at the cathode is removed via evaporation by the under-saturated oxidant stream, and the removal rate can be controlled by adjusting the inlet air humidity and flow rate through the flow field.

2. Mathematical Description

Based on the DNS model described in the previous sections, the following assumptions are additionally made for the modeling of water transport:

- water is in the gas phase even if the water vapor concentration slightly exceeds the saturated value (i.e., slight over-saturation is allowed);
- equilibrium between the water in the electrolyte phase and in the gas phase is assumed and hence it is sufficient to consider water transport only through the gas phase;
- the electro-osmotic drag coefficient is constant at unity; and
- the net water transport coefficient from the anode to cathode is assumed constant.

Figure 19 describes various water transport mechanisms in both gas and electrolyte phases through the catalyst layer that are included in the present DNS model. Similar to the oxygen transport equation described earlier, the conservation equation for water vapor concentration, c_{H_2O}, through the random catalyst layer microstructure, generated in Section III, can be expressed as:

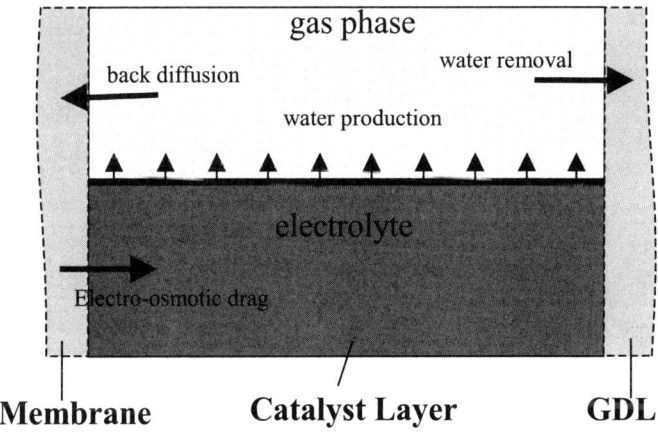

Figure 19. Schematic diagram of the water transport mechanisms inclued in the DNS model.

$$\nabla \cdot (D_{H_2O}^g \nabla c_{H_2O}) + a \int_\Gamma \frac{j}{2F} \delta(x - x_{interface}) ds = 0 \qquad (49)$$

where, $D_{H_2O}^g$ is the diffusion coefficient of water in the gas phase and definitions of the rest of the symbols remain the same as described earlier. Now, adopting the single-domain approach as detailed earlier, the water vapor diffusivity and source term in Eq. (49) can be expressed in terms of the discrete phase function, $f(i,j,k)$, respectively, as:

$$D_{H_2O}(i,j,k) = \\ D_{H_2O}^g \cdot [1 - f(i,j,k)] \cdot [2 - f(i,j,k)] \cdot [3 - f(i,j,k)]/6 \qquad (50)$$

$$\begin{aligned}
S_{H_2O}(i,j,k) &= \frac{i_0}{2Fc_{H_2O,ref}^g}[1 - f(i,j,k)]c_{H_2O}(i,j,k) \cdot \\
&\quad \{f(i-1,j,k)\exp[\frac{\alpha_c F}{RT}\phi_e(i-1,j,k)]/\Delta x \\
&\quad + f(i+1,j,k)\exp[\frac{\alpha_c F}{RT}\phi_e(i+1,j,k)]/\Delta x \\
&\quad + f(i,j-1,k)\exp[\frac{\alpha_c F}{RT}\phi_e(i,j-1,k)]/\Delta y \\
&\quad + f(i,j+1,k)\exp[\frac{\alpha_c F}{RT}\phi_e(i,j+1,k)]/\Delta y \\
&\quad + f(i,j,k-1)\exp[\frac{\alpha_c F}{RT}\phi_e(i,j,k-1)]/\Delta z \\
&\quad + f(i,j,k+1)\exp[\frac{\alpha_c F}{RT}\phi_e(i,j,k+1)]/\Delta z\}
\end{aligned} \qquad (51)$$

The water vapor diffusivity, $D_{H_2O}^g$, also depends on the temperature and pressure similar to the treatment of oxygen diffusivity as described by Eq. (41) and $D_{H_2O,0}^g$ is the reference water vapor diffusivity in the gas phase. The proton conductivity in the electrolyte phase has been correlated by Springer et al.[26] from the experiments as:

$$\kappa_0(\lambda) = 100 \exp[1268(\frac{1}{303} - \frac{1}{T})](0.005139\lambda - 0.00326) \text{ S/m} \qquad (52)$$

where, the water content in the membrane, λ, depends on the water activity, a, in the gas phase according to the following fit of the experimental data:

$$\lambda = \begin{cases} 0.043 + 17.81a - 39.85a^2 + 36.0a^3 & for \quad 0 < a \leq 1 \\ 14 + 1.4(a-1) & for \quad 1 < a \leq 3 \end{cases} \quad (53)$$

The water activity, a, is defined as:

$$a = \frac{c_{H_2O}}{c_{H_2O}^{sat}} \quad (54)$$

where, $c_{H_2O}^{sat}$ is the saturation concentration of water vapor corresponding to the fuel cell operating temperature. The saturation pressure of water vapor is only a function of temperature, which has been formulated by Springer et al.[26] as follows:

$$\begin{aligned} \log_{10} p^{sat} = &-2.1794 + 0.02953(T - 273.15) \\ &- 9.1837 \times 10^{-5} (T - 273.115)^2 \\ &+ 1.4454 \times 10^{-7} (T - 273.15)^3 \end{aligned} \quad (55)$$

where the pressure is in bar. Substitution of Eq. (53) into Eq. (52) provides the dependence of proton conductivity on water activity. Thus a water concentration distribution will cause the proton conductivity of the electrolyte phase to vary at every point within the catalyst layer.

Similar to the boundary conditions for charge and oxygen transport, a symmetry boundary condition is applied for the water conservation equation in y and z directions of the computational domain. At the CL-membrane interface, a net water transport coefficient, α, is employed to account for the net water flux across the membrane. It combines the electro-osmotic drag and back diffusion effects, and can be expressed as:

$$N_{w,net} = \alpha \cdot \frac{I}{F} = N_{w,drag} - N_{w,dif} \quad (56)$$

where, $N_{w,dif}$ is the water flux through the membrane due to back diffusion from the cathode side to the anode side. In the present study, α is assumed to be constant although it depends on the reaction rate and humidity conditions at anode and cathode inlets. Thus the boundary condition at the CL-membrane interface is given by

$$\frac{\partial c_{H_2O}}{\partial x}\bigg|_{x=x_L} = -N_{w,net} / D^g_{H_2O} \tag{57}$$

At the boundary connected to the GDL, i.e., at the CL-GDL interface, the water vapor concentration can be calculated from the water concentration at the channel inlet, $c_{H_2O,inlet}$, after the mass transport resistance through the GDL is accounted for and is given by:

$$c_{H_2O}\big|_{x=x_L} = c_{H_2O,inlet} + N_w\big|_{x=x_L} \cdot \frac{\Delta X_{GDL}}{D^{g,eff}_{H_2O,GDL}} \tag{58}$$

The water vapor profile is assumed to be linear across the GDL. The water flux through the GDL is the sum of the net flux across the membrane and the water production rate in the catalyst layer, $N_{w,prod}$, and can be expressed as:

$$N_w\big|_{x=x_L} = N_{w,net} + N_{w,prod} = (\alpha + 0.5) \cdot \frac{I}{F} \tag{59}$$

In Eq. (58), the inlet water concentration, $c_{H_2O,inlet}$, is calculated from the inlet air humidity and fuel cell operating pressure and temperature. Similar to the treatment in Section III, the water transport in the gas channel is assumed constant, which corresponds to a relatively large stoichiometric flow rate. From the inlet relative humidity, represented by RH, water vapor concentration of the humidified air is calculated by:

$$c_{H_2O,inlet} = RH \cdot c^{sat}_{H_2O} \tag{60}$$

All the parameters used in this section remain the same as those summarized in Table 2 in Section III.

3. Results and Discussion

(i) Inlet-Air Humidity Effect

Fuel/oxidant inlet humidity designs play an important role in the balance of water production and removal. Without a proper control, imbalance of water may result in either dehydration of the polymer membrane or flooding of the air cathode. Low-humidity is always desirable from the standpoint of external humidifiers with reduced size and cost.

Figure 20 displays several polarization curves of the cathode for the relative humidity ranging from 5% to 100%. There are three distinct characteristics corresponding to the three different regimes, kinetic control regime, ohmic control regime, and mass transport control regime, respectively. First, within the kinetic regime, the cathode performance is about the same and the only loss is kinetic activation loss. The only factor influencing the kinetic loss among these three cases is the oxygen concentration inside the catalyst layer. Calculations show that the oxygen concentrations are 7.5 mol/m^3, 9.2 mol/m^3 and 10.7mol/m^3 for 100%, 50%, and 5% inlet

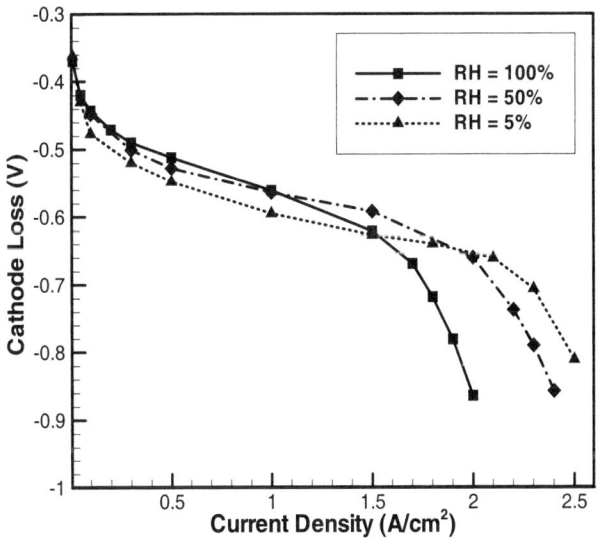

Figure 20. Polarization curves for various inlet humidity.

air humidity, respectively. These differences only bring about minor variations to the kinetic loss. According to the Tafel slope of 70 mV/dec, the case with 100% air humidity has 10 mV more kinetic loss than that with 5% relative humidity.

Secondly, low humidity tends to extend the ohmic control regime, postponing the occurrence of the mass transport limitation. This is the salient advantage of low-humidity operation. At 100% humidity, the oxygen concentration near the membrane end is as low as 0.2 mol/m^3 and the region is about to be depleted of oxygen. From the polarization curve, it can also be seen that the curve begins to fall off at the current density greater than 1.5 A/cm^2.

Thirdly, the disadvantage of low-humidity operation is its larger voltage loss in the ohmic control regime though the regime itself is enlarged. This is because of the low proton conductivity associated with partially hydrated electrolyte phase, leading to the increased ohmic loss. Figure 21 shows that the reaction zone shifts towards the back end of the catalyst layer (close to the membrane) with lowering of the inlet humidity. Apparently, this is due to

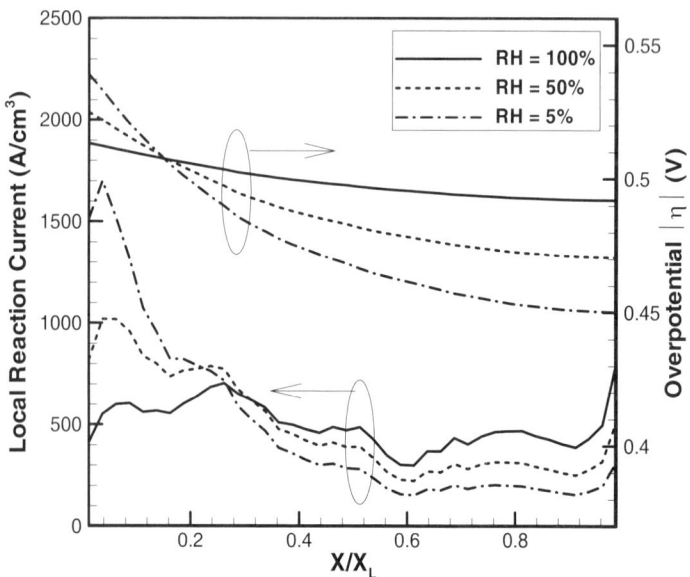

Figure 21. Overpotential and reaction current distributions for different inlet humidity.

poorer proton conductivity or higher ohmic resistance in the electrolyte phase and results in a much lower surface overpotential at the front end of the catalyst layer, making the overpotential distribution more nonuniform. In order to compensate for lower reaction current produced near the front end of the catalyst layer, the back end must provide higher reaction current as the average current density is fixed. This leads to a higher surface overpotential needed at the back end of the layer, representing more total voltage loss of the cathode in the low-humidity case.

In summary, 50% appears to be the optimal relative humidity in the catalyst layer configuration modeled here. It has a large range of ohmic control at a minimum expense of kinetic loss. It should be noticed that since the present model only considers the water transport in the gas phase, another significant advantage of low-humidity operation to alleviate cathode flooding cannot be demonstrated although it is widely recognized in practical applications.

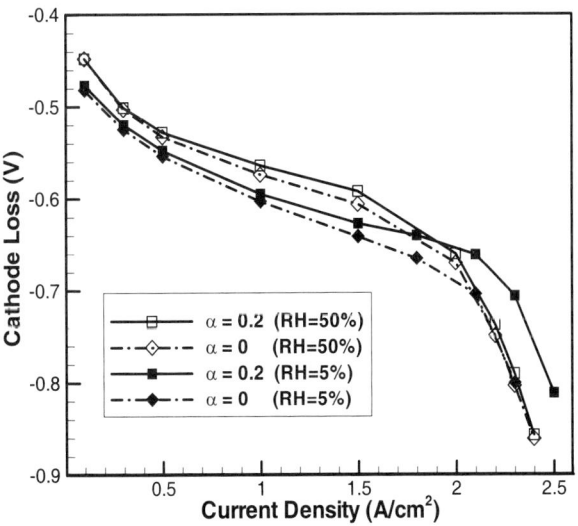

Figure 22. Polarization curves with different net water transport coefficients at the inlet humidity of 50% and 5%.

(ii) Water Crossover Effect

In the present DNS model, the net water flux through the polymer membrane to the cathode catalyst layer is quantified by the net water transport coefficient as given by Eq. (57). The value of this coefficient reflects the contribution of water transferred from the anode to the cathode. It depends on the humidity condition on the anode side. Low value of α indicates strong back diffusion from the cathode, which means a relatively dry condition in the anode. When α is close to its highest value, 1.0, it means the water flux due to electro-osmotic drag is dominant and the back diffusion is negligible. Therefore, by studying the effect of the net water transport coefficient, we can understand how important the anode gas humidification is to the cathode catalyst layer performance.

Two cases of the net water transport coefficient, 0.2, and 0, are studied under two different cathode inlet humidities, i.e., 50% and 5%. The polarization curves of the cathode are plotted in Figure 22 for all the four operating conditions. First, it shows that under 50% cathode inlet humidity, the value of α has almost no influence on the cathode voltage loss. The greatest difference of about 15 mV occurs at 1.5 A/cm^2. At operating current densities greater than 1.5 A/cm^2, the cathode performance almost has no change, which indicates the cathode side is already humid enough due to the large water production rate even if there is no water transported from the anode. However, with the cathode inlet humidity of 5%, α value becomes important especially at large current densities. The possible explanation is that the cathode is largely dry and hence any water supply from the anode is very helpful for increasing the proton conductivity of the electrolyte.

In summary, The DNS results clearly show that at 5% RH there is a considerable increase in the cathode voltage loss as α is reduced to zero. Again, this is due to the lower water activity, leading to the lower proton conductivity of the electrolyte phase. The lower proton conductivity would push the reaction zone to the back end, consuming more oxygen there. The larger consumption rate of oxygen, in turn, reduces the oxygen concentration there, which eventually requires a larger overpotential to drive the reaction. This series of consequences leads to the total cathode voltage loss to increase greatly with α being zero. The parametric study clearly suggests that at low cathode inlet humidity, restraining the back diffusion of water to the anode side is important and can significantly improve the cathode performance with better-hydrated electrolyte phase.

(iii) Optimization of Catalyst Layer Compositions

Besides the inlet humidity and water transport effects on the cathode performance, optimal design of the catalyst layer compositions is also of great practical interest. A series of simulations for various combinations of pore and electrolyte volume fractions were carried out to investigate the electrode composition effect. The predicted polarization curves under different compositions of pore and electrolyte phases are displayed in Figure 23. It can be seen that the greater the porosity, the larger the mass transport limiting current density. This is quite straightforward, as the oxygen transport would benefit from a large number of pores.

Figure 24 shows the local reaction current and overpotential distributions at a current density of 0.5 A/cm^2. With the porosity of 0.34 and electrolyte volume fraction of 0.32, the resulting small amount of "transport" pores limits the back end of the catalyst layer being accessed by oxygen. The increasing loss is then due to the ionic current passing through the back part of the catalyst layer without reaction (the resistance is much lower to pass current through the electronic phase after charge transfer reaction). As the porosity is increased, between 0.36 and 0.4, there is only a slight difference, which is due to the larger reaction sites of porosity 0.4. At the porosity of 0.6, Figure 24 shows that most of the reaction is concentrated at the back end, an indication that the proton conductivity is very poor. This is obviously because the electrolyte fraction has been reduced to 0.2 (the mixed electrolyte/electronic phases occupy ($1-\varepsilon_g$)), a typical design of the catalyst layer.

In summary, by assuming a sufficiently large electronic conductivity of the Pt/C phase and a fixed electronic phase volume fraction, it is important to select an appropriate pore to electrolyte volume ratio in order to achieve the best performance of the cathode. In general, both the pore and electrolyte volume fractions should be larger than the percolation threshold. Under this prerequisite, further increasing the porosity and electrolyte volume fraction will relax the mass transport limitation and reduce the ohmic drop, respectively. Hence, there always exists an optimal composition design for the best trade-off. According to the present simulation results of the DNS model, the optimal compositions are

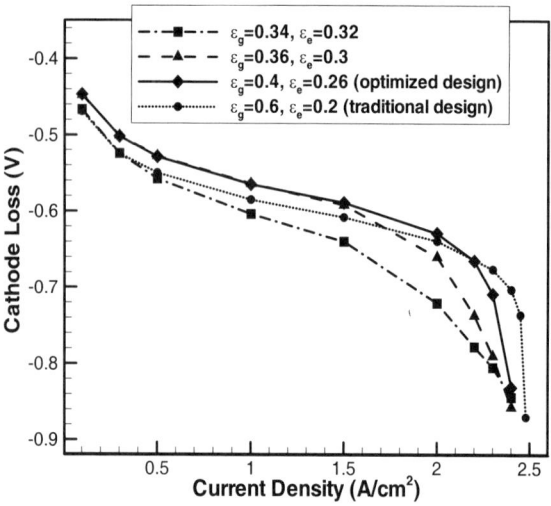

Figure 23. Polarization curves for different combinations of void and electrolyte volume fractions.

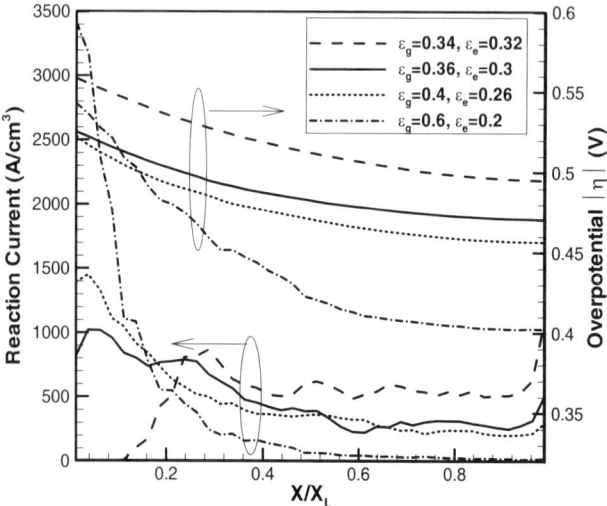

Figure 24. Local reaction current and overpotential distributions at 0.5 A/cm^2 for different combinations of void and electrolyte volume fractions.

0.4 for the porosity and 0.26 for the electrolyte volume fraction. In comparison, the traditional design with the electrolyte volume faction of 0.2 gives high ionic resistance. On the other hand, a relatively large porosity does not benefit the cathode performance much other than slightly delaying the mass transport control regime.

V. 3-D CORRELATED MICROSTRUCTURE

In the Section III, the generated 3-D microstructure only portrays the random nature of the porous media with an arbitrarily chosen porosity. However, the disorder in terms of randomness incorporated into the structure does not directly relate to the statistical inputs from an actual catalyst layer structure. This limitation is circumvented in the present section through the stochastic generation of a 3-D correlated microstructure.

1. Stochastic Generation Method

The process of generating a three-dimensional random porous medium with a given porosity and a given correlation function has been detailed by Adler.[18] The principle of this numerical reconstruction method is composed of two major steps. The first step involves the experimental measurement of any salient geometry features. Different features can be chosen for various materials. In most studies, the porosity and correlation function of pore spaces are selected. For this purpose, a representative portion of a SEM or TEM image of the catalyst layer is digitized to obtain an array of grayscale values for each pixel. Then using a computer program the image is converted into a two-dimensional solid map with artifacts removed by certain filters. Subsequently, statistical data such as correlation function and porosity are collected. The second step is the reconstruction process. Random samples of porous media are generated in such a way that, on average, they possess the same statistical properties as the real samples that they are assumed to mimic. Once these samples are generated, all transport processes can be studied at least in principle. For example, in the DNS model, using the phase function, a single set of governing equations can be derived and solved to analyze transport of reactants and product in both phases.

The stochastic method is based on the idea that an arbitrarily complex pore structure can be described by the values of a phase function $\vec{Z(r)}$, described by Eq. (28) at each point $\vec{r}$ in the porous medium. For a statistically homogeneous porous medium, the resulting microstructure can be described fully, albeit implicitly, by the first two moments of the phase function, namely porosity, ε and two-point autocorrelation function, $R_z(\vec{u})$ which are represented by Eqs. (29) and (30) respectively.

In general, the stochastic method creates a realization of the 3-D porous medium in terms of a binary, discrete population $Z(x)$, which takes only two-values 0 and 1, by transforming a Gaussian set $X(x)$ of standard, normal variates x. The final 3-D binary image represents a porous rock of prescribed porosity and autocorrelation function. This statistics-based reconstruction method was originally developed in two-dimensions by Joshi[27] and extended to three dimensions by Quiblier.[28] Adler et al.[29] applied it to the reconstruction of Fontainbleau sandstone. Ioannidis et al.[30] modified this method slightly by using Discrete Fourier Transform. In our study, we employed a simplified version, by Bentz et al.,[31] of the approach outlined by Quiblier.[28] The simplification of this method, over the approach utilized by Quiblier,[28] comes from the fact that there is no need to handle a large system of nonlinear equations in order to compute the matrix of filtering coefficients and this is numerically superior since no inversion is required.

Starting with the 2-D TEM image of an actual catalyst layer, as shown in Figure 25(a), the stochastic reconstruction technique first computes the two-point autocorrelation function from the binarized (pore/solid) 2-D image in Figure 25(b), generated from the original 2-D image in Figure 25(a). Using this correlation function and a nominal porosity of 0.6 as inputs, it finally produces the 3-D correlated microstructure representation of the catalyst layer, as shown in Figure 26, after passing it through the structural designation loop mentioned in Section III, which assigns the corresponding "transport" and "dead" pore and electrolyte phases.

2. Governing Equations, Boundary Conditions and Numerical Procedure

The governing equations for the conservation of charge, oxygen and water vapor remain the same as furnished in the previous sections. The symmetry boundary condition is imposed in the y and z directions of the computational domain, as shown in Figure

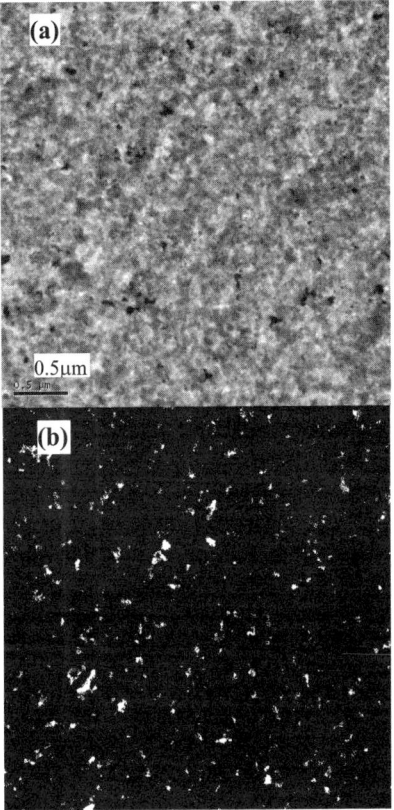

Figure 25. (a) Original 2-D TEM image of the CL, (b) Binary (pore/solid) 2-D image as input to the microstructure reconstruction model.

26. The rest of the boundary conditions are identical to the ones imposed for the 3-D random CL microstructure, as detailed in Sections III and IV. The conservation equations (16), (17) and (49) were solved using the commercial CFD software Fluent.$^{®32}$ The User Defined Functions (UDF) capability was deployed to customize the source terms given by Eqs. (38), (39) and (51) for modeling the electrochemical reactions at the phase interface as well as to solve the set of governing equations for DNS. Convergence was considered achieved when the relative error, for each scalar, between two successive iterations reached 10^{-6}. The

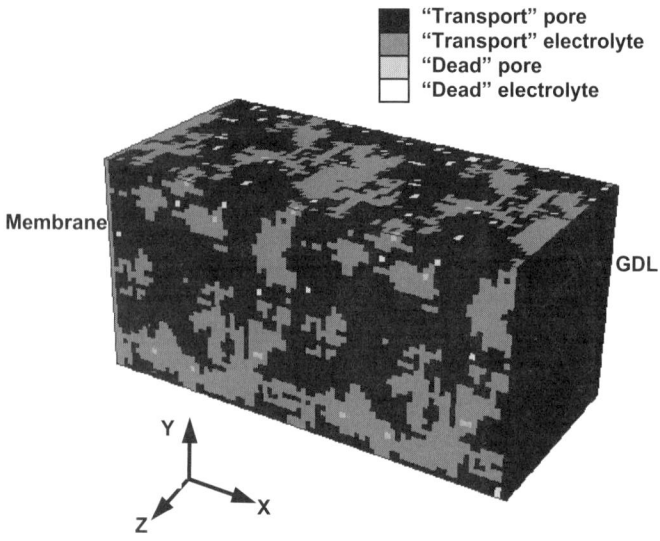

Figure 26. 3-D correlated microstructure of the catalyst layer with nominal porosity of 0.6.

Table 3
Property Data for the DNS Calculations with the 3-D Correlated Microstructure.

Parameter	Value
Oxygen diffusivity in air, $D^g_{O_2}$ (m^2/s)	9.5×10^{-6}
Water vapor diffusivity in air, $D^g_{H_2O}$ (m^2/s)	1.28×10^{-5}
Oxygen diffusivity in helox, $D^g_{O_2}$ (m^2/s)	2.0×10^{-5}
Water vapor diffusivity in helox, $D^g_{H_2O}$ (m^2/s)	3.3×10^{-5}
Pressure at the gas channel inlet, p (kPa)	200
Operating temperature, T (°C)	70
GDL thickness, ΔX_{GDL} (μm)	290
GDL porosity, ε_{GDL}	0.6
GDL tortuosity, τ_{GDL}	1.5
Nominal porosity of catalyst layer, ε_g	0.6

same numerical procedure was also employed for the DNS calculations with the regular and random 3-D microstructures in the previous sections. The property data used for the DNS calculations using the correlated microstructure are summarized in Table 3. In the present study for the correlated microstructure, the number of cells within the computational domain in the x, y and z directions are 100×50×50, respectively, leading to an average pore size of 0.1 μm and is found to be sufficient.

3. Results and Discussion

Figures 27, 28 and 29 show the comparison of cross-section averaged reaction current, cathode overpotential and oxygen concentration profiles at an average current density of 0.6 A/cm^2 with air as the oxidant along the thickness of the CL between the DNS and 1-D macrohomogeneous models, respectively. Different Bruggeman factors have been attempted. In the case of the reaction current (Figure 27) and cathode overpotential (Figure 28) distributions, the DNS result exhibits good agreement with the 1-D macrohomogeneous model with the Bruggeman factor of 3.5. However, Figure 29 shows that the factor of 4.5 gives a better match for the oxygen concentration profile. This feedback about

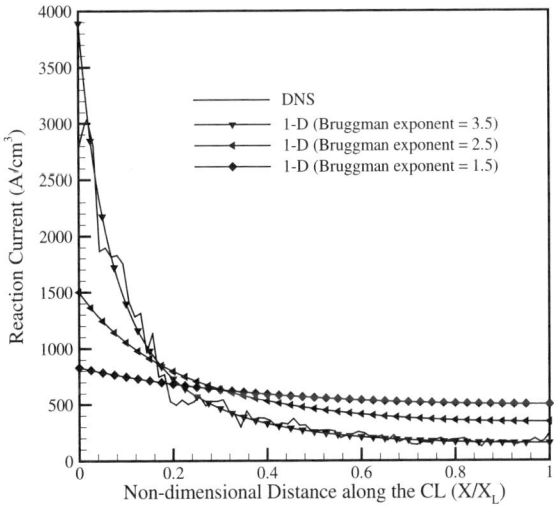

Figure 27. Local reaction current distribution along the thickness of the CL.

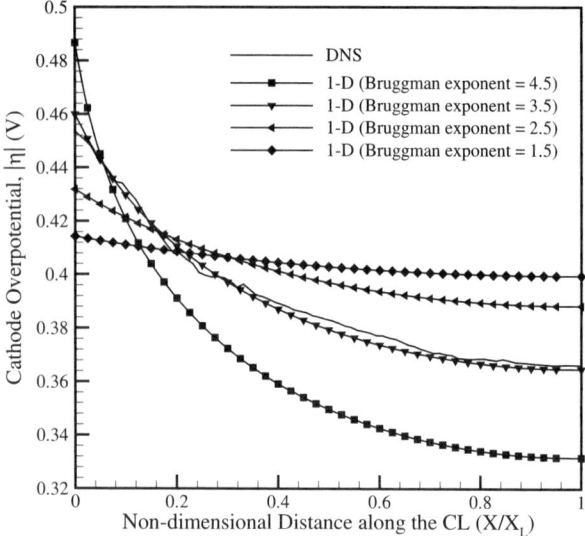

Figure 28. Local cathode overpotential distribution along the thickness of the CL.

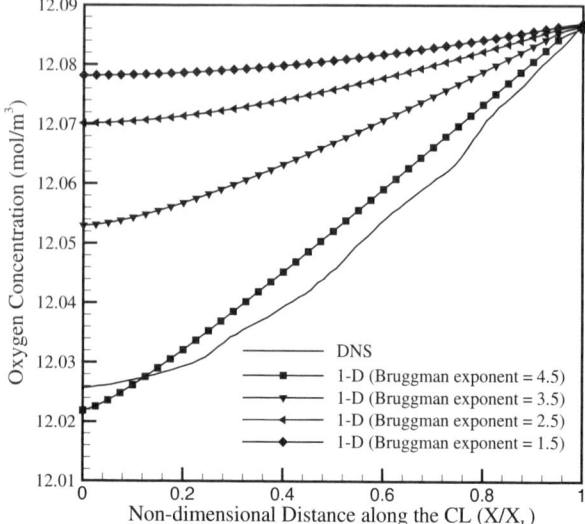

Figure 29. Local oxygen concentration distribution along the thickness of the CL.

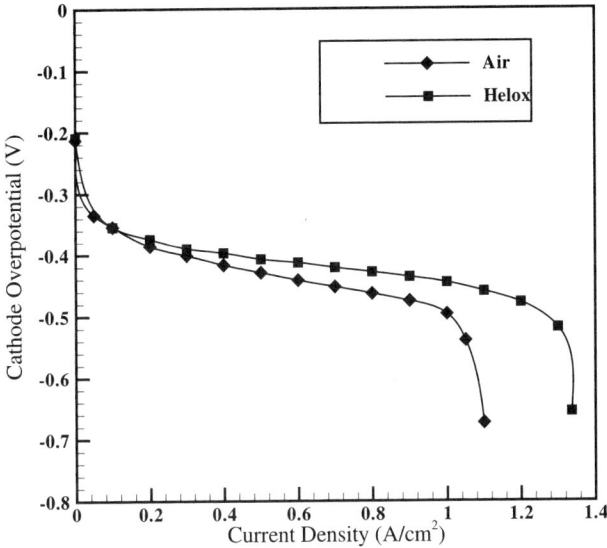

Figure 30. Polarization curves for 100% RH air and helox as oxidants.

the Bruggeman factor from the DNS data is very useful and can be used in the full-scale computational fuel cell dynamics (CFCD) models to effectively model the porous media. It is also important to note that the high reaction current in the 15-20% of the region near the membrane, as is evident in Figure 27, could be attributed to the limited ionomer conductivity resulting from low electrolyte phase volume fraction, estimated to be approximately 11%, which was calculated, based on the method outlined by Gastieger,[17] from the mass loading data of the respective constituent phases i.e., Pt, carbon and Nafion,® used in the recipe for the catalyst layer preparation.

Figure 30 shows the polarization curves with air and helox as oxidants under 100% RH inlet condition. From, Figure 30, it is evident that the cell performance is greatly improved when operated with helox as compared to with air due to the reduction in oxygen transport resistance. Higher performance is expected for helox, since oxygen diffusivity is almost two-times higher in He background (as in helox) as compared to in N_2 background (as in air). Also, as expected, for fully-humidified helox, DNS calculations predict a higher value of the limiting current density than that for fully-humidified air.

VI. CONCLUSIONS

In this chapter, the progress recently made in the pore-scale modeling of the PEFC catalyst layer is described. The DNS method, presented in the chapter, can directly model the transport and electrochemical reactions at the pore level of the catalyst layer with detailed morphology of the electrochemically active surface, ionomer network and pore tortuosity. It, thus, helps to investigate the microstructural influence of electrodes and permit optimization of morphology, composition and operating conditions to achieve high performance. The DNS model also has the capability to allow for virtual assessment of Pt/C ratio, solvents used in catalyst layer fabrication, Nafion® content etc. to maximize catalyst utilization and minimize g Pt/kW. To sum up:

- The DNS model provides a fast screening tool for optimizing catalyst layer compositions and structures.
- It provides an alternative approach to simulate fuel cell catalyst layers that is otherwise difficult to be characterized experimentally.
- It establishes a fundamental database to corroborate empirical correlations for effective parameters used in macroscopic models based on computational fluid dynamics.
- The DNS method establishes a science-based approach instead of Edisonian approach in providing inputs for development of novel recipes for next-generation, high-performance catalyst layers.

ACKNOWLEDGEMENTS

Financial support from NSF and industrial sponsors of ECEC is gratefully acknowledged. PPM wants to express sincere thanks to Mr. D. P. Bentz at National Institute of Standards and Technology (NIST), USA for many fruitful discussions during the stochastic reconstruction model development for the 3-D, correlated microstructure from TEM images, and Dr. X. G. Yang at ECEC for providing the TEM images.

REFERENCES

[1] S. Gottesfeld and T. A. Zawodzinski, in *Advances in Electrochemical Science and Engineering*, Vol. 5, Ed. by C. Tobias, Wiley and Sons, New York, 1997.
[2] T. E. Springer, T. A. Zawodinski, and S. Gottesfeld, *J. Electrochem. Soc.* **138** (1991) 2334.
[3] D. M. Bernardi and M. W. Verbrugge, *J. Electrochem. Soc.* **139** (1992) 2477.
[4] V. Gurau, H. Liu, and S. Kakac, *AIChE J.* **44** (1998) 2410.
[5] S. Um, C. Y. Wang, and K. S. Chen, *J. Electrochem. Soc.* **147** (2000) 4485.
[6] T. E. Springer and S. Gottesfeld, in *Modeling of Batteries and Fuel Cells*, Ed. by R. E. White, Electrochem. Soc. Proc., Pennington, NJ, 1991, p. 197.
[7] M. L. Perry, J. Newman, and E. J. Cairns, *J. Electrochem. Soc.* **145** (1998) 5.
[8] M. Eikerling and A. A. Kornyshev, *J. Electroanal. Chem.* **453** (1998) 89.
[9] C. Y. Wang, *Chem. Rev.* **104** (2004) 4727.
[10] A. Z. Weber and J. Newman, *Chem. Rev.* **104** (2004) 4679.
[11] L. Pisani, M. Valentini, and G. Murgia, *J. Electrochem. Soc.* **150** (2003) A1558.
[12] G. Wang, PhD Dissertation, Deptartment of Mechanical and Nuclear Engineering, The Pennsylvania State University (2003).
[13] J. Stumper, H. Haas, and A. Granados, *J. Electrochem. Soc.* **152** (2005) A837.
[14] T. Baritaud, T. Poinsot and M. Baum, *Direct Numerical Simulation for Turbulent Flows*, Publisher, Paris. Ed. Technip (1996).
[15] M. Sahraoui and M. Kaviany, *Int. J. Heat Mass Transfer* **37** (1994) 2817.
[16] S. V. Patankar, *Numerical Heat Transfer and Fluid Flow*, Hemisphere, Washington DC (1980).
[17] H. A. Gasteiger, W. Gu, R. Makharia, M. F. Mathias, and B. Sompalli, in *Handbook of Fuel Cells—Fundamentals, Technology and Applications*, Vol. 3, Ed. by W. Vielstich, A. Lamm, and H. A. Gasteiger, Wiley, Chichester, 2003, Ch. 46.
[18] P. M. Adler, *Porous Media: geometry and transports*, Butterworth-Heinemann, Stoneham, MA (1992).
[19] R. B. Bird, W. E. Stewart, and E. N. Lightfoot, *Transport Phenomena*, John Wiley & Sons Inc., New York (1960).
[20] S. Um, and C. Y. Wang, *J. Power Sources* **125** (2004) 40.
[21] S. Dutta, S. Shimpalee, and J. W. Van Zee, *J. Appl. Electrochem.* **30** (2000) 135.
[22] S. Dutta, S. Shimpalee, and J. W. Van Zee, *Int. J. Heat Mass Transfer* **44** (2001) 2029.
[23] T. Berning, D. M. Lu, and N. Djilali, *J. Power Sources* **106** (2002) 284.
[24] S. Mazumder and J. V. Cole, *J. Electrochem. Soc.* **150** (2003) A1503.
[25] T. A. Zawodzinski, J. Davey, J. Valerio, and S. Gottesfeld, *Electrochimica Acta*, **40** (1995) 297.
[26] T. E. Springer, T. A. Zawodzinski, and S. Gottesfeld, *J. Electrochem. Soc.* **138** (1991) 2334.
[27] M. Joshi, Ph.D. Dissertation, U. of Kansas, Lawrence, Kansas (1974).
[28] J. A. Quiblier, *J. Coll. Interf. Sci.* **98** (1984) 84.
[29] P. M. Adler, C. G. Jacquin, and J. A. Quiblier, *Int. J. Multiphase Flow* **16** (1990) 691.
[30] M. Ioannidis, M. Kwiecien, and I. Chatzis, *SPE Petroleum Computer Conference*, Houston, 11-14 June (1995).
[31] D. P. Bentz, and N. S. Martys, *Transport in Porous Media* **17** (1995) 221.
[32] Fluent 6.1 UDF Manual, Fluent Inc., NH, USA.

Index

1-D model, DMFC, 265
2-D model, DMFC, 273
3-D model, DMFC, 273
Accelerators, 30
AFCs, 128
Alkali conducting polymer membranes, 162
Alkaline conducting membrane, 183
Alkaline methanol fuel cells, 184
Alloys in bipolar plates, 20
Alternative electrode materials, 102
Alternative hypothesis, 46
Alumina membranes, 80
Alumina template, 117
  home grown, 80
Aluminum coupling, 25
Analysis of covariance, applications, 53
ANCOVA, 37, 53
  batch electrolyzer, 58
  CRE, 53
  flow electrolyzer, 58
  high conversion chemical reactor, 58
  RBE, 56
  two concomitant variables, 58
ANCOVA with velocity, 53
ANCOVA with velocity and pressure drop, 58
Anion exchange membrane, 183, 184
Anode,
  honeycomb, 118

Anode catalysts, SPE-DMFCS, 140
ANOVA, 37
  latin squares, 51
  one-way classification, 40
  three-way classfication, 49
  two factor classification, 46
  two-way classification, 46
ANOVA-related random effects, 66
Autocorrelation function, 306, 307, 334
Bartlett test, 71
Binary catalysts, 178
Bipolar plate, 1, 3
  carbon, 18
  cost, 13
  water management, 11
  features, 8
  materials and processes, 15
  parts of, 4
Bistructural catalyst, 154
Body of statistics, 37
Bottleneck pores, 80
Boxes, 8
Canonical variate analysis, 200, 216
Capacity ratio, 99
Capillary saturation, 211
Carbon as bipolar plate, 18
Carbon dioxide
  and methanol transport, 206
Carbon paper plates, 19
Catalyst evaluation, 143

Catalyst layer, 287
Catalyst layers, modeling of, 287
Catalyst materials, SPE-DMFCS, 140
Catalysts, 178
Catalysts supported on carbon, 287
Cathode Li-ion battery, 83
Cathodic process, 60
Caustic soda production as an example, 43
Cell models, 210
Cell stack, tolerances, 9
Cell-voltage behaviour, empirical model, 198
Charge/discharge reaction, Li-ion battery, 83
Charge-storage ability, mechanism, 100
Chemical grafting, 33
Cladding of bipolar plates, 22
Coatings for bipolar plates, 21
Colloidal crystallization, 81
Combinatorial synthesis, 143
Completely randomized experiment, 42
Computation of power, 63
Conducting membranes
 alkali, 162
 composite, 158
 grafted, 159
 methanol permeability, 160
 proton conductivity, 160
 radiation grafted, 162
 sulphonation of polymers, 157
 temperature, 160
 polyvinylalcohol, 159
Conducting polymers, 149
Conductive fillers, 33
Conductive polymer coatings, 24
Constant variance, 71
Contrast, 40
Contrasts analysis, 69
Control electrode, 92
Control electrodes, fabrication, 85
Corrosion resistant materials, 21
Cost of metal bipolar plates, 16
Covariance analysis for a two-factor, 60
Covariates, 58
CRE approach, 43
Crosslinkers, 31
Crossover
 CO_2, effect of, 258
 magnitude of, 253
Crossover, methanol, 253
Current collector, 77
CVD, 22
Cycle life,
 Li-ion battery, 102
Cyclic voltammogram and Li-ion batteries, 88
Deintercalation of Li ion, 88
Diffusion coefficient, 208
Direct methanol fuel cells, 229
DMFC, 127
 anode catalysts, 173
 anodes, 145
 applications, 172
 catalysts, 178
 components, 194
 electrode reactions, 137
 feed, 163

Index

hydrogen peroxide, 185
interest in, 163
low temperature performance, 172
mathematical modelling, 192
MEA, 163
membrane, 202
membranes, 178
mixed reactant, 186
modeling of, 264
modelling of the dynamic behaviour, 215
performance, 163
portable devices, 172
potential distribution model, 212
power density, 166, 171
simulation, 194
stack development, 166
stack performance, 175
stoichiometry, 171
temperature, 165
DMFC cell models
single phase operation, 211
two-phase operation, 211
DMFC electrode modelling, 209
assumptions, 210
DMFC model
two-phase model, 211
DNS model, 288
Dynamic behaviour of the DMFC, 215
Dynamic model, applications, 215
Dynamics and modelling, 214
Effective diffusion coefficient, 203
Electrical conduction in bipolar plate, 10
Electrocatalysts
novel systems for DMFC, 145
Electrocatalysts by electrodeposition, 142
Electrochemical characterization of electrodes, 87
Electrochemical investigations of Sn-base anode, 95
Electrode fabrication, 84, 103
Electrode morphology, 111
Electrode reaction mechanisms, 132
Electronic conductivity Li-ion battery, 101
Electro-osmotic water transport, 137
Electropolymerisation, 147
Empirical models, 198
Energy requirement for an electrolytic process, 47
Epoxies, 29
Etching solution, 79
External manifolds, 8
Fast CVD of bipolar plates, 23
F-distribution, 69
Fe-TMPP catalyst, 182
Fixed-effect based analysis, 66
Flame spraying of bipolar plates, 22
Flow field, 5
Forming cost of bipolar plates, 16
Frame, 8
F-statistic, critical values, 39
F-test, 39

Fuel cells, 127
Galvanic corrosion, 13
Gas diffusion layer, 208
Gas diffusion layers, 3, 4
GDL, electronic conduction, 10
General method, 77
  Li-ion batteries, 77
Grafoil or Graflex exfoliated graphite in bipolar plates, 19
Graft initiator, 31
Grafting, 24
Grafting, chemical, 27
Graphite in bipolar plates, 15
Gravimetric capacities, 109
Hierarchical classification, 66
Highly significant error, 39
Honeycomb carbon
  electrochemical characterization, 121
  preparation, 118
Honeycomb carbon framework, 117
Hydrogen PEMFCs, 173
Hydrogen peroxide in fuel cells, 184
Inductively coupled plasma, 93
Initiator, 31
Intercalation, 76
Intercalation of Li ion, 88
Internal manifolding, 8
Intrinsically corrosion resistant metals, 21
Ionomer, 287
Iron tetramethoxyphenylporphyrin, 152
Jet spray, 22
Large area filtered arc deposition (LAFAD) methods, 22
Laser ablation, 22
Li^+ diffusion distance, 117
$LiFePO_4$ cathode, 102
LiFePO4 electrode
  carbon analysis, 106
  electrochemical studies, 104
  electrochemistry of, 107
  imaging, 104
  volumetric capacity, 109
Li-ion batteries
  electrochemical characterization, 87
  electrodes, 76
Li-Ion Battery, 75
Li-ion battery cathodes
  electronic conductivity, of cathode 101
Linear combination of treatment, 40
Liquid-feed DMFC, 214
Mass transport and gas evolution, 205
Materials of bipolar plates, 15
Mathematical modelling of DMFC, 192
MEA, nafion
  problems, 2
MEAs
  high temperature, 2
  problems, 2
Membrane
  acid-base, 182
  acid-doped PBI, 179
  alkaline conducting, 183
  anion exchange, 183
  cross-linked copolymers, 182
  in DMFC, 178

nano-porous, 179
transport, 202
Membrane electrode
assemblies, 2
technologies for, 178
Membrane materials, DMFC, 156
Membranes, high temperature, 248
Metal bipolar plates, 19
Metal forming processes, 20
Metallic corrosion as an example, 45
Metals in bipolar plates, 15
Methanol
  anodic oxidation, 132
Methanol anodic oxidation
  alkaline electrolytes, 137
  catalysts, 134, 140
  electrolyte, 136, 137
  mechanism, 136
  pH, 137
  promoter, 133
Methanol crossover, 140, 149, 157, 162, 179, 202, 253
  MR-DMFC, 192
  Nafion membranes, 149
  parameters affecting, 203
Methanol crossover
  and fuel cell performance, 203
  membranes, 261
Methanol diffusion coefficients, 203
Methanol oxidation
  anode kinetics, 232
  catalysts, 233
  electrocatalytic polyaniline, 148
  low temperature, 195
  mechanism, 195
  reaction mechanism, 232
  sputter deposited anodes, 149
  supports, 144
Methanol Oxidation, 195
Methanol oxidation catalysts
  metal mini-mesh electrodes, 145
  polymer films, 147
Methanol transfer by diffusion, 205
Microporous polymeric filtration membranes, 78
Microstructure,
  correlated, 333
  random, 305, 306, 307
  regular, 290, 293
Microwire electrode, 99
Mixed-potential effects, 260
Mixed-reactant fuel cell, 185
Molded graphite in bipolar plates, 18
Monopolar cells, 10
MR-DMFC
  methanol crossover, 192
  performance, 189
  selective electrodes, 190
  cathode selectivity, 188
Nafion membrane
  and Teflon, 157
Nafion membrane electrolyte, 156
Nafion membrane modified, 179
Nafion membranes
  plasma etching, 157
Nanocomposite of $LiFePO_4$/carbon, 102
Nanocomposites
  DMFC, 145

Nanoelectrodes
  applications, 97
  low-temperature performance, 97
Nanomaterials, 123
  parameters related to, 77
Nanomaterials in Li-ion battery electrode design, 75
Nanoporous polymeric filtration membranes, 78
Nanoscopic particles electrode
  low-temperature performance, 99
Nanosphere template, 81
Nanostructured anodic electrodes, 91
Nanostructured cathodic electrode, 83
Nanostructured electrode fabrication, 84
Nanostructured electrodes and thn-film electrodes, 112
  avantages, 77
Nanostructured electrodes, 92
Nanostructured V_2O_5 electrodes, 97
Nanotechnology
  advantages, 102
Nanowire electrode, 99
Niobium in bipolar plates, 20
Null hypothesis, 46
Numerical simulation, 285
One-dimensional model, 211
One-dimensional potential distribution model, 197
One-way classification test, 63
Optimization of catalyst layer, 331
Oxygen cathodic reduction, 139
  mechanism, 139
Oxygen electrocatalysts
  types, 149
Oxygen reduction
  poisoning, 151
Oxygen reduction
  catalysts, 152, 247
  C-supported Pt-alloys, 151
  cyclic voltammograms, 155
Oxygen reduction catalysts, 149
Oxygen reduction reaction on Pt, 150
Oxygen-reduction
  cathode polarization curves, 189
PAFCs, 128
PEFC, 285
PEM fuel cell,
  bipolar plate technology, 1
PEMFCs, 128
Permeability of methanol, 203
Permeation rates, 203
Plasma polymerization, 178
Plating of bipolar plates, 21
Platinum and platinum alloy catalyst performance, 240
Polarisation curve
  SPE-DMFC, 131
Polybenzimidazole/H_3PO_4, 3
Polycarbonate template membrane, 76, 98
Polyimides, 30
Polymer electrolyte, 161
Polymer electrolyte fuel cell, 285
Polymer matrix, 29, 33

Polymer matrix materials, 30
Porosity, 306
Port, 6
Port bridge, 6
Potential distribution model, 212
Power of test, 40
Promoters, 144
  quaternary, 144
  ternary, 144
Proton-conducting membranes, 161
Pt/C, 155
Pt-based catalysts
  dispersion of, 146
Pt-Co/C, 155
Pt-Fe/C, 155
Pt-Ru, 178
Pt-Ru alloy
  preparation, 142
Pt-Ru anodes
  co-catalysts, 143
Pt-Ru catalyst, 141
  and transition metal DMFC, 144
  oxides, 144
Pulsed laser deposition, 24
Punching and stamping method, 20
P-value, 39
Pyrolised transition metal macrocycles, 150
Quaternary catalyst, 179
Quaternary promoters, 144
Randomized Block Experiment, 42
Randomized blocks, 67
Rate capabilities, 114
Rate capabilities of Li-ion battery electrode, 89
RBE example, 43, 45

Reactive ion beam-assisted electron beam-physical vapor deposition, 23
Reactive plasma spraying, 23
Resistivity
  bipolar plate, 11
RuSe catalyst, oxygen reduction, 153
RuSe/C catalyst, 154
Seals, 7
Separator plate, 4
Significant error, 39
Silane processes, 33
Silenes, bonding mechanism, 27
Single phase flow, 212
Single-phase one-dimensional model for the DMFC, 213
Size of observations, 64
Sn-based anodes, 91
  rate capability, 96
Sn-base anodes
  fabrication, 92
Sn-based electrode
  advantages/disadvantages, 96
SnO_2 control electrode, 92
SnO_2 electrode
  structural investigation, 93
Solid-polymer electrolyte, 137
Solid polymer electrolyte DMFCs, 183
Solid polymer electrolyte fuel cells
  and mass transport, 205
Solvent, 31
SPD-DMFC
  thermodinamics, 129
Specific capacity, 83

Specific energy requirement, 47
SPE-DMF
 half cell, 191
SPE-DMFC, 128
 conducting membrane, 156
 development of, 140
 disadvantages, 131
 dynamic response, 211
 electrode reaction materials for, 140
 mechanism, 132
 methanol crossover, 157
 operating principles, 129
 static response, 211
Sputtering, 178
Stack Hydraulic, 216
Stainless steels in bipolar plates, 20
Statistical analysis, 38
Statistical experiment, 42
Statistical techniques and economy, 71
Stochastic generation, 333, 334
Structural Investigations, 93
Structural investigations of materials, 93
Structural investigations of materials in Li-ion batteries, 86
Surface activation, 28
Synthetic method, Li-ion batteries, 77
Template synthesis, 76, 123
Templates, 78
Template-synthesis materials, 91
Template-synthesis methods, 103
Template-synthesized electrodes
 volumetric energy density, 110
Template-synthesized nanostructured
 Sn-base anode, 97
Template-synthesized nanostructured electrodes, 77
Template-synthesized nanostructures, 79
Ternary catalyst, 148
Ternary promoters, 144
Thermal laser assisted CVD of bipolar plates, 23
Thermal Management MEA, 9
Thermal models, 216
Thermal vapor deposition of bipolar plates, 22
Thermally activated CVD of bipolar plates, 23
Three dimensional modelling, 213
TIVO, 111
Tolerances in
 cell stack, 9
Track-etch membranes, 78
Track-etch method, 78
Track-etch polymer template, 81
Transfer of methanol in DMFC, 205
Transport of methanol, 203
Transport of water, 203
Treatment population size, 68 68
Two-dimensional modelling, 213
Two-factor analysis, 65

Index

Two-way classification, example of, 47
Type I error, 39
Type II error, 40, 62, 66
Vapor deposition plasma, 22
Vapour-feed DMFC, 208
Variance and covariance in electrochemistry, 37
Volumetric capacity, 109
Volumetric energy density, 110
Volumetric rate capability, 115
Water management in bipolar plates, 11
Water removal, 12
Water transport in DMFC, 204
Zeolite catalysts, 145

Printed in the United States of America.